Application of Empirical, Analytical, and Numerical Approaches in Mining Geomechanics

Application of Empirical, Analytical, and Numerical Approaches in Mining Geomechanics

Guest Editors

Mohammad H.B. (Farzine) Nasseri
Bibhu Mohanty
Kamran Esmaeili

MDPI

Basel • Beijing • Wuhan • Barcelona • Belgrade • Novi Sad • Cluj • Manchester

Guest Editors
Mohammad H.B. (Farzine) Nasseri
Department of Civil and Mineral Engineering
University of Toronto
Toronto, ON
Canada

Bibhu Mohanty
Department of Civil and Mineral Engineering
University of Toronto
Toronto, ON
Canada

Kamran Esmaeili
Department of Civil and Mineral Engineering
University of Toronto
Toronto, ON
Canada

Editorial Office
MDPI AG
Grosspeteranlage 5
4052 Basel, Switzerland

This is a reprint of the Special Issue, published open access by the journal *Mining* (ISSN 2673-6489), freely accessible at: https://www.mdpi.com/journal/mining/special_issues/mining_geomechanics.

For citation purposes, cite each article independently as indicated on the article page online and as indicated below:

Lastname, A.A.; Lastname, B.B. Article Title. *Journal Name* **Year**, *Volume Number*, Page Range.

ISBN 978-3-7258-6666-3 (Hbk)
ISBN 978-3-7258-6667-0 (PDF)
https://doi.org/10.3390/books978-3-7258-6667-0

Contents

About the Editors

Mohammad H.B. (Farzine) Nasseri

Mohammad H.B. (Farzine) Nasseri holds a Ph.D. in experimental rock mechanics. He has published more than sixty papers in peer-reviewed international journals and conferences including nature and science journals. In addition to these achievements, he has presented his research findings at international conferences and numerous invited talks during his academic career at the department of Civil and mineral Engineering, University of Toronto, where he has worked since 1999.

His journey has taken him through the University of Toronto's Civil and Mineral Engineering department, where his expertise in rock mechanics flourished. Teaching and designing new courses at the undergraduate and graduate level for more than 28 years, managing industrial projects, setting up advanced rock mechanical laboratories at UofT and all over the world, and spearheading fracture dynamics research encapsulates his most pivotal contributions. His background in structural geology has assisted him in many collaborative projects with nuclear waste repositories and mining, civil, and petroleum clients. He has made headway in characterizing petro-fabric structures and unraveling fracture propagation, guided by a synergy of problem-solving, concise communication, and incisive data analysis. These efforts converge on his personal goals and enlighten engineering applications with profound geological insights.

He has been retired from his position since October 2021. He is now active as an expert and a consultant in the rock mechanics field.

Bibhu Mohanty

Bibhu Mohanty previously held the NSERC Industrial Research Chair in Rock Dynamics and Fragmentation at the Department of Civil and Mineral Engineering, University of Toronto, Canada. He received his Ph.D. in Physics from the University of Toronto. Previously, he taught at McGill University, Montreal, and he has also served as a Science and Technology Associate at Orica, a leading manufacturer of explosives and accessories. In addition to his research activities, he consults regularly on blast- and explosives-related issues in the mining industry, as well as on explosion hazards and blast-resistant and containment structures.

Understanding the dynamic fracture behavior of rock is a key step in quantifying the response of rock mass to high-energy transient loads such as in the drilling and crushing of rock and in fragmentation due to explosive action. This is integral to all civil and mining excavation activities and determines the safety, economic success, and viability of these operations.

The objectives of his research group are to identify the key variables in this complex system and to determine their relative importance through appropriate parametric studies. Direct experimental and systematic verification of theoretical predictions remains a cornerstone of this attempt to develop reliable "building blocks," which can then be used to develop predictive models for the fracture and fragmentation of rock.

Kamran Esmaeili

Kamran Esmaeili leads the work undertaken at the Mine Modeling and Analytics lab. He holds the Robert Smith Chair in Geomechanical Mine Design and Analysis. Over several years of industrial work experience and academic research, he has contributed to several important fields such as rock slope stability, geomechanical underground mine design, numerical modeling in geomechanics, blasting and rock fragmentation, material handling in underground mines, UAV application in mining and data analytics, and machine learning applied in mining and geoscience. Dr. Esmaeili's work experience includes several years of engineering experience and working as a consultant for various companies involved in the mining industry.

His research aims to explore the development of predictive analytics and smart modeling solutions for the mining industry, enabling better mine design and continuous mining process control and optimization. His ongoing research projects fall within the following three key interrelated areas: the development of innovative methods for the real-time acquisition of high-resolution mining data, the digital transformation of mining data, data integration, analytics, and visualization; the development of predictive 3D spatial models of rock mass properties (i.e., geomechanical, structural and geometallurgical models) based on advanced statistical and geostatistical techniques; and the investigation of the influence of rock mass heterogeneity and anisotropy on rock responses when subjected to different static and dynamic loading conditions (i.e., the influence of heterogeneity and anisotropy on rock mass strength, fragmentation, and stability).

mining MDPI

Article

Required Plug Strength for Continuously Poured Cemented Paste Backfill in Longhole Stopes

Murray Grabinsky [1,*], Will Bawden [2] and Ben Thompson [3]

[1] Department of Civil & Mineral Engineering, Faculty of Applied Science and Engineering, University of Toronto, Toronto, ON M5S 1A4, Canada
[2] Bawden Engineering, Toronto, ON M4S 3H7, Canada; wbawden@mdt.ca
[3] Mine Backfill Geomechanics Inc., London, ON N6A 2X1, Canada; ben.thompson@minebackfill.ca
* Correspondence: murray.grabinsky@utoronto.ca

Abstract: Continuously poured paste backfill dramatically improves underground mining efficiency through reduced stope cycle time and simplified logistics. For longhole stopes, a backfill "plug" is poured to a few meters above the undercut brow and must gain sufficient strength to prevent failure through the plug when the "main" pour begins. A novel, rational engineering design approach that determines the required plug strength is developed. The potential failure mechanism during continuous pouring is identified and the theoretical solution and its numerical validation/calibration are discussed. Four field case histories are then used, three of them involving continuous pours, to demonstrate the theoretical solution's validity in back-analysis. These case studies are unique in the extent and quality of total stress and water pressure measurements made throughout backfilling. Additionally, comprehensive laboratory data are available to characterize strength development during binder hydration in the first few days, which are critical to the back-analyses. Results indicate that continuous backfilling is feasible with reasonably attainable backfill strengths at most mines. However, mines must undertake comprehensive early strength laboratory testing, and must carry out field measurements during the pour to ensure the placed backfill behaviour is consistent with the analysis assumptions.

Keywords: cemented paste backfill; continuous pouring; Prandtl analysis; cohesion; strength analysis; field instrumentation

1. Introduction

Ongoing underground mining and progressive orebody extraction can generate dangerous stress concentrations leading to rockbursts, especially in deep orebodies. Backfill provides global support to the host rock and as such it is integral to mining strategies that mitigate rockburst hazard. Cemented paste backfill (CPB) is increasingly favored over other backfilling methods (rock fill and hydraulic fill) due to shortest filling time, uniform as-placed engineered properties, and tight filling potential of the excavated volume. For longhole stopes, a CPB containment structure (usually a shotcrete barricade or rammerjam rock berm) is constructed in the undercut and then a CPB "plug" is poured to a few meters above the undercut brow. This CPB plug must develop sufficient strength to protect the containment structure from the further effects of the "main" pour in the remainder of the stope's volume. Ideally, the required CPB plug strength will be realized during the plug pour so that once the CPB reaches the desired plug height the backfill plant can change to the main pour mix design (assuming this is different from the plug pour mix design) and continuously deliver CPB until the entire stope is filled.

No rational process exists to determine required CPB plug strength before starting the main pour (including for the most desirable scenario, a continuous pour). Some mines continuously pour CPB using empirically developed procedures [1], but these do not provide insight into the factors controlling CPB plug stability and cannot be universally applied. One well-cited industry handbook suggests the CPB should attain a "strength" (presumably unconfined compressive strength, *UCS*) 150 kPa [2], but no explanation or substantiating case histories are provided. Whether mines use *UCS* = 150 kPa or some other target strength, there are trade-offs to how such strength is achieved. Some mines use higher binder content in the CPB plug to realize higher early strength, while others use a "cure period" after the plug pour (in some cases, several days) prior to backfilling the remainder of the stope. Regardless, if the target CPB strength is higher than necessary, either too much binder is being consumed or the stope cycle time is unnecessarily long, respectively.

The implications of insufficient CPB plug strength leading to barricade failure include fatality, damaged plant, and production delays. Four CPB failures reported by [3] and five by [4] highlight various potential contributing factors, but in all cases the failure mechanism through the CPB plug is washed out once the barricade is breached, making it impossible to identify progressive failure details. There is, therefore, urgent need to develop rational engineering design procedures (laboratory test methods, analysis methods, and field monitoring methods) to continuously pour CPB safely and economically.

The problem just described falls into the "data-limited, understanding-limit" class considered by [5], where numerical modeling should be used to explore potential failure mechanisms, understand the impact of different material behaviour assumptions, and identify the data and understanding further needed to better analyze the problem. An important step in this process is making appropriate simplifying assumptions: Start with the simplest model that seems to capture essential system behaviour, and only add complexity as needed. Therefore, field data that motivates the developed system idealization, and which facilitates analytical and numerical solutions is initially considered. Subsequently the solution is applied to four case histories where sufficient field monitoring and laboratory testing data exists to reasonably assess the predictive solution's validity. Steps that other mines can take towards rationally engineering continuously poured CPB are then discussed.

2. Predictive Model Development, Validation, and Calibration

The simplifying assumptions used in developing the analytic solution are considered first. These motivate an analytic solution based in part on the Prandtl solution for footing bearing capacity in undrained clay. Numerical analysis is then used to validate and calibrate the analytic solution. Finally, a design procedure based on the solution is proposed.

2.1. *Effective Stress Development during Field Pours*

The CPB literature typically cites several examples of field monitoring, but most [6–10] do not collect data frequently enough, do not combine total stress and pore water pressure measurements, or do not have instrumentation in the appropriate backfill plug location to be relevant to the continuous CPB pour analyses considered here. The field monitoring studies undertaken by the authors at Williams, Kidd, and Cayeli mines [11,12] involved deploying up to five instrumentation clusters throughout the stope (two in the undercut and three throughout the height), with each cluster containing a minimum three orthogonal total earth pressure cells (TEPCs), a pore water pressure transducer (PWP, or piezometer), and a thermistor (TEMP) and electrical conductivity (EC) transducers. Figure 1 shows an example instrumentation cluster and deployment at Kidd mine. The TEMP and EC transducers help identify onset of significant binder (Portland cement and cement extenders, including fly ash and granulated ground blast furnace slag) hydration reactions. Additional TEPC and PWP transducers were installed on the barricade. Data was logged every five minutes throughout pouring, and every 15 minutes after pour completion.

(**a**) (**b**)

Figure 1. University of Toronto fieldwork instrumentation: (**a**) cluster with TEPCs and PWP (photo credit: Ben Thompson while a Research Associate with University of Toronto); (**b**) clusters deployed in stope at Kidd mine (photo credit: Ryan Veenstra while a PhD candidate at the University of Toronto).

Ref. [13] describe a similar approach at Casa Berardi mine, but using two clusters with TEPCs and PWPs. [14] give results from Kanowna Bell mine using one TEPC oriented for vertical total stress, and one PWP.

Published field monitoring results from the above mines were digitized in approximating linear segments (estimated precision generally ± 0.5 h (hour) and ± 5 kPa) for the instrument clusters closest to the stope's middle base, in the CPB plug. As will be shown later, this is the critical location for formation of a potential failure surface through the plug, which could then lead to barricade failure. Figure 2 shows field test results in terms of vertical effective stress development with respect to time after the CPB covers the transducers, for up to the subsequent 48 h (two days), which is generally sufficient time to complete the plug pour. Initial zero effective stress periods in which the CPB remains in an essentially fluid state ($\sigma_v' = \sigma_h' = PWP$) ranges between ~3 h (i.e., for Williams) and ~14 h. Horizontal effective stresses for the same mines and time periods remain below 25 kPa (see the previously cited studies for full monitoring results). Typical CPB rise rates range from 0.25 m/h to 0.40 m/h, so a ~5 m high undercut would fill in 12–20 h. For most of the mines shown in Figure 2, very little effective stress develops in either vertical or horizontal directions during this initial CPB plug pour period.

In these field studies the main CPB pours featured higher rise rates or lower binder contents, and the monitoring results indicated zero effective stress over much longer time periods, typically to the end of backfilling (see previously cited studies). For example, Figure 2 shows at Kidd the zero effective stress period within the CPB plug persisted for ~7 h, while in the main pour (which had half the binder concentration compared to the plug) the zero effective stress period existed for greater than 48 hours after the transducers in the main CPB pour were covered [11].

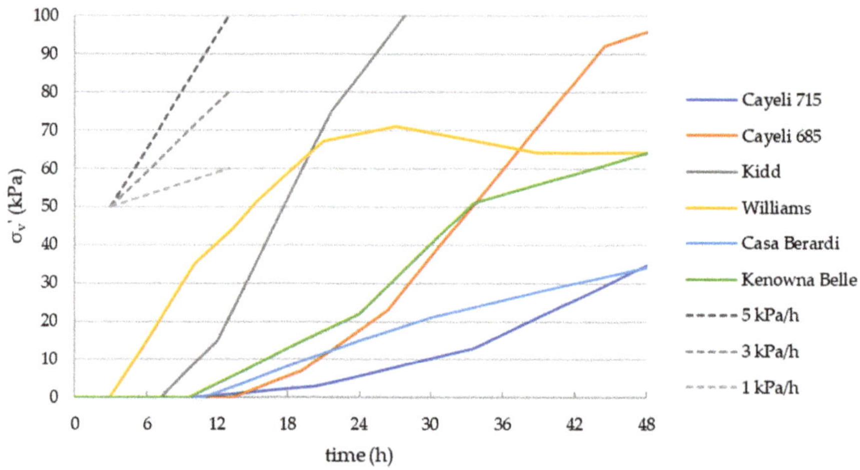

Figure 2. Vertical effective stress development at the indicated mines (referenced with respect to time when backfill covers the TEPC and PWP transducers). Adapted from data presented in [11–14].

2.2. *Strength Development during Field Pours*

The field monitoring results reported by [11,12] show increasing TEMP measurements during the zero effective stress periods, which indicate binder hydration is occurring and so the CPB plug is gaining strength. [15] used direct shear tests to determine strengths at 4 h and at 1, 2, 4, 7 days and beyond. The CPB plug pours often take less than 24 h and occur under essentially zero effective stress which means material friction cannot contribute to backfill strength. Therefore, only material cohesion data is considered (Figure 3).

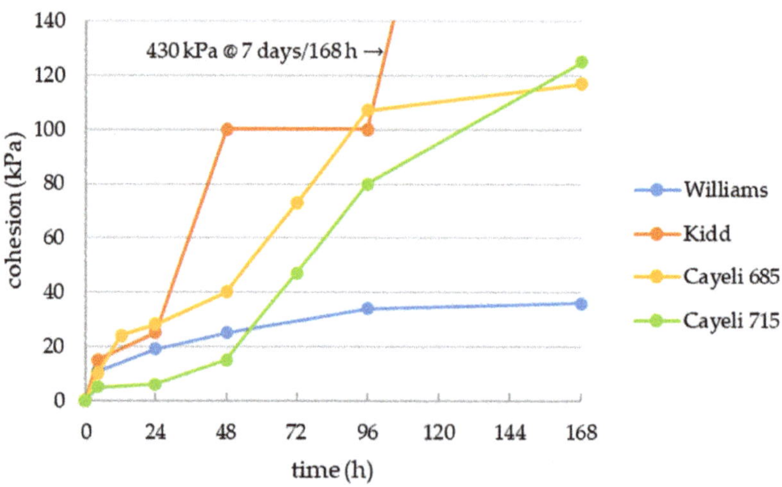

Figure 3. Cohesion development with curing time for Williams, Kidd, and Cayeli mines. Adapted from data presented in [15].

The Williams binder is a blend of 50 wt% Portland Type I (GU) cement and 50 wt% Type C fly ash and shows relatively continuous cohesion development with cure time. In contrast, the Kidd binder is a blend of 10 wt% cement and 90 wt% ground blast furnace slag, and the Cayeli binder is CEM V from a Turkish manufacturer which is equivalent to

60 wt% cement and 40 wt% extenders. These binders show cohesion developing in stages, with significant increases after one day and again after four days for Kidd, and after two days for Cayeli. These stages probably correspond to the secondary reactions of the slag and other extenders, although this was not investigated in detail.

Although others have considered CPB strength increases due to self-weight consolidation during deposition [16–20], the slow start to effective stress development (Figure 2) compared to material cohesion development (Figure 3) suggests minimum potential for self-weight consolidation and associated strength gain at the studied mines. Indeed, [21,22] show that the as-placed bulk properties at Williams, Kidd, and Cayeli mines are the same as the as-mixed properties, except that occluded air was mixed into the paste during deposition which had the effect of slightly increasing void ratio and decreasing degree of saturation. Similar results are reported by [13] at Casa Berardi where cores from the CPB plug had void ratios similar to the as-mixed design. Therefore, the strengths shown in Figure 3 should be considered indicative of CPB curing under field conditions.

2.3. Preliminary Analytical Model for CPB Plug Strength

A vertical cross section through a typical stope and undercut are illustrated in Figure 4, with indicated dimensions that will be used in formulating the analytical model. With reference to this figure, the following simplifying assumptions are made:

- The undercut has square cross section with height Hu, and the distance between the backfill barricade and the undercut brow is Lu;
- The undercut walls are rough so that shear occurs through the CPB across asperities, rather than along the CPB-rock interface;
- The CPB plug has height Hp with height above the undercut brow Hb;
- The height of main pour at any given time is Hm above the plug, and the main pour exerts fluid-like pressure on the plug γHm where γ is the backfill's unit weight;
- The total backfill height at any time is $Ht = Hp + Hm$;
- The average CPB plug undrained cohesion c at the end of the plug pour must be sufficient to create a "self-supporting" condition, such that the CPB plug no longer relies on the barricade for stability; and
- The average CPB plug undrained cohesion c must continue to increase sufficiently during the main pour to support the surcharge γHm and prevent failure through the CPB plug which would otherwise then exert further pressure on the barricade and potentially fail it.

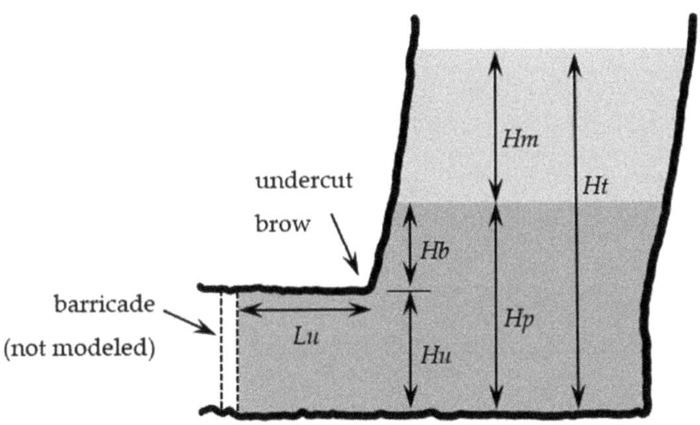

Figure 4. Idealized plug and main pour geometry and dimensions for developing an analytic solution.

In terms of barricade design, a worse-case loading scenario assumes the binder does not hydrate during the plug pour, in which case the barricade must safely resist the full slurry pressure head (i.e., γHb at the top of the barricade, increasing linearly to γHp at the base of the barricade). If this happens, or if the self-supporting strength has not yet developed, then the pouring must stop upon plug completion and a cure period be used to allow for CPB plug strength gain before starting the main pour. Pressure monitoring at the barricade and within the CPB plug are therefore essential requirements for making informed decisions to ensure pouring continuously can be performed safely.

The idealization of the CPB plug shown in Figure 5 incorporates these assumptions and shows the postulated failure mechanism through the plug, induced by the main pour. The out-of-plane extent of the failure mechanism is limited by the undercut width (assumed same as undercut height, Hu, although this could be generalized). Therefore, the failure zone above the brow (Zone 1 in Figure 5) has plan dimensions $Hu \times \sqrt{2}Hu$.

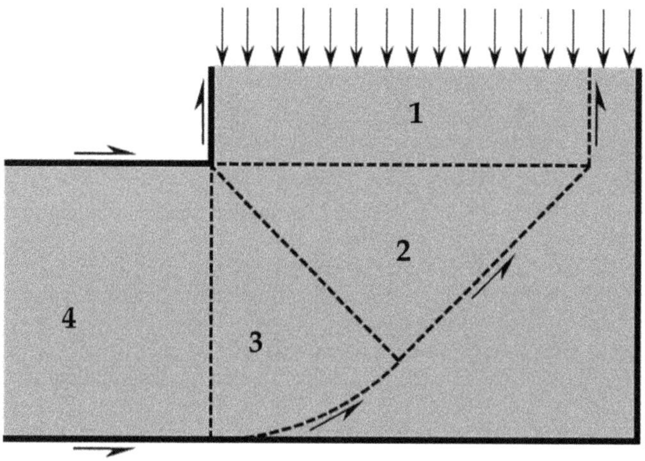

Figure 5. Simplified geometry and postulated plug failure mechanism defined as four discrete zones.

Note that if the distance from the undercut brow to the opposing stope wall is less than $\sqrt{2}Hu$, then the postulated failure mechanism cannot be fully developed. This would imply an increased plug stability, although the issue is not further investigated here. Zones 2 and 3 can be considered a $\frac{1}{2}$-Prandtl mechanism as used in analyzing shallow foundation bearing capacity for footings in undrained clay, for which the analysis when the out-of-plane footing dimension is finite follows Skempton's method [23].

Table 1 shows each zone's contribution to driving and resisting effects, with approximations noted (~). The $\frac{1}{2}$-Prandtl mechanism's self-weight driving effect assumes moment equilibrium about its rotation center, the brow. The $\frac{1}{2}$-Prandtl mechanism's resistance arises from Skempton's simplifications regarding theoretical bearing capacity factor $(2 + \pi) \approx 5$ and out-of-plane shape factor 1.2; thus, the bearing capacity factor becomes 6, and for a $\frac{1}{2}$-mechanism is 3.

Table 1. Driving and resisting effects for zones in Figure 5. Geometric parameters are defined in Figure 4.

Zone	Driving Effect	Resisting Effect
plug's top surface	γHm	n/a
above brow, Zone 1	γHb	~4 $c\ Hb/Hu$
$\frac{1}{2}$-Prandtl, Zones 2 and 3	~0.65 γHu	~3 c
undercut, Zone 4	n/a	4 $c\ Lu/Hu$

Some quantities approximated; see text.

Equation (1) expresses the above in limiting form: driving effects = resisting effects:

$$\gamma(Hm + Hb + 0.65Hu) = c\left(4\frac{Hb}{Hu} + 3 + 4\frac{Lu}{Hu}\right) \tag{1}$$

2.4. Numerical Validation/Calibration

Three-dimensional (3D) nonlinear analyses were carried out using Rocscience finite element software (RS3) to validate the postulated failure mechanism, and to calibrate the coefficients in Equation (1). Additional calibration details are given in [24]. Figure 6 shows a sample result on a cross section through mid-undercut, and the emerging $\frac{1}{2}$-Prandtl mechanism can be identified in the zones of high plastic shear strain. Note that a barricade is *not* included in the model, as previously explained. The critical pressure on the plug's top surface was evaluated by incrementally increasing this pressure and monitoring (i) the out-of-balance force; (ii) the displacement at mid-point on the free face of the material in the undercut; and (iii) the evolution of plastic shear strain. The result shown in Figure 6 is very close to the ultimate applied pressure, and on the next pressure increment the plastic strain fully develops in the $\frac{1}{2}$-Prandtl zones and instability ensues. This result implies there will be very little warning prior to an actual failure in an underground mine, which is consistent with observations from previous failures [3,4]. The 3D models also confirmed that the failure mechanism is limited in the out-of-plane direction by the lateral extents (i.e., width) of the undercut.

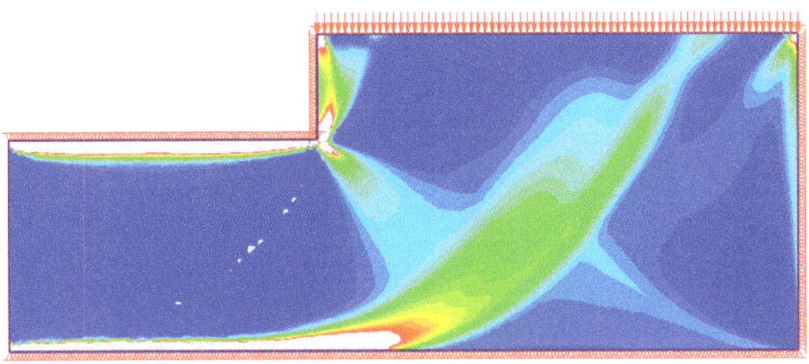

Figure 6. Sample failure mechanism from numerical analysis (plastic shear strain contours, dark blue being elastic, with increasing plastic strain magnitude to the red and white zones).

The coefficients in Equation (1) were tested using parametric studies to ensure agreement between numerical results and the predictive equation's results. For example, a model with no material above the brow, $Hb = 0$, was used with varying distances Lu, thereby determining that the $\frac{1}{2}$-Prandtl zone's self-weight effect should be 0.55 (not 0.65). Note that this self-weight term is irrelevant in the footing analysis for a full Prandtl mechanism, considered by Skempton, due to the mechanism's symmetry about the rotation point, i.e., the footing's edge which is analogous to the brow in the undercut. Also note that the mechanism in the material above the brow shown in Figure 6 is not entirely vertical as assumed in Figure 5, however, this did not significantly affect the coefficients used in Table 1. The remaining coefficients were deemed appropriate. Therefore, Equation (2) shows the calibrated design equation that will be used (in modified forms) for subsequent back-analyses of field data:

$$\gamma(Hm + Hb + 0.55Hu) = c\left(4\frac{Hb}{Hu} + 3 + 4\frac{Lu}{Hu}\right) \tag{2}$$

2.5. Non-Uniform Strength Effects

At the completion of the CPB plug pour, the backfill at the plug's top surface will still be fluid while layers deeper into the plug have had longer to cure and should have increasing strength with depth. This was tested using a model like that shown in Figure 6, except that the cohesion varied linearly from 1 kPa at the plug's top surface to a maximum value at the plug's bottom surface. The cohesion at height $\frac{1}{2}Hu$ was then used in a uniform strength model, with similar ultimate capacity predicted. Therefore, the reference cohesion value used in Equation (2) should be based on cohesion developed at $\frac{1}{2}Hu$. The implication of this is that the "cure time" (Figure 3) needs to be referenced with respect to the time when the backfill reached $\frac{1}{2}Hu$, and not with respect to when pouring started.

2.6. Recommended Analysis Procedure

The first analysis step determines if there is sufficient cohesion at elevation $\frac{1}{2}Hu$ at the end of the CPB plug pour to make the plug self-supporting, i.e., as if the barricade could be removed and the CPB plug would not fail (although any mining operation would be imprudent to venture removing the barricade at this point). Under these circumstances $Hm = 0$ and Equation (2) can be rearranged to obtain Equation (3) for the limiting self-supporting cohesion, c_{ss}, normalized as follows:

$$\frac{c_{ss}}{\gamma Hu} = \frac{0.55 + \frac{Hb}{Hu}}{3 + 4\frac{Hb}{Hu} + 4\frac{Lu}{Hu}} \tag{3}$$

When the main pour starts and as it continues, the CPB plug's strength must continue to increase to support the additional load on the plug's top surface arising from the main pour's pressure. The second analysis stage therefore considers evolving CPB plug strength and imposed plug surcharges. The analysis procedure involves computing the maximum main pour height Hm, max supportable for the available cohesion developed at $\frac{1}{2}Hu$. This can then be compared to the planned Hm, or actual Hm if used in a back-analysis. Rather than defining a conventional Strength Factor (e.g., available-cohesion/required-cohesion) it is recommended instead to use an "excess fill height capacity", $Hm, maxHm$, because techniques exist to determine Hm during the main pour. The time variable t is referenced with respect to start of plug pour. The CPB plug's cure time is $(t - t_{ref})$ where t_{ref} is when the plug pour reaches $\frac{1}{2}Hu$. The time-varying quantities of interest during the main pour are cohesion $c\left(t - t_{ref}\right)$, maximum supportable main pour height $Hm, max(t)$, and actual main pour height $Hm(t)$. $Hm, max(t)$ depends on cohesion developed greater than c_{ss}. Therefore, Equation (4) shows Equation (3) rearranged to incorporate this increasing strength effect.

$$Hm, max(t) = \frac{1}{\gamma}\left(c\left(t - t_{ref}\right) - c_{ss}\right)\left(3 + 4\frac{Hb}{Hu} + 4\frac{Lu}{Hu}\right) \tag{4}$$

The maximum total backfill height with time is then simply $Ht, max(t) = Hp + Hm, max(t)$. The next section uses Equation (3) in the case study back-analyses to determine each CPB plug's self-supporting condition, and then uses Equation (4) to assess ongoing CPB plug stability during the main pours.

3. Back-Analyses of Field Case Histories

The distinction between plug and main pours depends on the mine, with variations including (i) use a cure period after plug pour to allow for plug strength gain before main pour, and (ii) use higher binder content in the plug to achieve higher early strength and enable continuous pouring. It must be noted that each case study involved multiple measurement locations within the backfill and at the barricade, and the pressures were monitored in real time to ensure continuous pouring was safe.

3.1. Case History Analysis Parameters

Deciding appropriate Hu, Hb, Lu values requires judgment, since real mine geometries are never exactly as assumed in the preceding analysis. When in doubt, parametric studies are needed to determine the computed $Hm, max(t)$ sensitivity to assumed parameter values. [11,12] provide fuller information related to each case study, but Figure 7 illustrates the most problematic geometry of these, Cayeli Stope 685. This involved a continuous pour with binder transition at $Hp = 9.0$ m. By choosing $Lu = 1.8$ m and determining the overlying brow location (which is an approximation made from a cavity monitoring survey) it is estimated $Hu = 5.5$ m and $Hb = 3.5$ m for a combined $Hp = 9.0$ m. Table 2 gives the geometric parameters for all case studies.

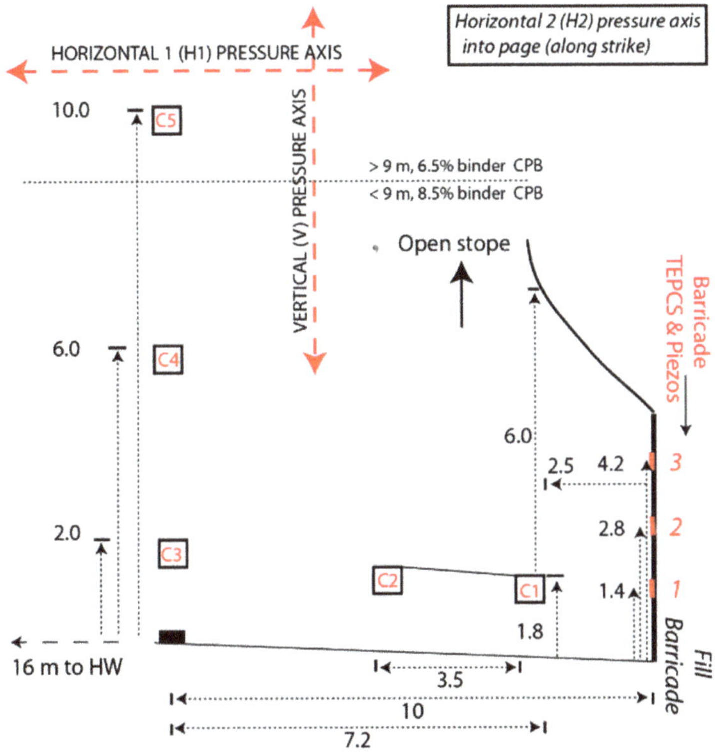

Figure 7. Undercut geometry at Cayeli Stope 685; dimensions in meters. From [11].

Table 2. Geometric parameters for all case studies (all dimensions in m, see Figure 4).

Mine/Stope	Lu	Hu	Hb	Hp	Hm_{final}	Ht_{final}
Cayeli 685	1.8	5.5	3.5	9.0	7.4	16.4
Cayeli 715	1.8	4.5	2.0	6.5	8.5	15.0
Kidd 67-SL1	6.75	4.4	2.0	6.4	25.6	32.0
Williams L70-5	9.3	4.5	1.5	6.0	49.0	55.0

The distinctions between the plug and main pours in the case studies are as follows: Cayeli 685, Kidd 67-SL1, and Williams L70-5 use higher binder content in the CPB plug and a continuous pour; Cayeli 715 uses the same binder throughout and featured a cure period after the CPB plug pour.

The analysis also requires backfill unit weight, backfill rise rate in the plug pour rr_{plug}, time t_{ref} when the backfill reaches height $\frac{1}{2}Hu$, time t_{plug} when the plug pour finishes, time t_{main} when the main pour starts (=t_{plug} if a continuous pour), backfill rise rate in the main pour rr_{main}, and time t_{final} when the stope is filled. Table 3 gives values for these parameters. The times referenced are based on field recorded filling data and are actual times, and the rise rates represent averages in the plug and main pours. According to the idealized geometry used, $t_{ref} = \frac{1}{2}Hu/rr_{plug}$, although this is an approximation because the actual undercut geometry varies from the idealization.

Table 3. Fill analysis parameters for all case studies, with values determined from [11,12].

Mine/Stope	γ	rr_{plug}	t_{ref}	t_{plug}	t_{main}	rr_{main}	t_{final}
	(kN/m^3)	**(m/h)**	**(h)**	**(h)**	**(h)**	**(m/h)**	**(h)**
Cayeli 685	21.4	0.23	9	39	39	0.24	70
Cayeli 715	22.4	0.33	8	20	93	0.39	115
Kidd 67-SL1	20.6	0.36	5	18	18	0.31	101
Williams L70-5	19.8	0.33	7	18	18	1.00	67

3.2. Self-Supporting CPB Plug Strength Requirements

In terms of material strength, only undrained cohesion can be relied upon under zero effective stress conditions. An assumption used in some backfill design methods is that the material's cohesion equals one-half the unconfined compressive strength, $c = \frac{1}{2}UCS$. However, this assumption can be confused with the undrained shear strength of unstructured clay which may be written $s_u = \frac{1}{2}UCS$, with the corresponding undrained friction angle $\phi_u = 0$. While unstructured clays do not dilate when sheared, CPB (which is predominantly silt) does dilate and so the $\phi_u = 0$ assumption is invalid for CPB; the failure envelopes in the drained and undrained states are virtually identical for CPB. This has important implications that are considered in the Discussion section, and an appropriate relationship between UCS, c, and ϕ is considered in this section. For the remainder of the article the terms cohesion (c) and friction (ϕ) are used with the understanding that the values for these parameters could be determined from drained tests, or from undrained tests with pore water pressure measurements, with identical results.

Mines inevitably assess strength using UCS, which is a strength parameter that depends on both cohesion and friction. [15] determined that for curing times up to 48 h, which is longer than all the plug pour times in Table 3, the friction angle ϕ varied between 35° and 37°. Equation (5) expresses the relation between these parameters, and for the given range of ϕ a reasonable approximation is $UCS = 4c$, as shown in Figure 8 for tests on Williams CPB. This provides a straightforward basis to convert the self-supporting cohesion from Equation (3) to a self-supporting unconfined compressive strength, $UCSss$.

$$UCS = \frac{2c\ \cos\phi}{1 - \sin\phi} \tag{5}$$

The dimensionless parameters in Equation (3) facilitate general consideration of CPB plug strength requirement for a range of undercut geometries, shown graphically in Figure 9. Note that this is the equation for limiting strength (Strength Factor = 1.0) for a self-supporting plug, and the strength must continue to increase to support the main pour as considered in subsequent analysis. Figure 9 considers the plausible extreme range of Hb/Hu and Lu/Hu and indicates the more usual design range for these parameters. The four case histories are also shown.

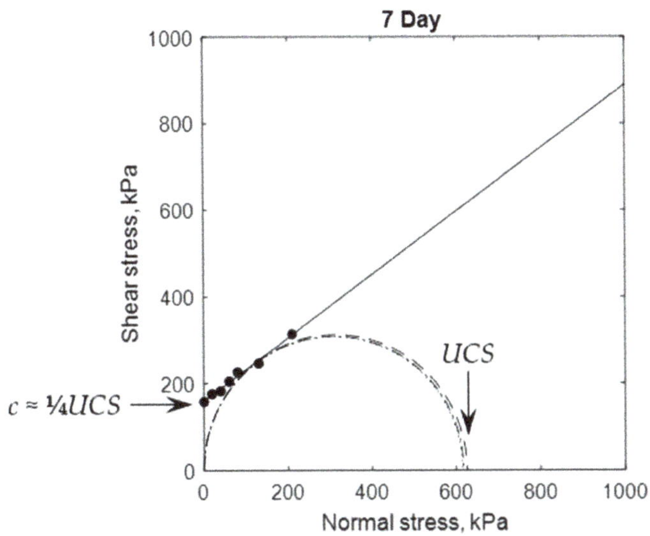

Figure 8. Relationship between cohesion, friction angle, and *UCS*, shown in Mohr's stress space. Dashed circles are from *UCS* tests, solid dot markers are from direct shear tests [25].

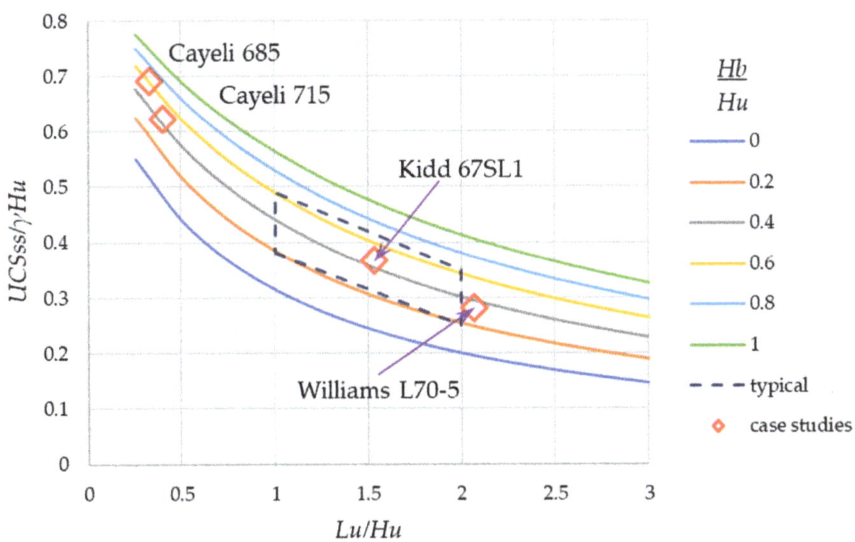

Figure 9. Self-supporting plug strength UCS_{ss} based on plug geometry and assuming $UCS = 4c$ ($\varnothing \approx 37°$). Contours based on Equation (3), mine's values from Table 2.

Table 4 shows available cohesion (interpolated from Figure 3) and the required self-supporting cohesion and corresponding UCS_{ss}. When comparing the available cohesion to the required self-supporting cohesion, only Cayeli 715 did not meet the indicated strength requirement at the end of the plug pour. However, as shown in Table 2 this was the only case study not involving a continuous pour. The difference between Cayeli 685 and 715 backfill performance is apparently due to differences in tailings mineralogy (Cayeli separates the tailings into "clastic" and "non-clastic" streams that are processed separately to maximize recovery). The delayed binder hydration reaction is reflected in the slow cohesion development shown in Figure 3. For sake of comparison, Table 4 also shows the Strength Factor at the end of the plug pours, defined as the ratio of available cohesion to that required for the self-supporting condition, $SF = c\left(t_{plug} - t_{ref}\right)/c_{SS}$. For the successful continuous pours, the required plug UCS ranges from 24 kPa (Williams) to 80 kPa (Cayeli 685). The implications of these strength requirements are considered in the Discussion section.

Table 4. Developed cohesion at elevation $\frac{1}{2}Hu$ at end of plug pours versus self-supporting strength requirements (cohesion and UCS). The Cayeli 715 case is italicized as the only considered non-continuous pour. c_{ss} values determined using Equation (3).

Mine/Stope	$t_{plug}-t_{ref}$	$c(t_{plug}-t_{ref})$	c_{ss}	SF	UCS_{ss}
	(h)	(kPa)	(kPa)		(kPa ϕ = 36°)
Cayeli 685	30	31	20	1.6	80
Cayeli 715	*12*	*5*	*16*	*0.3*	*64*
Kidd 67-SL1	13	20	8	2.5	32
Williams L70-5	11	14	6	2.3	24

3.3. Main Pour Back Analyses to Determine Continuous Pour Viability

Figure 10 shows maximum main pour heights computed using Equation (4) and actual main pour heights from the field monitoring data for Kidd and Cayeli 685 stopes; and Figure 11 shows the same for Williams and Cayeli 715 stopes. To use Equation (4) in calculating the supportable main pour heights, cohesion development with time data shown in Figure 3 was moved forward in time by t_{ref} (Table 3) to reflect the representative cohesion developed at $\frac{1}{2}Hu$ with respect to pour time. The linearization in the $Hm, max(t)$ curves reflect the frequency of cohesion measurements (see Figure 3); more frequent cohesion determinations would be desirable. The calculation sequence can start at $t = 0$, but results $Hm, max(t) < 0$ mean the plug has yet to achieve self-supporting strength.

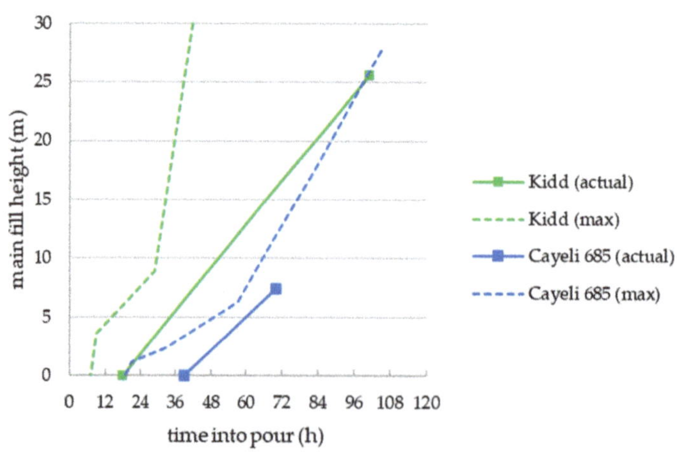

Figure 10. Main pour analysis for Kidd and Cayeli 685 stopes. *Hm, max(t)* values determined using Equation (4).

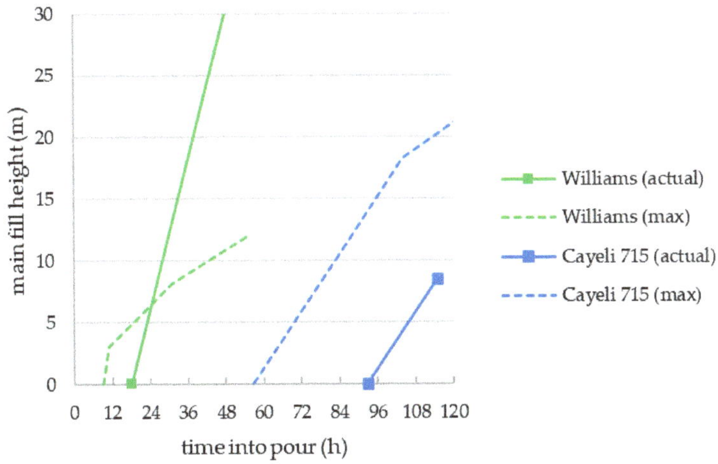

Figure 11. Main pour analysis for Williams and Cayeli 715 stopes. *Hm, max*(*t*) values determined using Equation (4).

Figure 10 shows the successful continuous pours at Kidd and Cayeli 685 stopes. At Kidd, the as-placed main pour height was always less than the maximum supportable height by at least 5 m. The sudden increase in capacity at 29 h is due to the cohesion increase seen in Figure 3 at 24 h (the time difference being t_{ref} to reflect cohesion acting at $\frac{1}{2}Hu$). In contrast, the analysis for Cayeli 685 suggests at 57 h there was only ~1.5 m extra main fill capacity, however by this time, close to the end of pour, effective stresses developed within the CPB plug (see Figure 2 for vertical effective stress; full stress results are given in (Thompson, et al., 2012)). These effective stresses mobilize the frictional strength components which contribute significantly to CPB plug stability.

Figure 11 shows the Cayeli 715 main fill height analysis with the actual fill height reflecting the three-day CPB plug cure period routinely used at the mine for this backfill type. However, the analysis in Figure 11 indicates the main pour could have safely started at least one shift (12 h) earlier based on the limiting strength having been reached before 60 h. Figure 10 also shows the Williams pour and the analysis indicates that CPB plug instability should be expected ~6 h after the start of the main pour (~t = 24 h). However, there are two factors mitigating potential failure. Firstly, the undercut's height was 4.5 m and the distance from its brow to the opposite stope wall was about 6 m (see [12] for detailed geometry). Compared to the mechanism shown in Figures 5 and 6, the proximity of the opposite wall could interfere with development of the assumed failure mechanism. The other cases studied had significantly larger distances to opposing stope walls. Secondly, and likely more significantly, Williams CPB was the fastest of the cases studied to develop effective stresses, which started ~3 h after CPB reached the transducers in the CPB plug location (which was about 3 h into the pour (not shown)), and Figure 2 shows that at 21 h (24 h into the pour) there was about 68 kPa vertical effective stress. At 24 h pour time, or 17 h cure time with respect to the reference location, there was already ~16 kPa cohesion, equivalent to $UCS = 64$ kPa, and even greater vertical effective stress could have been supported by the CPB plug with the developing effective horizontal stresses. Therefore, CPB plug strength was developing fast enough to withstand the effective stresses applied to the plug backfill. To understand backfill pour stability with fast effective stress development requires much more sophisticated analyses, such as the techniques demonstrated in [9,14,15,26–30]. These sophisticated analysis techniques are not warranted for routine designs.

3.4. Final CPB Plug Strength Requirement

The analysis presented herein can also be applied to calculate the final CPB plug strength required at the completion of backfilling, UCS_{final}. To determine UCS_{final} for a final main fill height Hm_{final} the corresponding required cohesion is found by rearranging Equation (2) to obtain:

$$c_{final} = \frac{\gamma\left(Hm_{final} + Hb + 0.55Hu\right)}{4\frac{Hb}{Hu} + 3 + 4\frac{Lu}{Hu}} \quad (6)$$

Finding the corresponding UCS at these later cure times may need to account for a changing friction angle. For Williams CPB [15] found that the friction angle remained ~37° throughout curing up to 28 days, and so $UCS = 4c$. However, the Cayeli 715 CPB had a progressively reducing friction angle to ~28° by 7 days, the Cayeli 685 CPB's friction angle progressively decreased to about ~27° by four days, and the Kidd CPB's friction angle started decreasing after 4 days and fell to ~26° by seven days. From Equation (5), $UCS = 3.25c$ when the friction angle is ~27°. Therefore, for these other mines the equivalent UCS_{final} is between $3.25c_{final}$ and $4c_{final}$. Table 5 shows the ranges of these required strengths along with the equivalent cure time at which these strengths would have to be achieved.

Table 5. CPB plug strength requirements at end of backfilling (Strength Factor = 1). The equivalent curing time ($t_{final} - t_{ref}$) for Cayeli 715 includes the plug cure period and would be about 36 h less if the pour had been continuous. c_{final} values determined using Equation (6).

Mine/Stope	Hm_{final}	c_{final}	UCS_{final}	UCS_{final}	$t_{final}-t_{ref}$
	(m)	(kPa)	(kPa, ϕ = 27°)	(kPa, ϕ = 36°)	(h)
Cayeli 685	7.4	43.5	141	174	61
Cayeli 715	8.5	45.6	148	182	107
Kidd 67-SL1	25.6	56.5	184	226	96
Williams L70-5	49.0	83.2	n/a	333	60

Note that the strengths shown in Table 5 assume no effective stress development in the CPB plug, but this assumption is increasingly tenuous with longer cure times (see Figure 2) and so these unfactored strength estimates are conservative from a safety perspective. This is particularly the case for Williams, as discussed in the previous section, and the assessed strength 333 kPa is unrealistically high for that operation. Field monitoring in the CPB plug is required to demonstrate when, and the extent to which effective stresses develop for a given mining operation. For mines embarking on stress measurement campaigns with the view to achieving continuous pours, a starting point for consideration (assuming site specific assessment) would require the CPB plug to achieve an appropriately factored UCS_{final} before starting the main pour (i.e., to ensure stability throughout the main pour), and the mine could then work towards achieving a plug strength UCS_{ss} with a reasonable strength factor to facilitate continuous pouring.

4. Discussion: Application at Other Mine Sites

Notwithstanding the apparent agreement between the proposed analysis methods and results from the four cases studied, the potential consequences of a plug failure mean that future applications of the presented framework must be approached cautiously. This section discusses some requirements in applied design.

Many mines collect UCS data at 3, 7, 14, 28 days (sometimes less frequently, and sometimes for longer cure periods if cement extenders such as slag and fly ash are used, or if there is concern about long-term strength deterioration due to sulfate attack, for example). However, the cases studied demonstrate that the critical time for continuous pour analysis is before three days. Furthermore, cohesion cannot be determined from

UCS unless ϕ is also known (or assumed). Therefore, mines interested in performing rational continuous pour analysis must consider alternate strength testing programs.

Figure 9 shows that a single empirically derived CPB plug strength is not universally applicable. More importantly, it demonstrates the sensitivity of required strength to (i) the setback from the undercut's brow to the barricade, Lu; and (ii) to the height of plug backfill above the brow Hb. It also shows how the relationships between these parameters scale with the undercut's absolute height, Hu. Of particular note, the required strength is sensitive to Lu: many operations want to minimize Lu so that they do not have to subsequently mine out the paste in the undercut when developing the adjacent stope, but this has consequences for increasing binder content (and strength development with time) if a continuous pour is to be achieved.

No empirically derived CPB plug strength is widely accepted, but one well-cited design handbook recommends the following:

> *"In bulk mining stopes, the paste fill should be placed approximately 1 m above the draw point brow and permitted to cure to approximately 150 kPa strength before filling the remainder of the stope."*
>
> ([2], p.92)

Note that this recommendation does not indicate *where* in the plug this strength should be achieved, nor does it account for the effects of the main pour height, or different main pour rise-rates on the CPB plug. Given that the 150 kPa strength is a general recommendation, it is presumably empirically validated under normal operating conditions and deemed to be safely conservative, although to an unknown extent. As an aside, note that the suggested 1 m height above the brow is significantly less than the corresponding Hb values shown in Table 2.

For comparison with the design approach for UCS_{ss} suggested here, consider the UCS = 150 kPa heuristic in the context of Figure 9. If a reasonably conservative Strength Factor = 2.0 is assumed so that the limiting UCS_{ss} is 75 kPa, and assuming a large Hu = 7.5 m and a typical $\gamma = 20$ kN/m^3, then $UCS_{ss}/\gamma Hu = 0.5$ and this corresponds to a range of Lu/Hu from 0.4 to 1.3 (see Figure 9), which many operations would consider unsafely close to the brow. (For reference, note that Kidd's $Lu/Hu = 1.5$ is achieved using a frame that remotely pushes the barricade framework into place prior to shotcreting.) Most operations have more conservative operating parameters (smaller Hu, larger Lu/Hu) and so would have Strength Factors > 2.0 if they strictly adhered to the recommended heuristic. On the other hand, in comparison with the UCS_{final} values shown in Table 5, the recommended UCS = 150 kPa strength before starting the main pour falls within the range of calculated UCS_{final} values.

As mentioned, mines wishing to use the analysis approach described in this work need to carry out backfill strength testing at frequent intervals in the first few days representing the duration of CPB plug pour and start of main pour. These tests must focus on determining cohesion, either using direct shear tests combined with UCS as shown in Figure 8 (the preferred method), or by combining UCS and confined triaxial tests with confining stresses up to $\frac{1}{2}$ UCS.

In addition, mines should determine typical effective stress development rates by carrying out field stress measurements prior to attempting continuous pours. This instrumentation should include PWP and vertical TEPC transducers at minimum, but preferably two orthogonal horizontal TEPCs as well, since this combination provides more reliable interpretation of effective stress development [11,12]. The instrumentation cage shown in Figure 1 is robust and straightforward for underground construction crews to install, as demonstrated by [13]. The transducers have internal thermistors to capture TEMP data that is helpful to ascertain when binder hydration becomes significant at the measurement point. The mid-point at the base of the stope, ~1.5 m off the stope floor is an ideal measurement point since this is where a potential failure mechanism would be expected to form (Figure 6). If potential rock fall prior to stope filling is a concern, then the cage can also be placed at the undercut's brow although its proximity to blasted host rock provides better drainage potential and so the PWPs may be lower than at mid-stope and, therefore, the interpreted effective stresses higher, which could lead to unsafe estimates of plug stability. When continuous pouring attempts begin, the same field monitoring strategies can be used to verify that the backfill behaviour corresponds to the analysis assumptions. While not strictly part of the CPB plug design, barricade monitoring is also strongly recommended to ensure safe barricade loading limits are adhered to throughout the pour. It should be cautioned that barricade pressure measurements of effective stress likely are not representative of the effective stress state within the plug; [11] reported that effective stress at the barricade developed significantly more quickly than at the brow (C1 location, Figure 7). As such, barricade pressure measurements alone are not recommended for verifying the conditions of this analysis.

The approach considered here only addresses plug stability under pressures arising from the main pour. Other design considerations may also need to be addressed. One of these is liquefaction resistance, for which an often-referred to heuristic is $UCS = 100$ kPa [31]. However, [32] found that after 12 h the Golden Giant CPB (similar to Williams CPB, because both operations mine the same orebody) could not be liquefied using the highest cyclic loading possible from the available MTS load frame. Similar results have been found by [33] for Kidd tailings after 7 h binder cure time. As mentioned, the barricade should be designed to withstand the full slurried CPB pressure during the plug pour, and in 12 h only the top 3 m would be prone to liquefaction (assuming deposition rate 0.33 m/h), so the liquefaction concern is overstated in this specific design case.

Finally, the strength assumption stated in Section 3.2 requires discussion in the context of the proposed plug strength analysis method. The common undrained strength assumption, that $\phi = 0$ and therefore the undrained cohesion is $\frac{1}{2}UCS$, only applies to materials that do not dilate during shearing (i.e., unstructured clays). [34] used miniature tensiometers installed in the base platen of a triaxial cell to monitor a CPB's pore water pressure and suction changes during UCS tests, as well as in non-backpressure saturated consolidated-undrained (CU) tests. The test results showed how dilation potential during shear induced more than 60 kPa incremental suction and therefore increased the effective stress, leading to measured shear resistance larger than for the assumed confining stress level. [35] showed the same effects in conventional backpressure saturated undrained (CU) and drained (CD) tests on Williams materials, results for which are shown in Figure 12. Further, [35] demonstrated that the failure envelopes determined from drained and undrained testing were virtually identical (angles within about 1°) for unamended tailings (shown in Figure 12) as well as for the mine's CPB. Similar results were obtained by [33] for tests on Kidd mine materials including (i) mine tailings alone, (ii) tailings blended with esker sand, and (iii) the mine's CPB (tailings, sand, and binder).

Some authors [36–38] have tested CPB under undrained, but non-backpressure saturated conditions and without using pore water pressure measurements. In these cases, the induced suction during shearing at low confining stress levels leads to increases in effective confining pressure (to an unmeasured degree) and, therefore, artificially high interpretations of mobilized shear resistance, as shown in [34]. Studies on Kidd tailings [33] and Williams tailings [35] found backpressures of at least 400 kPa were required for saturation. Therefore, tests by [36–38] at higher confining pressure levels would result in closer to saturated conditions. However, as shown in Figure 12, the mobilized shear resistance in undrained testing is not as high as in drained testing. The combined result of over-interpreted shear resistance at low confining stresses and under-interpreted shear resistance at high confining stresses leads to lower estimates of friction angle and higher estimates of cohesion, potentially resulting in the interpreted friction angle being zero and undrained cohesion being $\frac{1}{2}UCS$. An extreme example is shown by some of the test results using Kidd CPB in [38] where stress deviator decreases with increasing confining stress, implying a negative friction angle in total stress space. This result is inconsistent with results of [33] which used backpressure saturation and pore water pressure measurements when testing the same Kidd CPB. The authors in [36] note that results from tests without backpressure saturation and pore water pressure measurements must be used carefully in design and could lead to overestimating backfill strength in some cases. Estimating plug strength requirements is such a case, where using $c = \frac{1}{2}UCS$ would reduce the calculated strength (UCS) requirement by a factor of two. If the recently placed backfill remains close to saturated, as was determined from field tests by [21,22], then $c = \frac{1}{4}UCS$ must be used. It is therefore recommended that the assumption $c = \frac{1}{4}UCS$ continue to be used until it can be demonstrated that greater estimates of cohesion are warranted.

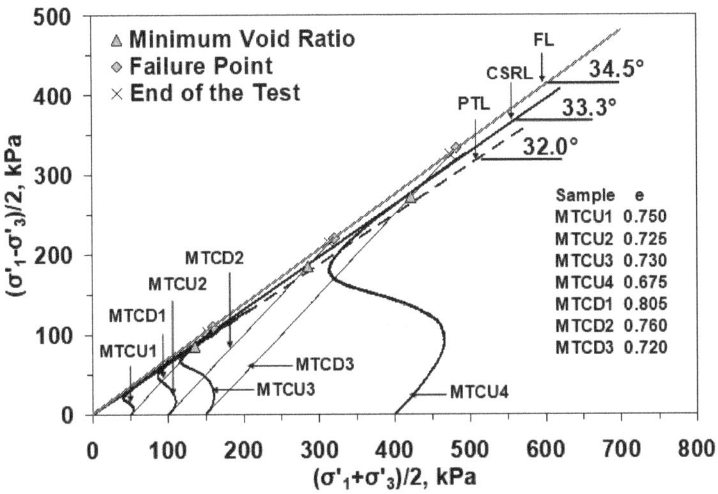

Figure 12. Triaxial test effective stress paths for Williams Mine Tailings in undrained (CU) testing (MTCUn, curved stress paths) and drained (CD) testing (MTCDn, linear stress paths). FL is failure line for drained tests. CSRL is constant stress ratio line for undrained tests (interpreted as FL in the undrained condition). PTL is the phase transition line for undrained tests (PT, phase transition, is the point in undrained testing where incremental pore water pressure changes from positive to negative and corresponds to the onset of dilation in drained tests). Figure from [35].

5. Conclusions

The proposed CPB plug stability assessment method is the first attempt to provide a framework for rational engineering analysis of CPB strength required for continuous pouring. The case studies used to compare against the developed approach's predictions are unique in terms of (i) the extent and quality of field monitoring data available, and (ii) the frequency and type of laboratory strength testing carried out as part of a larger previous research program at these mine sites. However, mines have the capacity to carry out field instrumentation of the kind illustrated in the case studies, albeit in a focused way at select stope locations. Consulting companies should generally have the equipment for the needed strength testing procedures, and with further research some improvement and simplification of these test methods may result in equipment that also can be used routinely at mine sites along with slump/viscometer testing, UCS testing, etc.

The proposed analysis method correctly assessed the continuous pour potential for Kidd and Cayeli 685 stopes. It also correctly assessed that the Cayeli 715 stope required a cure period after completing the CPB plug pour, and the results indicated that the CPB plug cure duration used by the mine was likely conservative and could have been shortened by at least one shift (12 h). For Williams mine the analysis results correctly indicated the potential for continuous pour, but then incorrectly indicated that the CPB plug strength was insufficient ~6 h into the main pour. However, in this case the narrow stope geometry relative to the undercut height, and the quickly developing CPB strength relative to the developing effective stress imposed by self-weight helped to mitigate against failure, and a continuous pour was successful. It will be important in future to determine similar mining scenarios where the CPB plug strength analysis is overly-conservative from a strength perspective.

The self-supporting CPB plug strength assessment (Equation (3), Figure 9) suggests that the recommended $UCS = 150$ kPa heuristic likely over-estimates required CPB self-supporting plug strength (in the context of a continuous pour) for most backfilling cases, even if a reasonable Strength Factor is applied to the analysis. The backfilling cases studied indicate that if the self-supporting strength has been achieved then the remainder of the pour is likely to be successful, although this cannot be assured, and Equation (4) should still be used as a check. The final case study, Williams, shows the limitations of the analysis approach when the stope geometry infringes on development of the potential failure surface, or when effective stress development occurs. Even in these cases, however, the initial CPB plug strength assessment (Equation (3), Figure 9) will give a reasonable upper-bound strength estimate that in many cases will be lower than $UCS = 150$ kPa.

If the suggested analysis and design approach is used at other mines that are willing to share their experience, the rational engineering approach recommended here can be improved and confidence will increase in carrying out continuous pours of cemented paste backfill under properly controlled conditions.

Author Contributions: All authors contributed to the iterative refinement of postulated failure mechanism and to recommended design procedures for other mines. M.G. developed the analytic solution and oversaw numerical calibration. W.B. contributed to contextualization of results in an operating mine environment. B.T. oversaw the original field instrumentation at Williams, Kidd, and Cayeli mines and contributed to field data processing and interpretation. M.G. provided first manuscript draft and colleagues contributed to reviewing and editing. All authors have read and agreed to the published version of the manuscript.

Funding: This original work received no external funding, however it arose out of a multi-year research project originated by the authors at the University of Toronto involving partnership with Williams, Kidd, and Cayeli mines (and their parent companies at the time, Barrick, Xstrata, and Inmet) and the Natural Sciences and Engineering Research Council (NSERC) Canada.

Institutional Review Board Statement: Not applicable.

Informed Consent Statement: Not applicable.

Data Availability Statement: All data supporting reported results has been included.

Acknowledgments: A preliminary version of this work appears as a conference paper at Minefill 2020 (postponed to the last week of May 2021). Flac3D models presented in that work were carried out by Jeff Oke while at Mine Design Engineering (MDE) in Kingston, Ontario, under supervision of the first author. Compared to the conference paper, the work here provides additional case studies and a more practical recommended analysis and design approach.

Conflicts of Interest: The authors declare no conflict of interest.

References

1. Li, J.; Ferreira, J.; Le Lievre, T. Transition from discontinuous to continuous paste filling at Cannington Mine. In Proceedings of the 11th International Symposium on Mining with Backfill, Perth, Australia, 20–22 May 2014; Australian Center for Geomechanics: Perth, Australia, 2014.
2. Potvin, Y.; Thomas, E.; Fourie, A. *Handbook on Mine Fill*; Australian Centre for Geomechanics: Perth, Australia, 2005; ISBN 0-9756756-2-1.
3. Yumlu, M.; Guresci, M. Paste backfill bulkhead failures and pressure monitoring at Cayeli mine. *CIM Bull.* **2007**, *100*, 1001–1010.
4. Revell, M.; Sainsbury, D. Paste bulkhead failures. In Proceedings of the 9th International Symposium on Mining with Backfill, Montreal, QC, Canada, 29 April–2 May 2007; Canadian Institute of Mining: Montreal, QC, Canada, 2007; p. 10.
5. Starfield, A.; Cundall, P. Towards a methodology for rock mechanics modelling. *Int. J. Rock Mech. Min. Sci. Geomecahnics Abstr.* **1988**, *25*, 99–106. [CrossRef]

6. Hassani, F.; Fotoohi, K.; Doucet, C. Instrumentation and backfill performance in a narrow vein gold mine. *Int. J. Rock Mech. Min. Sci.* **1998**, *35*, 106. [CrossRef]
7. Hassani, F.; Ouellet, J.; Servant, S. In situ measurements in a paste backfill: Backfill and rock mass response in the context of rockbursts. In Proceedings of the 17th International Mining Congress and Exhibition of Turkey, IMCET 2001, Ankara, Turkey, 19–22 June 2001; Chamber of Mining Engineers of Turkey: Ankara, Turkey, 2001.
8. Belem, T.; Harvey, A.; Simon, R.; Aubertin, M. Measurement and prediction of internal stresses in an underground opening during its filling with cemented fill. In Proceedings of the 5th International Symposium on Ground Support in Mining and Underground Construction, Perth, Australia, 28–30 September 2004; Taylor & Francis Group: London, UK, 2004; pp. 619–630.
9. Doherty, J.; Hasan, A.; Suazo, G.; Fourie, A. Investigation of some controllable factors that impact the stress state in cemented paste backfill. *Can. Geotech. J.* **2015**, *52*, 1901–1912. [CrossRef]
10. Seymour, J.B.; Raffaldi, M.J.; Abraham, H.; Johnson, J.C.; Zahl, E.G. Monitoring the in situ performance of cemented paste backfill at the Lucky Friday mine. In Proceedings of the 12th International Symposium on Mining with Backfill, Minefill 2017, Denver, CO, USA, 19–22 February 2017; Society for Mining, Metallurgy, and Exploration, Inc.: Englewood, CO, USA, 2017; pp. 19–22.
11. Thompson, B.; Bawden, W.F.; Grabinsky, M. In situ measurements of cemented paste backfill at the Cayeli Mine. *Can. Geotech. J.* **2012**, *49*, 755–772. [CrossRef]
12. Thompson, B.; Grabinsky, M.; Bawden, W. In situ monitoring of cemented paste backfill pressure to increase backfilling efficiency. *CIM J.* **2011**, *2*, 199–209.
13. Alcott, J.; Dallaire, D.; Belem, T. Pastefill optimization at Hecla Quebec's Casa Berardi mine. In Proceedings of the 53rd US Rock Mechanics/Geomechanics Symposium, New York, NY, USA, 23–26 June 2019.
14. Helinski, M.; Fahey, M.; Fourie, A. Numerical modeling of cemented mine backfill deposition. *J. Geotech. Geoenviron. Eng.* **2007**, *133*, 1308–1319. [CrossRef]
15. Veenstra, R.L. A Design Procedure for Determining the In Situ Stresses of Early Age Cemented Paste Backfill. Ph.D. Thesis, University of Toronto, Toronto, ON, Canada, 2013. Available online: http://hdl.handle.net/1807/36027 (accessed on 19 December 2018).
16. Yilmaz, E.; Belem, T.; Benzaazoua, M. Effects of curing and stress conditions on hydromechanical, geotechnical and geochemical properties of cemented paste backfill. *Eng. Geol.* **2014**, *168*, 23–37. [CrossRef]
17. Yilmaz, E.; Benzaazoua, M.; Belem, T.; Bussière, B. Effect of curing under pressure on compressive strength development of cemented paste backfill. *Miner. Eng.* **2009**, *22*, 772–785. [CrossRef]
18. Fahey, M.; Helinski, M.; Fourie, A. Development of specimen curing procedures that account for the influence of effective stress during curing on the strength of cemented mine backfill. *Geotech. Geol. Eng.* **2011**, *29*, 709–723. [CrossRef]
19. Walske, M.; McWilliam, H.; Doherty, J.; Fourie, A. Influence of curing temperature and stress conditions on mechanical properties of cementing paste backfill. *Can. Geotech. J.* **2015**, *53*, 148–161. [CrossRef]
20. Cui, L.; Fall, M. Mechanical and thermal properties of cemented tailings materials at early ages: Influence of initial temperature, curing stress and drainage conditions. *Constr. Build. Mater.* **2016**, *125*, 553–563. [CrossRef]
21. Grabinsky, M.; Bawden, W.F.; Simon, D.; Thompson, B.T.; Veenstra, R.L. In situ properties of Cemented Paste Backfill from three mines. In Proceedings of the 66th Canadian Geotechnical Conference, GeoMontreal 2013, Montreal, QC, Canada, 29 September–3 October 2013; p. 8.
22. Grabinsky, M.; Simon, D.; Thompson, B.D.; Bawden, W.F.; Veenstra, R.L. Interpretation of as-placed cemented paste backfill properties from three mines. In Proceedings of the Mine Fill 2014, Perth, Australia, 20–22 May 2014; Australian Centre for Geomechanics: Perth, Australia, 2014. Available online: https://papers.acg.uwa.edu.au/p/1404_28_Grabinsky/ (accessed on 20 March 2020).
23. Skempton, A.W. The bearing capacity of clays. In Proceedings of the Building Research Congress, London, UK, 11–20 September 1951; Institution of Civil Engineers: London, UK, 1951; pp. 180–189. [CrossRef]
24. Grabinsky, M.; Thompson, B.; Bawden, W. *Evaluating Cemented Paste Backfill Plug Strength and the Potential for Continuous Pouring*; Minefill 2020–2021; Taylor & Francis OA: Katowice, Poland, 2021.
25. Pan, A. Mechanical Properties of Cemented Paste Backfill under Low Confining Stress. Master's Thesis, University of Toronto, Toronto, ON, Canada, 2019. Available online: http://hdl.handle.net/1807/98310 (accessed on 9 April 2021).
26. Helinski, M.; Fahey, M.; Fourie, A. Behavior of cemented paste backfill in two mine stopes: Measurements and modeling. *J. Geotech. Geoenviron. Eng.* **2010**, *137*, 171–182. [CrossRef]
27. El Mkadmi, N.; Aubertin, M.; Li, L. Effect of drainage and sequential filling on the behavior of backfill in mine stopes. *Can. Geotech. J.* **2013**, *51*, 1–15. [CrossRef]
28. Cui, L.; Fall, M. Multiphysics modeling of arching effects in fill mass. *Comput. Geotech.* **2017**, *83*, 114–131. [CrossRef]
29. Witteman, M.; Simms, P. Unsaturated flow in hydrating porous media with application to cemented mine backfill. *Can. Geotech. J.* **2017**, *54*, 835–845. [CrossRef]
30. Shahsavari Goughari, M. Effects of Transient Hydro-mechanical Cemented Paste Tailings Properties on One-dimensional Deposition Behaviour. Ph.D. Thesis, University of Toronto, Toronto, ON, Canada, 2020. Available online: http://hdl.handle.net/1807/103748 (accessed on 9 April 2021).

31. Le Roux, K.; Bawden, W.; Grabinsky, M. Liquefaction analysis of early age Cemented Paste Backfill. In Proceedings of the Eighth International Symposia on Mining with Backfill, Minefill 2004, Beijing, China, 19–21 September 2004; The Nonferrous Metals Society of China: Beijing, China, 2004; pp. 233–241.
32. Le Roux, K. In Situ Properties and Liquefaction Potential of Cemented Paste Backfill. Ph.D. Thesis, University of Toronto, Toronto, ON, Canada, 2004.
33. Abdelaal, A.G. Early Age Mechanical Behavior and Stiffness Development of Cemented Paste Backfill with Sand. Ph.D. Thesis, University of Toronto, Toronto, ON, Canada, 2011. Available online: http://hdl.handle.net/1807/31668 (accessed on 9 April 2021).
34. Simms, P.; Grabinsky, M. Direct measurement of matric suction in triaxial tests on early-age cemented paste backfill. *Can. Geotech. J.* **2009**, *46*, 93–101. [CrossRef]
35. Saebimoghaddam, A. Liquefaction of early age cemented paste backfill. Ph.D. Thesis, University of Toronto, Toronto, ON, Canada, 2010. Available online: http://hdl.handle.net/1807/24868 (accessed on 13 March 2020).
36. Belem, T.; Fourie, A.B.; Fahey, M. Time-dependent failure criterion for cemented paste backfills. In Proceedings of the 13th International Seminar on Paste and Thickened Tailings, Paste 2010, Toronto, ON, Canada, 3–6 May 2010; Australian Center for Geomechanics: Perth, Australia, 2010; pp. 147–162. [CrossRef]
37. Rankine, R.M.; Sivakugan, N. Geotechnical properties of cemented paste backfill from Cannington Mine, Australia. *Geotech. Geol. Eng.* **2007**, *25*, 383–393. [CrossRef]
38. De Araujo, E.E.B.; Simon, D.; de Franca, F.A.N.; de Freitas Neto, O.; dos Santos Junior, O.F. Shear strength of a Cemented Paste Backfill submitted to high confining pressure. *Appl. Mech. Mater.* **2017**, *858*, 219–224. [CrossRef]

mining

MDPI

Article

A Statistically Based Methodology to Estimate the Probability of Encountering Rock Blocks When Tunneling in Heterogeneous Ground

Maria Lia Napoli [1,*], Monica Barbero [1] and Roberto Fontana [2]

[1] Department of Structural, Geotechnical and Building Engineering, Politecnico di Torino, C.so Duca degli Abruzzi 24, 10124 Torino, Italy; monica.barbero@polito.it

[2] Department of Mathematical Sciences, Politecnico di Torino, C.so Duca degli Abruzzi 24, 10124 Torino, Italy; roberto.fontana@polito.it

* Correspondence: maria.napoli@polito.it

Abstract: Strong rock blocks embedded in a weaker soil matrix are found in many geological units. When tunneling in ground containing cobbles and boulders, extremely challenging conditions can be encountered. Such inconveniences may be avoided by means of appropriate tunneling methods and cutterhead designs, which require the content, frequency, and size of rock blocks to be predicted as accurately as possible. Several approaches have been developed to estimate the block fraction of heterogeneous geomaterials for excavation. However, the estimation of cobble–boulder quantities both all along the tunnel and only partially embedded within the tunnel face remains a critical issue. This study develops a methodology for the estimation of the probability of encountering blocks partially or totally contained within the tunnel excavation area, wherein the area of intersection with the tunnel face is greater than the given critical values. For this purpose, a statistical approach has been implemented in a Matlab routine. The potential of this code is that it provides extremely useful and statistically based information that can be used for making a more rational choice regarding tunneling technique and in terms of designing a suitable cutterhead in order to avoid technical problems during tunnel excavations in heterogeneous ground. The executable code is provided.

Keywords: block-in-matrix; heterogeneous ground; tunneling; cutterhead design; statistical simulation; executable code

1. Introduction

Heterogeneous geological formations, such as glacial tills, conglomerates, flysches, breccias, melanges, alluviums, and talus deposits, are widespread all over the world and are characterized by strong rock blocks of variable dimensions embedded in a weaker soil matrix [1–4]. Due to their considerable spatial, lithological, and mechanical variability, the characterization, design, and construction concerning these geomaterials are extremely challenging tasks. Since the 1990s, by performing many laboratory and in situ tests, as well as numerical analyses, significant advancements have been made in understanding how these heterogeneous formations behave [5–11]. However, improvements are still needed, mainly to reduce the geotechnical and construction risks associated with the presence of rock inclusions in the different engineering works.

Unexpected and expensive difficulties can arise when tunneling in these geomaterials due to the mixed-face conditions [4,12]. Cobbles and boulders can cause technical problems at the heading and excavation chamber, and of the mucking system of a TBM if cutting tools are not capable of comminuting large blocks to small cobble or gravel sizes [13–15]. In fact, the content, distribution, dimension, lithology, abrasivity, and strength of cobbles and boulders can induce, among other problems, face instabilities, extraordinary high strains and stresses on the tunnel lining, obstructions or damage to cutter housings, and more rapid wear of cutters, with consequent schedule delays and costs increments [16–21].

The risks associated with tunneling in these heterogeneous geological formations mainly depend on the cobble–boulder characteristics but also on the site conditions (i.e., tunnel depth, surface access constraints, settlement limits, etc.), the size and type of the machine (i.e., TBM with or without face access), and the strength and stiffness of the matrix. Different strengths and stiffnesses of the matrix can significantly influence the effectiveness of the cutter type and the breaking mechanism when the cutter penetrates the block (i.e., classical chip formation, boulder plucking by adhesive bond or matrix-bearing capacity failure through going boulder fracturing) [13,14,22,23].

In order to limit the adverse impact of the cobbles and boulders on tunneling, an appropriate TBM type and an adequate cutterhead design (i.e., opening size, shape, etc.) are necessary. This requires careful geologic and subsurface investigations to predict the rock block content, lithology, frequency, and size. Large diameter borings, boulder volume surveys, geophysical methods, geological maps, test pits, and excavations may help in assessing boulder and cobble quantities, as well as their characteristics [2,22,24,25]. Boulder and cobble volume ratios (i.e., the total volume of rock blocks divided by the total excavation volume) may also be evaluated by means of semi-empirical correlations of geologic and volumetric data from previous excavation works in nearby areas (with the same geologic settings). Fractal block-size distributions have been observed by several authors for fractured rock masses, melange formations, and similar rock–soil mixtures containing a few large blocks and a greater number of smaller rock inclusions [26–28]. A few probabilistic methods have also been proposed [24,29,30]. Frank and Chapman (2005) developed a mathematical method for the prediction of rock block quantities and frequencies, which characterize the soil to be excavated. In particular, an exponential distribution of the form $N = C/V^d$ was proposed, with N representing the number of clasts of a given size, C a constant dependent on the sample size data, V the size of the blocks being counted, and d a constant correlated with the clast size distribution. Napoli et al. [30] developed a 3D statistical approach to provide an uncertainty factor to adjust the estimated block quantity as a function of the size of the outcrop area investigated. These methods represent extremely valid and useful tools for the estimation of boulder and cobble quantities contained in an entire rock mass.

However, the prediction of the number and position of blocks of different clast sizes that could be encountered during underground excavation works is of utmost importance. In particular, the estimation of boulders that are near the tunnel perimeter or only partially embedded within the tunnel face (i.e., protruding rock blocks) represents a critical issue [2,17,18,24]. In fact, these blocks are much more difficult to cut and more likely to be pushed aside or plucked, and may cause severe impacts, such as significant settlements, mucking system damage, sinkholes, high contact stresses at the cutterhead-ground interface causing lining damage, boring machine stuck, and obstruction or deflection of a TBM shield. Moreover, they may also produce excessive torque and thrust demand and a significant wear (or breakage) of the cutterhead tools, with consequent lower cutting efficiency, more frequent cutter-change intervention intervals, safety risks, and higher costs.

The aim of this study is to develop a statistically based methodology to estimate the probability of encountering blocks partially or totally contained within both the whole excavation area and the lateral extremities of the tunnel when excavating through heterogeneous ground. The blocks to be considered are those for which the area of intersection (with the tunnel face) is greater than a given critical value. This value must be chosen on the basis of the current project characteristics (i.e., geological–geotechnical soil properties, tunnel diameter, etc.) and must correspond to boulder dimensions that may cause technical problems during the excavation work. The possibility of coming across such boulders should be considered when choosing the excavation method as well as for designing an appropriate cutterhead.

For this purpose, a Matlab routine implementing a statistical approach was written. The code allows for different ground conditions to be simulated in order to take the inherent variability of these geological deposits into account. It is based on algorithms that, among others, (i) generate many tunnel configurations composed of populations of 2D circular blocks with random dimensions and positions within a control area containing the tunnel geometry in order to take the inherent spatial and dimensional variability of the heterogeneous formations into account; (ii) check if the blocks generated intersect—totally or partially—the tunnel face; (iii) identify the blocks that are located at the lateral distance furthest from the center of the tunnel; (iv) determine the intersection area of each block and their equivalent diameter; and (v) calculate the average number and probability of encountering rock blocks of different dimensions—up to six size classes—during the excavation work. Although based on conceptually quite simple operations, the code developed in this paper provides (in a very short time) extremely useful and statistically based information that geopractitioners can use for making a more rational choice of tunneling technique and for designing a proper cutterhead in order to avoid damage to cutting tools, obstructions, etc., during tunnel excavations in heterogeneous ground.

The executable open-source code, named PBE_vers1.2, is provided for further research on this topic.

2. The PBE Code

In order to statistically model the spatial and dimensional variability inherent in heterogeneous formations, a specific Matlab code, performing Monte Carlo simulations, was implemented to generate a great number of boulder–ground configurations. As illustrated in Figure 1, each configuration is characterized by a control area of dimension BxH, represented by the square window, containing the blocks and the tunnel section (i.e., the circular region in the center of the control area). The dimension of the control area can be set each time according to the tunnel diameter. A ratio of at least 5 between the side of the control area and the tunnel diameter is suggested to ensure that the control area is representative of the real in situ geological conditions.

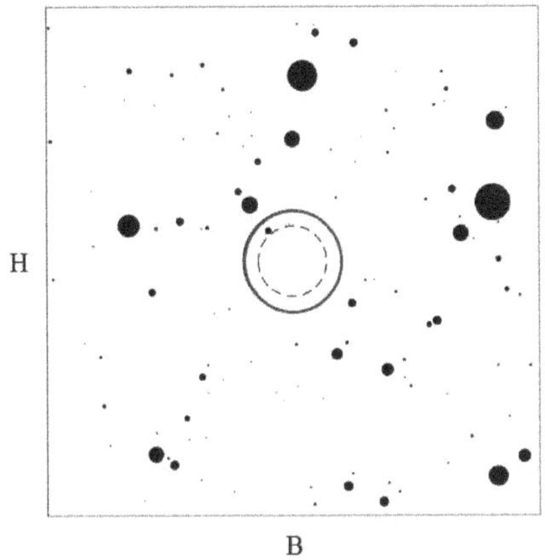

Figure 1. An example of a tunnel configuration in a block-in-matrix formation with a VBP equal to 2%. The length of the control area was set at 5 times greater than the tunnel diameter, indicated with the continuous circle. The dotted line indicates the inner boundary of the tunnel sub-area, where blocks are more difficult to cut and may cause severe impacts, such as higher tool wear.

The tunnel has a circular shape and its center can be located anywhere within the control area. This allows for the modelling of different geological ground properties in the tunnel section. For example, if blocks are expected to occur only in half of the tunnel section, the center of the tunnel can be positioned on one control area boundary (Figure 2).

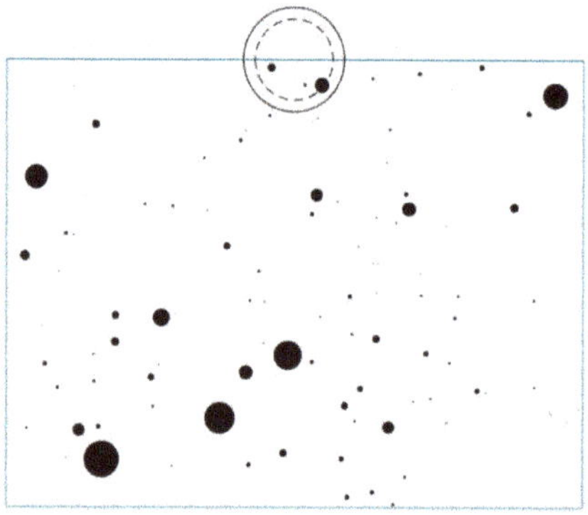

Figure 2. An example of a tunnel partially located outside the rectangular control area to simulate different geologic units being present in a tunnel section.

For the estimation of cobble–boulder quantities at the lateral distance furthest from the center of the tunnel, the thickness of an internal circular crown must be set. This value represents the distance of the inner boundary of the circular crown from the tunnel perimeter, defining the position of the dotted line in Figure 1. The probability of encountering blocks located inside this tunnel sub-area or that extending past the perimeter (i.e., protruding blocks) is extremely important for the reasons highlighted above.

The clasts, represented by circles of variable sizes, are located randomly within the control area. Their number depends on the boulder and cobble volume ratio expected. For each configuration, the code randomly generates n diameters (d) extracted from a population distributed according to the cumulative distribution function of Equation (1), as shown in Figure 3 [31,32], until the block content requested as input is achieved:

$$F(d) = \left(a^{1+D} - d^{1+D}\right) / \left(a^{1+D} - b^{1+D}\right) \quad (1)$$

where D is the fractal dimension and a and b are the minimum and maximum expected clast dimensions [31].

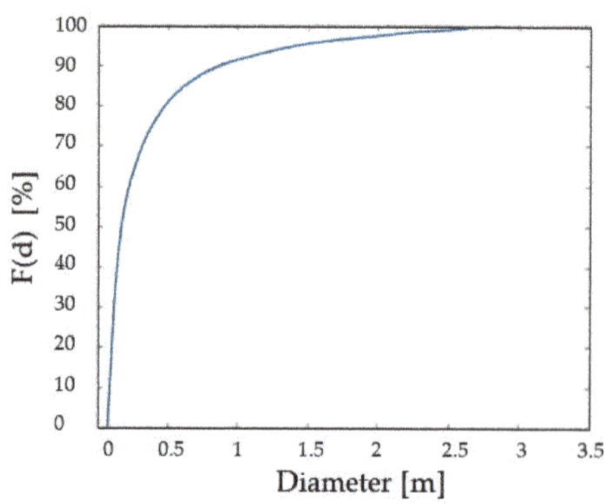

Figure 3. An example of a cumulative distribution function (CDF) obtained for $D = -1.75$, $a = 0.075$ m, and $b = 3$ m.

Such a distribution properly reflects the grain size distribution of geological units with a block-in-matrix fabric containing a few large boulders and increasing numbers of smaller blocks [33,34]. The parameters *a*, *b*, and *D* of Equation (1) should be estimated on the basis of geological surveys, site observations, and field data, all of which should be as detailed as possible.

Once the *n* diameters are generated, the code locates the blocks randomly within the control area while avoiding block–block interpenetrations, as this would have no physical meaning. Furthermore, intersections between the blocks and control area boundaries are not allowed either as this would no longer reflect the requested rock content. To these aims, a minimum distance between two clasts and between the clasts and control area boundaries were set and equal to 10 cm.

A great number of simulations can be requested in order to achieve a statistical validity of the results. For each configuration generated, the code computes and returns the number and intersection area of all the blocks that are either entirely or partially contained within the tunnel section. Then, the code compares each intersection area with a threshold user-defined value corresponding to the minimum block dimension deemed a possible cause of obstruction or tool damage. If the intersection area of a block (either fully or partially contained within the excavation volume) is smaller than the minimum requested intersection area (i.e., the threshold value), the block is not considered problematic and is discarded from the subsequent analyses (Figure 4).

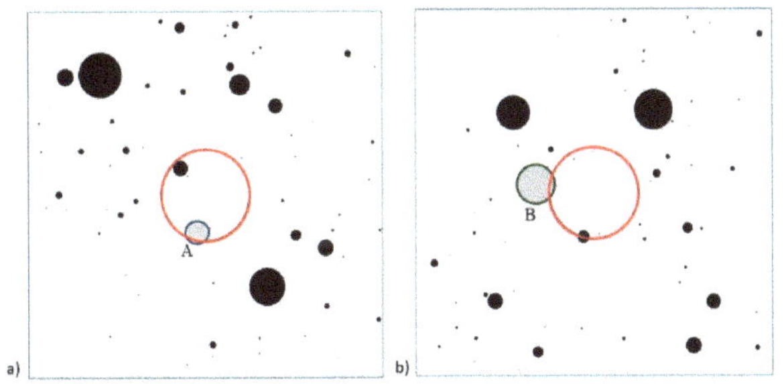

Figure 4. Two tunnel configurations with a protruding boulder (in light gray). (**a**) The protruding block A has an intersection area greater than the threshold value. Hence, the code classifies it as a possible cause of obstruction. (**b**) The protruding block B has an intersection area smaller than the threshold value. Since it does not represent a possible cause of obstruction or other technical problems, the code discards it.

Furthermore, in order to estimate the probability of encountering cobbles and boulders of different sizes during the excavation, six dimensional categories of intersecting rock blocks (i.e., six equivalent clast areas) can be set. The smallest category corresponds to the previously defined threshold value.

The estimated probability that a certain number, *n*, of rock blocks (with $n = 0, 1, 2, \ldots,$ 10, >10) belonging to a given size class can be encountered during the tunnel excavation is finally computed by dividing the number of configurations, in which *n* blocks of that size class were found for the total number of simulations performed. The potential of this new tool is discussed in the next section.

3. Application Example

In order to show the validity of the Matlab routine implemented in the free executable code PBE_vers1.2, the excavation of a circular tunnel in heterogeneous ground was simulated. The parameters required as input are listed in Table 1.

The values assigned in this example to the 16 input parameters can be found as default values in the executable code. The A_thr1 variable corresponds to the minimum requested intersection area (i.e., the minimum block dimension deemed a possible cause of technical problems). Its default value was set as equal to 177 cm^2, corresponding to an equivalent circular block fully encapsulated in the tunnel with a diameter of 15 cm, accord-

ing to [24]. The threshold areas listed in Table 1 define six size classes in terms of equivalent clast diameters: class 1: 15–30 cm (i.e., 177–707 cm^2); class 2: 30–50 cm (i.e., 707–1,963 cm^2); class 3: 50–75 cm (i.e., 1963–4418 cm^2); class 4: 75–100 cm (i.e., 4418–7854 cm^2); class 5:
100–150 cm (i.e., 7854–17,663 cm^2); and class 6: >150 cm (i.e., >17,663 cm^2). All the blocks with an intersecting area smaller than A_thr1 (i.e., 177 cm^2, corresponding to an equivalent circular block with a diameter of 15 cm) were not considered further in this study.

Table 1. Input parameters for the probabilistic analysis.

Parameter		Variable	Assigned Value
Length of the control area	[m]	B	32.5
Height of the control area	[m]	H	32.5
Tunnel radius	[m]	R_t	3.25
Circular crown thickness	[m]	R_ext	0.8
Tunnel x–y coordinates	[m]	coord_t	B/2 =16.25 H/2 =16.25
Threshold area class 1 (equivalent diameter of 0.15 m)	[m^2]	A_thr1	0.0177
Threshold area class 2 (equivalent diameter of 0.3 m)	[m^2]	A_thr2	0.0707
Threshold area class 3 (equivalent diameter of 0.5 m)	[m^2]	A_thr3	0.1963
Threshold area class 4 (equivalent diameter of 0.75 m)	[m^2]	A_thr4	0.4418
Threshold area class 5 (equivalent diameter of 1 m)	[m^2]	A_thr5	0.7854
Threshold area class 6 (equivalent diameter of 1.5 m)	[m^2]	A_thr6	1.7663
Block content	[-]	BC	0.02
Minimum expected clast dimension	[m]	a	0.075
Maximum expected clast dimension	[m]	b	3
Fractal dimension	[-]	D	−1.75
Number of configurations to generate	[-]	z	500

In total, 500 configurations were generated in this example in about 4 min. However, since the computation only takes a few minutes, many more configurations can easily be requested and obtained. Moreover, since uncertainties always exist in the determination of some of the input parameters (i.e., BC, a, b, and D), more than a single value may be assumed for each of them and a greater number of analyses can also be performed in a very short time. In this way, by averaging the data obtained, the user can obtain more reliable statistically based results.

Figure 5 shows five of the 500 configurations generated for the example considered, while the probability of finding n intersecting blocks greater than the threshold value, A_thr1, both within the tunnel and at the lateral distance furthest from the center of the cutterhead (i.e., inside the circular crown) is given in Table 2.

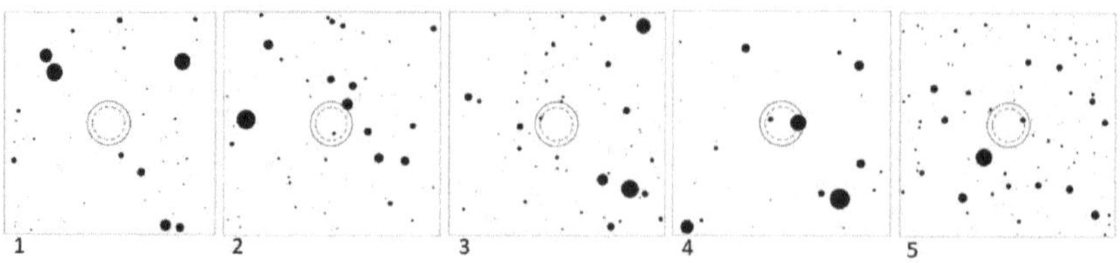

Figure 5. Five of the 500 configurations generated with the Matlab code. Different block sizes and positions are shown.

Table 2. Probability, P, of encountering *n* blocks (with *n* from zero to more than seven) with an intersection area greater than the threshold value A_thr1, equal to 0.0177 m^2. The average number of blocks corresponding to each equivalent clast diameter is also provided. The results are related to the entire tunnel section (table above) and to the circular crown (table below). These results are contained in the output text files "Probability" and "N_Average" of Table 3.

TUNNEL Equivalent clast diameter [m]	$P_{0\ blocks}$ [%]	$P_{1\ blocks}$ [%]	$P_{2\ blocks}$ [%]	$P_{3\ blocks}$ [%]	$P_{4\ blocks}$ [%]	$P_{5\ blocks}$ [%]	$P_{6\ blocks}$ [%]	$P_{\geq 7\ blocks}$ [%]	Average number of blocks
0.15–0.30	49.4	31.4	13.8	4.4	0.8	0	0.2	0	0.77
0.30–0.50	67.4	27.8	4.0	0.6	0.2	0	0	0	0.38
0.50–0.75	79.0	19.4	1.4	0.2	0	0	0	0	0.23
0.75–1.0	86.6	13.0	0.2	0.2	0	0	0	0	0.14
1.0–1.5	84.4	15.4	0.2	0	0	0	0	0	0.16
>1.5	90.8	8.2	1	0	0	0	0	0	0.10
CIRCULAR CROWN Equivalent clast diameter [m]	**$P_{0\ blocks}$ [%]**	**$P_{1\ blocks}$ [%]**	**$P_{2\ blocks}$ [%]**	**$P_{3\ blocks}$ [%]**	**$P_{4\ blocks}$ [%]**	**$P_{5\ blocks}$ [%]**	**$P_{6\ blocks}$ [%]**	**$P_{\geq 7\ blocks}$ [%]**	**Average number of blocks**
0.15–0.30	65.4	27.8	5.8	0.8	0.2	0	0	0	0.43
0.30–0.50	77.6	20.4	2.0	0	0	0	0	0	0.24
0.50–0.75	87.4	12.0	0.6	0	0	0	0	0	0.13
0.75–1.0	93.6	6.4	0	0	0	0	0	0	0.06
1.0–1.5	86.6	13.0	0.4	0	0	0	0	0	0.14
>1.5	97.6	2.4	0	0	0	0	0	0	0.02

Table 3. Outputs of the executable code. The output files with the "*" contain the results related to both the entire tunnel section and the circular crown.

Parameter		Output File Name
Total number of intersecting blocks	[-]	Total_IB *
Total intersection area	[m^2]	A_int *
Total number of intersecting blocks greater than the threshold value, A_thr1	[-]	N_crit_IB *
Intersection area of all the blocks greater than the threshold value	[m^2]	A_int_cr *
Probability of finding *n* intersecting blocks greater than the threshold value during the excavation	[%]	Probability * (Table 2)
Average number of intersecting blocks greater than the threshold value during the excavation	[-]	N_Average * (Table 2)
Graphical representation of the cumulative distribution function F(d)-d	[-]	CDF
Graphical representation of the configurations generated	[-]	Configuration 1 up to z (Figure 5)
List of input variables and values assigned	[-]	Info_viewer

What stands out in Table 2 is the high probability of encountering zero blocks (i.e., $P_{\_0\ blocks}$ was always greater than 49% and up to over 90%). This result is clearly related to the low BC set (i.e., 2%), which produced many configurations without blocks inside the tunnel section (e.g., configuration 1 of Figure 5).

Table 2 also shows that the probability of encountering a single boulder during tunneling ($P_{\_1\ block}$) is higher than (or at least equal to) the probability of encountering two ($P_{\_2\ blocks}$) or more blocks of the same clast size, regardless of the area examined (i.e., the entire tunnel face or a part of it).

Moreover, for a given number *n* of blocks, with $n \geq 1$, the highest probability is related to the smallest clast dimension (i.e., equivalent clast diameter in the range 0.15–0.30 m). Furthermore, in the proposed example, a very low probability of encountering blocks during the tunnel excavation was reached for *n* equal to four and six, while no configuration has more than six intersecting blocks (i.e., $P_{TUNNEL-n\geq 7\ blocks} = 0$). Conversely, the results obtained for the circular crown of the tunnel indicate that the probabilities of encountering two to four blocks are much lower than those related to the entire tunnel section. Moreover, no configuration had more than four blocks in the sub-area furthest from the center (i.e., $P_{CIRCULARCROWN-n>=5\ blocks} = 0$) and the probability of encountering more than two blocks was very low in the case related to only cobbles belonging to the smallest class dimension.

The executable code also generates other outputs, which are listed and described in Table 3 as text files or JPG images.

4. Discussion and Conclusions

When tunneling in soil containing cobbles and boulders, a number of unfavorable conditions may be encountered related to the size and type of the machine, as well as to the design of its cutterhead. Among other technical difficulties, damage to cutting tools, obstructions, lower penetration rates, and more rapid cutter wear, with consequent delays and extra tunneling costs, may often occur due to the presence of strong rock blocks.

This study provides a statistically based tool for the estimation of the probability of encountering a number n (with $n = 0, 1, 2, \ldots, 10, >10$) of cobbles and boulders during underground excavations. Specifically, blocks fully or partially located within the entire tunnel section or in a part of it at lateral distance furthest from the center of the cutterhead, have been considered. This information can be extremely useful for making a more rational choice of the tunneling technique and for designing a more suitable cutterhead.

In order to do this, a statistical method implemented in an executable free code is provided. A great number of boulder–ground configurations can be generated according to a given tunnel dimension and location within a control area of rectangular shape, block content, and clast-size distribution. This last input parameter can be estimated using one of the mathematical methods proposed in the literature. The area of intersection between the tunnel face and each block encountered is calculated and compared with a threshold area corresponding to the minimum block size deemed a possible cause of obstruction, tool damage, or other technical problems. Since problematic blocks can be manifold and depend on the excavation method and cutterhead design, up to six threshold values (i.e., size classes) can be set by the user. Finally, the probability of encountering one or more boulders of each size class, in part or in the whole excavation area, is provided. This information can be an extensive and useful tool to predict boulder occurrences all along the tunnel stretch to be excavated within a heterogeneous ground and close to its perimeters in order to appropriately choose the TBM type, including face access, cutterhead design, cutter types, and machine power (i.e., torque, thrust, and speed).

A limitation of the proposed tool is that only circular blocks are considered. Although they may well represent geological units, such as conglomerates and sedimentary melanges, most of the heterogeneous formations generally contain non-spherical rock inclusions. Hence, a future work could address the extension of the PBE code to simulate other block shapes, such as more realistic ellipses.

Author Contributions: Conceptualization, M.L.N. and M.B.; methodology, M.L.N.; software, M.L.N.; validation, M.L.N., M.B. and R.F.; formal analysis, M.L.N.; investigation, M.L.N. and M.B.; data curation, M.L.N.; writing—original draft preparation, M.L.N.; writing—review and editing, M.L.N., M.B. and R.F.; supervision, M.B. and R.F.; project administration, M.L.N. All authors have read and agreed to the published version of the manuscript.

Funding: This research study received no external funding.

Data Availability Statement: The data presented in this study are openly available on Mendeley Data at https://data.mendeley.com/datasets/jjgw7s8yvr/draft?a=d24acd93-5d7c-4bc3-b46a-e5f1280b76da (accessed on 7 September 2021)—doi:10.17632/jjgw7s8yvr.1.

Acknowledgments: The authors thank Eng. Alessandro Cresto from Herrenknecht Italia S.r.l. for the useful discussion on the topic of this research.

Conflicts of Interest: The authors declare no conflict of interest.

References

1. Jee, W.W.; Ha, S.G. Feasible Boulder treatment methods for soft ground shielded TBM. In *Underground Space—The 4th Dimension of Metropolises*; Taylor & Francis Group: Abingdon, UK, 2007; pp. 217–222.
2. Hunt, S.W. Tunneling in Cobbles and Boulders. In Proceedings of the Breakthroughs in Tunneling Short Course, Chicago, IL, USA, 14–16 August 2017; pp. 1–46.
3. Dong, W.; Pimentel, E.; Anagnostou, G. Experimental Investigations into the Mechanical Behaviour of the Breccias around the Proposed Gibraltar Strait Tunnel. *Rock Mech. Rock Eng.* **2013**, *46*, 923–944. [CrossRef]
4. Marinos, V. Tunnel behaviour and support associated with the weak rock masses of flysch. *J. Rock Mech. Geotech. Eng.* **2014**, *6*, 227–239. [CrossRef]
5. Montoya-Araque, E.A.; Suarez-Burgoa, L.O. Automatic generation of tortuous failure surfaces in block-in-matrix materials for 2D slope stability assessments. *Comput. Geotech.* **2019**, *112*, 17–22. [CrossRef]
6. Sonmez, H.; Ercanoglu, M.; Kalender, A.; Dagdelenler, G.; Tunusluoglu, C. Predicting uniaxial compressive strength and deformation modulus of volcanic bimrock considering engineering dimension. *Int. J. Rock Mech. Min. Sci.* **2016**, *86*, 91–103. [CrossRef]
7. Kalender, A.; Sonmez, H.; Medley, E.; Tunusluoglu, C.; Kasapoglu, K. An approach to predicting the overall strengths of unwelded bimrocks and bimsoils. *Eng. Geol.* **2014**, *183*, 65–79. [CrossRef]
8. Xu, Y.; Cai, M. Numerical study on the influence of cross-sectional shape on strength and deformation behaviors of rocks under uniaxial compression. *Comput. Geotech.* **2017**, *84*, 129–137. [CrossRef]
9. Coli, N.; Tanzini, M. Characterization of the "Chaotic Complex" (Northern Apennines, Italy). In *Rock Mechanics for Resources, Energy and Environment*; Taylor & Francis Group: London, UK, 2013; pp. 111–116.
10. Napoli, M.L.; Barbero, M.; Scavia, C. Effects of block shape and inclination on the stability of melange bimrocks. *Bull. Eng. Geol. Environ.* **2021**, *80*, 7457–7466. [CrossRef]
11. Napoli, M.L.; Barbero, M.; Scavia, C. Slope stability in heterogeneous rock masses with a block-in-matrix fabric. In Proceedings of the ISRM 2019, Foz do Iguassu, Brazil, 13–18 September 2019; pp. 3482–3489.
12. Kanji, M.A. Critical issues in soft rocks. *J. Rock Mech. Geotech. Eng.* **2014**, *6*, 186–195. [CrossRef]
13. Hunt, S.W.; Del Nero, D.E. Two Decades of Advances Investigating, Baselining and Tunneling in Bouldery Ground. In Proceedings of the World Tunneling Congress, Vancouver, BC, Canada; 2010; pp. 1–8.
14. Tarkoy, P.J. The boulder facts of life. In Proceedings of the World Tunneling Congress, Agra, India, 22–24 September 2008; pp. 25–28.
15. Medley, E.W. Bimrocks—Part 2: Case histories and Practical Guidelines. *Newsl. Hell. Soc. Soil Mech. Geotech. Eng.* **2007**, *8*, 26–31.
16. Barla, G.; Pelizza, S. TBM tunnelling in difficult ground conditions. In Proceedings of the GeoEng 2000, International Conference on Geotechnical & Geological Engineering, Melbourne, Australia, 19–24 November 2000.
17. DiPonio, M.A.; Chapman, D.; Bournes, C. EPB tunnel boring machine design for boulder conditions. In Proceedings of the Rapid Excavation and Tunneling Conference, Toronto, ON, Canada, 11 June 2007; pp. 215–228.
18. Gwildis, U.; Aguilar, J.; Mosavat, K. TBM Tool Wear Analysis for Cutterhead Configuration and Resource Planning in Glacial Geology. In Proceedings of the North American Tunneling Conference, Washington, DC, USA, 24–27 June 2018.
19. Moritz, B.; Grossauer, K.; Schubert, W. Short term prediction of system behaviour of shallow tunnels in heterogeneous ground. *Felsbau* **2004**, *22*, 44–52.
20. Button, E.A.; Riedmueller, G. Shallow Tunneling in a Tectonic Mélange: Rock Mass Characterization and Data Interpretation. In Proceedings of the 5th North American Rock Mechanics Symposium, Toronto, ON, Canada, 7–10 July 2002; pp. 1125–1132.
21. Li, C.; Zhang, D.; Du, S.-S.; Shi, B. Computed tomography based numerical simulation for triaxial test of soil–rock mixture. *Comput. Geotech.* **2016**, *73*, 179–188. [CrossRef]
22. Hunt, S.W. Managing Risk of Tunneling in Cobbly, Bouldery Ground. In Proceedings of the TAC 2014: Tunnelling in a Resource Driven World, Vancouver, BC, Canada, 26–28 October 2014.
23. Napoli, M.L.; Barbero, M.; Scavia, C. Tunneling in heterogeneous rock masses with a block-in-matrix fabric. *Int. J. Rock Mech. Min. Sci.* **2021**, *138*, 104655. [CrossRef]
24. Frank, G.; Chapman, D.R. A new model for characterizing the cobble and boulder fraction for soft ground tunneling. In Proceedings of the Rapid Excavation and Tunneling Conference, Seattle, WA, USA, 27–29 June 2005; pp. 780–791.
25. Goss, C.M. Predicting boulder cutting in soft ground tunneling. In *North American Tunneling*; Ozdemir, L., Ed.; CRC Press: Boca Raton, FL, USA, 2002.
26. Nagahama, H. Technical note fractal fragment size distribution for brittle rocks. *Int. J. Rock Mech. Geomech.* **1993**, *30*, 469–471. [CrossRef]
27. Riedmüller, G.; Brosch, F.J.; Klima, K.; Medley, E.W. Engineering Geological Characterization of Brittle Faults and. *Felsbau* **2001**, *19*, 13–19.
28. Medley, E.W. The Engineering Characterization of Melanges and Similar Block-in-Matrix Rocks (Bimrocks). Ph.D. Thesis, University of California, Berkeley, CA, USA, 1994.

29. Medley, E.W. Uncertainty in estimates of block volumetric proportions in melange bimrocks. In Proceedings of the International Symposium on Engineering Geology and the Environment, Athens, Greece, 23 June 1997; pp. 267–272.
30. Napoli, M.L.; Milan, L.; Barbero, M.; Scavia, C. Identifying uncertainty in estimates of bimrocks volumetric proportions from 2D measurements. *Eng. Geol.* **2020**, *278*, 105831. [CrossRef]
31. Napoli, M.L.; Barbero, M.; Ravera, E.; Scavia, C. A stochastic approach to slope stability analysis in bimrocks. *Int. J. Rock Mech. Min. Sci.* **2018**, *101*, 41–49. [CrossRef]
32. Barbero, M.; Bonini, M.; Borri-Brunetto, M. Three-Dimensional Finite Element Simulations of Compression Tests on Bimrock. In Proceedings of the 12th International Conference of International Association for Computer Methods and Advances in Geomechanics (IACMAG), Goa, India, 1–6 October 2008; pp. 631–637.
33. Wakabayashi, J.; Medley, E.D.; Wilcox, G.; Risden, C. *Tunnels through Fault Rocks and Tectonic Melanges: A Short Course for Engineering Geologists and Geotechnical Engineers;* Association of Engineering Geologist: San Francisco, CA, USA, 2002.
34. Medley, E.W. Orderly Characterization of Chaotic Franciscan Melanges. *Felsbau* **2001**, *19*, 20–33.

mining MDPI

Article

Thermoporoelastoplastic Wellbore Breakout Modeling by Finite Element Method

Lijing Zhang [1], Hua Zhang [2], Kezhen Hu [3], Zhuoheng Chen [3] and Shunde Yin [4,*]

1 School of Mathematics and Physics, University of Science and Technology Beijing, Beijing 100083, China; zhanglij@ustb.edu.cn
2 Department of Petroleum Engineering, University of Wyoming, Laramie, WY 82071, USA; zhanghua_cec@126.com
3 Geological Survey of Canada, Calgary, AB T2L 2A7, Canada; kezhen.hu@canada.ca (K.H.); zhuoheng.chen@canada.ca (Z.C.)
4 Department of Civil and Environmental Engineering, University of Waterloo, Waterloo, ON N2L 3G1, Canada
* Correspondence: shunde.yin@uwaterloo.ca

Abstract: Drilling a hole into rock results in stress concentration and redistribution close to the hole. When induced stresses exceed the rock strength, wellbore breakouts will happen. Research on wellbore breakout is the fundamental of wellbore stability. A wellbore breakout is a sequence of stress concentrations, rock falling, and stress redistributions, which involve initiation, propagation, and stabilization sequences. Therefore, simulating the process of a breakout is very challenging. Thermoporoelastoplastic models for wellbore breakout analysis are rare due to the high complexity of the problem. In this paper, a fully coupled thermoporoelastoplastic finite element model is built to study the mechanism of wellbore breakouts. The process of wellbore breakouts, the influence of temperature and the comparison between thermoporoelastic and thermoporoelastoplastic models are studied in the paper. For the finite element modeling, the D-P criterion is used to determine whether rock starts to yield or not, and the maximum tensile strain criterion is used to determine whether breakouts have happened.

Keywords: wellbore breakout; finite element method; Drucker-Prager criterion; the maximum tensile strain criterion; thermoporoelasticity; thermoporoelastoplasticity

1. Introduction

When a hole is drilled into the crust, the rock mass removed from the hole will not support the surrounding rocks, which leads to the stress concentration and redistribution. If the induced stresses exceed the rock strength, parts of rock fall from the wellbore wall, which is called a wellbore breakout [1,2]. Wellbore breakouts result in problematic wellbore stability and affects drilling efficiency, so the research of wellbore breakouts is important in petroleum engineering [3,4].

Numerous studies have been made to explain the mechanism of wellbore breakouts. Zoback found the analytical solution of initial breakout zone by incorporating the Mohr-Coulomb criterion to the Kirsch equation [2]. The studies of wellbore breakouts show that the wellbore breakouts occur by a series of successive spalls that result from shear failure subparallel to the direction of the local minimum principal stress [3,5–9]. In addition, some studies also show that a wellbore breakout can also influence the wellbore breakdown pressure [10]. To consider the effect of pore pressure and obtain effective stress distribution in a porous medium, Biot [11] proposed a linear poroelastic model based on Terzaghi's theory and introduced Biot's coefficient. Later, Fluid seepage and pressure diffusion between the wellbore and the formation were considered for both vertical and inclined wells [12–14]. Recently, numerical methods were also used to analyze the breakout mechanisms [15–19].

Most of these models do not consider the influence of plastic damage of the rock. For the plastic model, different yield criteria are used to analyze the character of rock and soil. The Mohr-Coulomb criterion is the most popular one, which relates the shearing resistance to the contact forces and friction [20]. The Drucker-Prager criterion is an extended version of the Von Mises criterion [21]. Whittle and Kavvadas [22] presented the MIT-E3 soil model to describing the behavior of overconsolidated clays that obey normalized behavior and are rate-independent. Akl and Whittle [23] analyzed horizontal wellbore stability in clay shale based on MIT-E3 soil model. Zhang and Yin [24] made a wellbore study using the tangent stiffness matrix method based on the Drucker-Prager criterion.

However, in reality, a wellbore breakout is a complex time-dependent process including initiation, propagation, and stabilization sequences [25]. The breakout initiation and its propagation are the result of stress concentration on the wellbore wall, further stress concentration at the tip of breakout, and the formation of a plastic zone around the tip [26]. Poroelastoplastic analysis of wellbore breakouts is still not well studied due to the complexity of the problem.

With the development of a deep well, consideration of the thermal effects is essential. Coupled thermal-hydraulic-mechanical processes play an important role in the stability of wells in thermal reservoir formations [27]. Lewis [28] used FE simulation to study thermal recovery processes and heat losses problems to surrounding strata. Aboustit et al. [29] used a general variational principle to investigate thermo-elastic consolidation problems, and Vaziri [30] also presented a fully coupled thermo-hydro-mechanical FE model.

In this paper, a fully coupled thermoporoelastoplastic finite element model is built to study the mechanism of wellbore breakouts. The Drucker-Prager criterion is used to determine whether rock starts to yield or not, and the maximum tensile strain criterion is used to determine whether breakouts happened.

2. Model Structure and Methodology

To simulate wellbore breakouts in plastic rock, the Drucker-Prager yield criterion and maximum tensile strain criterion are used to analyze the yield and failure of the rock. The finite element implementation of thermoporoelasticity is based on Biot's theory with the compressible fluid flowing through the saturated non-isothermal deformable porous medium.

2.1. Rock Failure Criteria

In this paper, the Drucker-Prager yield criterion and maximum tensile strain criterion are used to analyze the yield and failure of the rock. Figure 1 shows a simplified sketch for the relationship between stress and strain in the condition of yield and failure of the rock, where σ_{yield} and ε_{yield} are the yield stress and strain of rock, $\varepsilon_{breakout}$ is the maximum allowable tensile strain of rock.

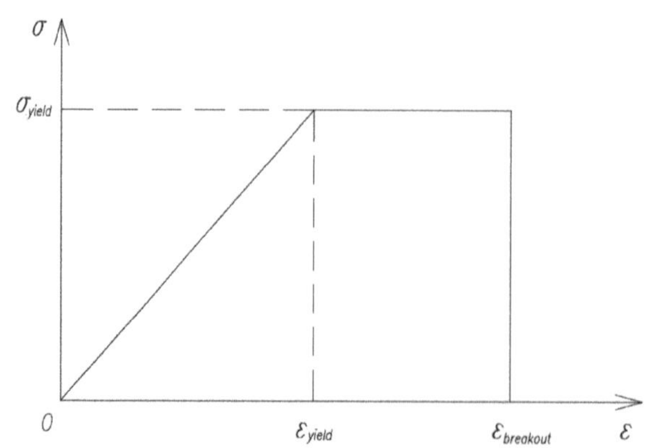

Figure 1. Sketch of the stress-strain relationship.

The Drucker-Prager yield criterion is an extended version of the Von Mises criterion. To obtain a smooth yield surface approximate to Mohr-Coulomb surface, Drucker and Prager [21] put forward the following yield criterion:

$$F = \alpha I_1 + \sqrt{J_2} - k_0 \tag{1}$$

where is F is the yield criterion, I_1 is the first stress invariant, J_2 is the second deviatoric stress invariant, and α, k_0 are the material constants.

Two of the most common approximations used are obtained by making the yield surfaces of the Drucker-Prager and Mohr-Coulomb criteria coincident either at the outer or inner edges of the Mohr-Coulomb surface. Coincidence at the outer edges is obtained when

$$\alpha = \frac{2sin\varphi}{\sqrt{3}(3 - sin\varphi)}, \; k_0 = \frac{6ccos\varphi}{\sqrt{3}(3 - sin\varphi)} \tag{2}$$

whereas, coincidence at the inner edges is given by the choice

$$\alpha = \frac{2sin\varphi}{\sqrt{3}(3 + sin\varphi)}, \; k_0 = \frac{6ccos\varphi}{\sqrt{3}(3 + sin\varphi)} \tag{3}$$

where c is the cohesion of the material; φ is the internal friction angle of the material.

The yield condition of rock can be determined by the following equation.

$$\begin{cases} F \leq 0 & no\ yield \\ F > 0 & yield \end{cases} \tag{4}$$

The failure (*breakout*) condition of rock can be determined by the following equation.

$$\begin{cases} \varepsilon_{max} \leq \varepsilon_{breakout} & no\ breakout \\ \varepsilon_{max} > \varepsilon_{breakout} & breakout \end{cases} \tag{5}$$

2.2. Finite Element Implementation

The general theory of thermoporoelasticity is based on Biot's theory. With the compressible fluid flowing through the saturated non-isothermal deformable porous medium, in the form of displacements, pressure and temperature as unknowns, the governing equations can be described as follows [31].

$$G\nabla^2 \boldsymbol{u} + (G + \lambda)\nabla div\boldsymbol{u} - \left(1 - \frac{K}{K_m}\right)\nabla \boldsymbol{p} - K\beta_s \nabla \boldsymbol{T} + \boldsymbol{f}^u = 0 \tag{6}$$

$$\left(1 - \frac{K}{K_m}\right)div\boldsymbol{u_t} + \left[\frac{1-\phi}{K_m} + \frac{\phi}{K_f} + \frac{1}{(3K_m)^2} i^T Di\right]\boldsymbol{p_t} + \frac{k}{\mu}\nabla^2 \boldsymbol{p} - \left[\phi\beta_f + (1-\phi)\beta_s + \frac{\beta_s}{9K_m} i^T Di\right]\boldsymbol{T_t} + \boldsymbol{f}^p = 0 \tag{7}$$

$$T\left[(1-\phi)\frac{\rho_s c_s}{K_m} + \phi\frac{\rho_f c_f}{K_f}\right]\boldsymbol{p_t} + \left[(1-\phi)\rho_s c_s + \phi\rho_f c_f - T(1-\phi)\rho_s c_s \beta_s - T\phi\rho_f c_f \beta_f\right]\boldsymbol{T_t} + \lambda_T \nabla^2 \boldsymbol{T} + \boldsymbol{f}^T = 0 \tag{8}$$

where K, K_f and K_m are the bulk moduli of the skeleton, fluid and matrix, respectively. k is the permeability of the porous medium, μ is the viscosity of the fluid, G and λ are Lamé constants. u, p and p_t denote the displacement of the porous medium, the pore pressure and its time derivative, ϕ is the porosity of the porous medium, λ_T is the thermal conductivity matrix of the porous media, T and T_t are the temperature and time derivative, $\rho_s c_s$ is the heat capacity of the solid phase, $\rho_f c_f$ is the heat capacity of the fluid phase, β_s is the thermal expansion coefficient of the matrix, and β_f is the thermal expansion coefficient of the fluid. Furthermore, i^T = [1, 1, 1, 0, 0, 0], and D is the elastic stiffness matrix, the subscript t denotes time derivative.

Using Galerkin finite element method to approximate the governing equations, the final form of the FEM solution is as follows.

$$\begin{bmatrix} M & -C_{sw} & -C_{sT} \\ 0 & H_{ww} & 0 \\ 0 & 0 & H_{TT} \end{bmatrix} \begin{Bmatrix} u \\ p \\ T \end{Bmatrix} + \begin{bmatrix} 0 & 0 & 0 \\ C_{sw}^T & R_{ww} & -C_{wT} \\ 0 & C_{Tw} & R_{TT} \end{bmatrix} \begin{Bmatrix} u_t \\ p_t \\ T_t \end{Bmatrix} = \begin{Bmatrix} f^u \\ f^p \\ f^T \end{Bmatrix} \tag{9}$$

where $[u, p, T]^T$, $[u_t, p_t, T_t]^T$ are the vectors of unknown variables and corresponding time derivatives, and $[f^u, f^p, f^T]^T$ is the vector for the nodal loads, flow source and heat sources. The explicit expressions of above matrixes are as follows.

$$M = \int B^T D B d\Omega \tag{10}$$

$$H_{ww} = \frac{k}{\mu} \int (\nabla \mathrm{N})(\nabla \mathrm{N})^T d\Omega \tag{11}$$

$$R_{ww} = \int \mathrm{N} \left[\frac{1-\phi}{K_m} + \frac{\phi}{K_f} + \frac{1}{(3K_m)^2} i^T D i \right] \mathrm{N}^T d\Omega \tag{12}$$

$$C_{sw} = \int \left(B^T i \mathrm{N} - B^T D \frac{i}{3K_m} \mathrm{N} \right) d\Omega \tag{13}$$

$$C_{sT} = \int B^T D i \frac{\beta_s}{3} \mathrm{N} d\Omega \tag{14}$$

$$C_{wT} = \int \mathrm{N} \left[(1-\phi)\beta_s - \phi\beta_f + \frac{\beta_s}{9K_m} i^T D i \right] \mathrm{N}^T d\Omega \tag{15}$$

$$C_{Tw} = \int \mathrm{N} \left[(1-\phi)\frac{\rho_s c_s}{K_m} - \phi \frac{\rho_f c_f}{K_f} \right] \mathrm{N}\overline{T}\mathrm{N}^T d\Omega \tag{16}$$

$$H_{TT} = \lambda_T \int (\nabla \mathrm{N})(\nabla \mathrm{N})^T d\Omega \tag{17}$$

$$R_{TT} = \int \left[(1-\phi)\rho_s c_s - \phi\rho_f c_f - (1-\phi)\rho_s c_s \beta_s \mathrm{N}\overline{T} - \phi\rho_f c_f \beta_f \mathrm{N}\overline{T} \right] \mathrm{N}^T d\Omega \tag{18}$$

To integrate the above equations with respect to time (θ method), the equation becomes:

$$\begin{bmatrix} M & -C_{sw} & -C_{sT} \\ C_{sw}^T & R_{ww} + \Delta t\theta H_{ww} & -C_{wT} \\ 0 & C_{Tw} & R_{TT} + \Delta t\theta H_{TT} \end{bmatrix} \begin{Bmatrix} \Delta u \\ \Delta p \\ \Delta T \end{Bmatrix} = \begin{Bmatrix} \Delta f^u \\ \Delta t\theta f_1^p + \Delta t(1-\theta) f_0^p - \Delta t H_{ww} p_0 \\ \Delta t\theta f_1^T + \Delta t(1-\theta) f_0^T - \Delta t H_{TT} T_0 \end{Bmatrix} \tag{19}$$

where $\begin{Bmatrix} \Delta u \\ \Delta p \end{Bmatrix}$ is the increment of displacement, pore pressure, $\begin{Bmatrix} \Delta f^u \\ \Delta f^p \end{Bmatrix}$ is the increment vector for the nodal loads and flow source, Δt is the increment of time, θ is the time integration coefficient, f_0^p is the initial vector for the nodal loads in a time step, and p_0 is the initial vector for the pore pressure in a time step.

If the stresses are in the plastic state at a special time t_a based on the yield criterion, Equation (19) becomes Equation (20) at the special time t_a [32].

$$\begin{bmatrix} M & -C_{sw} & -C_{sT} \\ C_{sw}^T & R_{ww} & -C_{wT} \\ 0 & C_{Tw} & R_{TT} \end{bmatrix}_a \begin{Bmatrix} \Delta u \\ \Delta p \\ \Delta T \end{Bmatrix}_a = \begin{Bmatrix} \Psi_n \\ 0 \\ 0 \end{Bmatrix} \tag{20}$$

$$\Psi_n = \Sigma_e \int B^T D^p \Delta\varepsilon d\Omega \tag{21}$$

$$D^p = \frac{D\left\{\frac{\partial Q}{\partial \sigma}\right\}\left\{\frac{\partial F}{\partial \sigma}\right\}^T D}{\left\{\frac{\partial F}{\partial \sigma}\right\}^T D\left\{\frac{\partial Q}{\partial \sigma}\right\} - A} \tag{22}$$

where Σ_e is the set of element nodal forces round the node, and Ψ_n is the unbalanced force. For an ideal elastoplastic body, $A = 0$.

In the process of iteration, the convergence condition can be described as:

$$norm\ (\Psi_n) < V \tag{23}$$

where $norm\ (\Psi_n)$ is the norm of vector Ψ_n, and V is the convergence precision.

For different iterative methods, the global stiffness matrixes are different, which are shown in Figure 2. For the constant stiffness method, the values in Equation (20) can be calculated based on Equations (10)–(18). For the tangent stiffness method, the explicit expressions of the values in Equation (20) are as follows.

$$M|_{t=t_a} = \int B^T D^{ep} B d\Omega \tag{24}$$

$$R_{ww}|_{t=t_a} = \int \mathrm{N}\left[\frac{1-\phi}{K_m} + \frac{\phi}{K_f} + \frac{1}{(3K_m)^2} i^T D^{ep} i\right] \mathrm{N}^T d\Omega \tag{25}$$

$$R_{TT}|_{t=t_a} = R_{TT} \tag{26}$$

$$C_{sw}|_{t=t_a} = \int \left(B^T i\mathrm{N} - B^T D^{ep} \frac{i}{3K_m} \mathrm{N}\right) d\Omega \tag{27}$$

$$C_{sT}|_{t=t_a} = \int B^T D^{ep} i \frac{\beta_s}{3} \mathrm{N} d\Omega \tag{28}$$

$$C_{wT}|_{t=t_a} = \int \mathrm{N}\left[(1-\phi)\beta_s - \phi\beta_f + \frac{\beta_s}{9K_m} i^T D^{ep} i\right] \mathrm{N}^T d\Omega \tag{29}$$

$$C_{Tw}|_{t=t_a} = C_{Tw} \tag{30}$$

$$D^{ep} = D - D^p = D - \frac{D\left\{\frac{\partial Q}{\partial \sigma}\right\}\left\{\frac{\partial F}{\partial \sigma}\right\}^T D}{\left\{\frac{\partial F}{\partial \sigma}\right\}^T D\left\{\frac{\partial Q}{\partial \sigma}\right\} - A} \tag{31}$$

where D^{ep} is elastic-plastic matrix [32,33]. For an ideal elastoplastic body, $A = 0$.

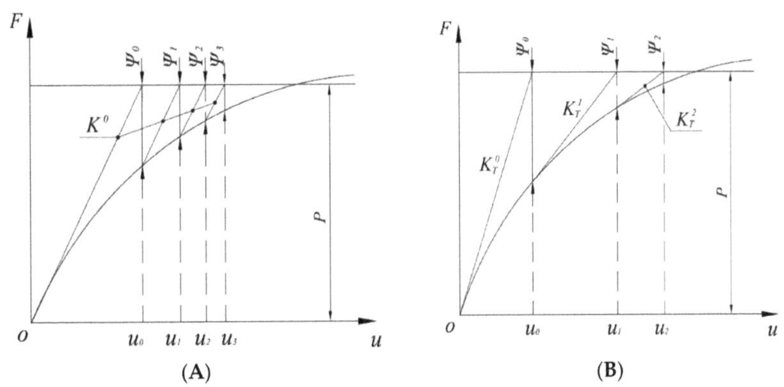

Figure 2. (**A**) Constant stiffness method, (**B**) Tangent stiffness method.

The tangent stiffness method is adopted in this paper.

3. Numerical Experiments

Thermoporoelastoplastic wellbore breakouts are simulated for a vertical well, and there are four parts in this section: the relationship between breakout shape and in situ stresses, verification of wellbore breakout process, influence of drilling fluid temperature, and a comparison between thermoporoelastic and thermoporoelastoplastic modeling.

3.1. Thermoporoelastoplastic Finite Element Modeling of Wellbore Breakouts

For a vertical wellbore shown in Figure 3 that is subjected to horizontal in-situ stresses σ_H and σ_h, the shape of breakouts ϕ_b and r_b can be acquired by finite element modeling, where r_b is the depth of breakouts, and ϕ_b is the width of breakouts (Figure 4).

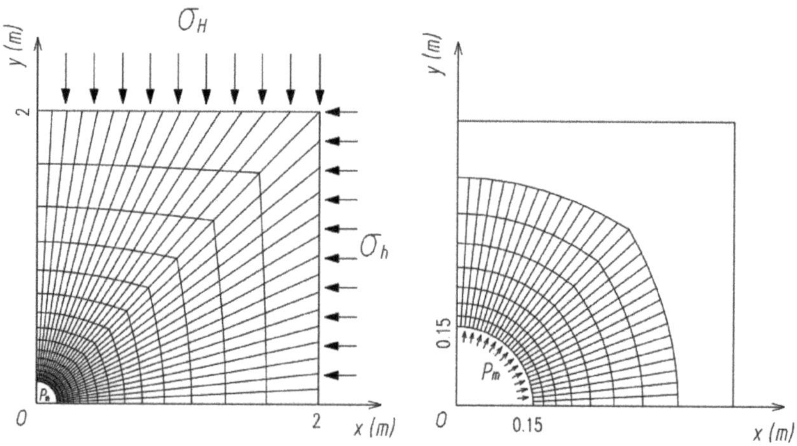

Figure 3. Mesh of finite element model.

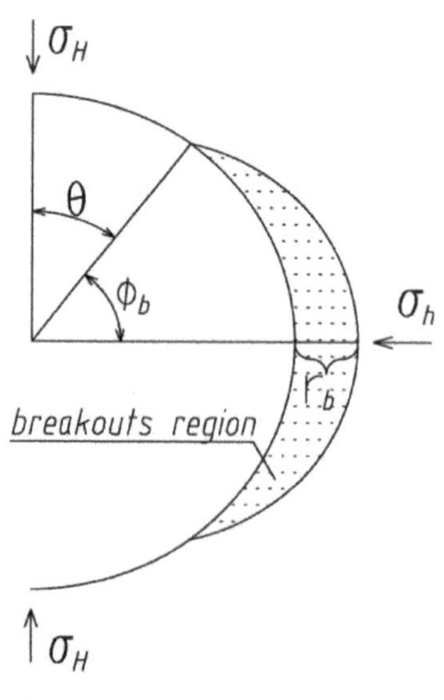

Figure 4. Schematic of a wellbore breakout shape.

The yield condition of rock is determined by Equation (4), and the failure (breakout) condition of rock is determined by Equation (5).

The geometric and mechanical parameters are shown in Table 1.

Table 1. Geometric and mechanical parameters.

Parameter	Value
Young's Modulus, E (MPa)	14,400
Poisson Ratio, ν	0.2
Cohesion, c (MPa)	6
Inner friction angle, φ (°)	35
Radius of well, R (m)	0.15
Bulk modulus of skeleton, K (MP_a)	8000
Bulk modulus of matrix, K_m (MP_a)	36,000
Bulk modulus of fluid, K_f (MP_a)	2250
Porosity of the porous medium, ϕ	0.19
Permeability of the porous medium, k (μm^2)	0.19
Viscosity of the fluid, μ $(MP_a s)$	10^{-9}
Thermal conductivity, λ_T $(Wm^{-1}K^{-1})$	2.5
Thermal expansion coefficient of solid, β_s (K^{-1})	2.1×10^{-5}
Thermal expansion coefficient of liquid, β_f (K^{-1})	2.0×10^{-4}
Specific heat of solid, $c_s(J\ kg^{-1}K^{-1})$	800
Specific heat of liquid, $c_f(J\ kg^{-1}K^{-1})$	4200
Maximum principal stress, σ_H (MP_a)	85~92.5
Minimum principal stress, σ_h (MP_a)	60~72.5
Vertical stress, σ_V (MP_a)	80
Drilling fluid pressure, P_m (MP_a)	45
Temperature difference between drilling fluid and surrounding, T (K)	−50
Maximum allowable tensile strain, $\varepsilon_{breakout}$	0.002

The results of the breakout shape are shown in Table 2. From Table 2, Figures 5 and 6, it can be seen that the relationship between breakout shape and in situ stresses is nonlinear. With relatively large σ_H and difference between σ_H and σ_h, as σ_h increases, the breakout width and depth increase. With relatively small σ_H and difference between σ_H and σ_h, as σ_h increases, the difference between σ_H and σ_h is smaller, the in situ stress is closer to symmetric in situ stresses state, and the breakout width and depth decrease.

Table 2. Result of the shape of wellbore breakouts r_b and ϕ_b by FEM.

No.	σ_h MPa	σ_H MPa	ϕ_b °	r_b mm	No.	σ_h MPa	σ_H MPa	ϕ_b °	r_b mm
1	85.0	60.0	23.3	53.5	13	90.0	67.5	33.2	112.7
2	85.0	62.5	22.9	51.2	14	90.0	70.0	33.3	115.1
3	85.0	65.0	22.1	48.3	15	90.0	72.5	33.7	118.8
4	85.0	67.5	21.2	44.1	16	90.0	60.0	34.3	120.9
5	85.0	70.0	20.4	40.3	17	90.0	62.5	35.0	124.0
6	85.0	72.5	19.3	33.4	18	90.0	65.0	35.8	129.6
7	87.5	60.0	29.7	90.6	19	92.5	67.5	36.2	132.2
8	87.5	62.5	29.8	90.1	20	92.5	70.0	36.4	136.1
9	87.5	65.0	29.9	89.3	21	92.5	72.5	36.9	139.8
10	87.5	67.5	30.0	88.9	22	92.5	60.0	37.5	144.6
11	87.5	70.0	30.2	88.2	23	92.5	62.5	38.2	149.3
12	87.5	72.5	30.4	85.2	24	92.5	65.0	39.3	154.6

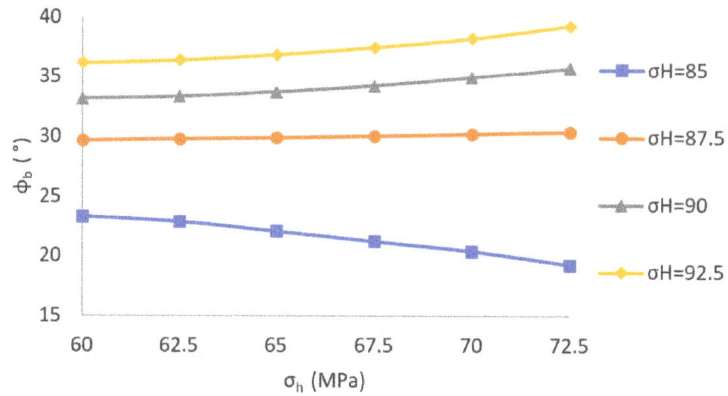

Figure 5. Breakout width for different in situ stresses.

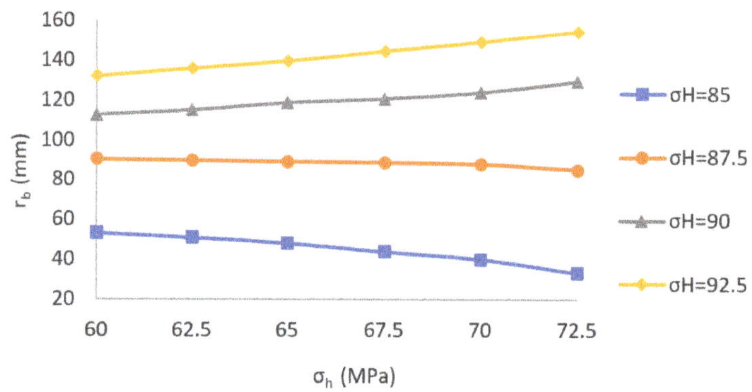

Figure 6. Breakout depth for different in situ stresses.

3.2. *Verification of Wellbore Breakout Process*

For a vertical well shown in Figure 5, associated with horizontal in situ stresses σ_H = 87.5 MPa, σ_h = 72.5 MPa and other parameters are listed in Table 1, the process of well breakout is shown in Table 3 and Figures 7–9.

Table 3. Data of effective principal stresses and breakout region in the process of breakouts.

Elem	Iteration1				Iteration2				Iteration3			
	$\sigma_{1'}$ (MPa)	$\sigma_{2'}$ (MPa)	$\sigma_{3'}$ (MPa)	ε_{max}	$\sigma_{1'}$ (MPa)	$\sigma_{2'}$ (MPa)	$\sigma_{3'}$ (MPa)	ε_{max}	$\sigma_{1'}$ (MPa)	$\sigma_{2'}$ (MPa)	$\sigma_{3'}$ (MPa)	ε_{max}
1	−19.64	−87.86	−54.31	0.00294	−20.56	−91.03	−55.97	0.00315	−28.26	−113.46	−66.27	0.00397
2	−19.86	−87.44	−54.21	0.00292	−20.79	−90.46	−55.86	0.00314	−28.98	−111.17	−65.55	0.00387
3	−20.29	−86.59	−54.00	0.00288	−21.28	−89.41	−55.64	0.00311	−30.35	−106.72	−64.05	0.00362
4	−20.93	−85.35	−53.70	0.00282	−21.97	−87.94	−55.30	0.00305	−31.98	−99.88	−61.76	0.00328
5	−21.73	−83.74	−53.31	0.00275	−23.08	−86.13	−54.89	0.00295	−33.25	−90.66	−58.60	0.00290
6	−22.67	−81.78	−52.84	0.00266	−24.60	−84.05	−54.28	0.00277	−33.25	−79.86	−54.76	0.00236
7	−23.71	−79.54	−52.31	0.00257	−26.56	−81.64	−53.33	0.00248	−29.37	−64.73	−48.08	0.00168
8	−24.85	−77.08	−51.74	0.00246	−27.60	−78.51	−51.49	0.00211	−25.69	−53.71	−43.49	0.00133
9	−26.13	−74.51	−51.08	0.00231	−27.75	−73.95	−50.08	0.00191	−23.57	−49.53	−40.89	0.00094
10	−27.67	−71.91	−50.22	0.00205	−27.14	−66.88	−48.84	0.00199	−23.10	−47.39	−40.83	0.00091

Table 3. *Cont.*

Elem	Iteration4				Iteration5				Iteration6			
	$\sigma_{1'}$ (MPa)	$\sigma_{2'}$ (MPa)	$\sigma_{3'}$ (MPa)	ε_{max}	$\sigma_{1'}$ (MPa)	$\sigma_{2'}$ (MPa)	$\sigma_{3'}$ (MPa)	ε_{max}	$\sigma_{1'}$ (MPa)	$\sigma_{2'}$ (MPa)	$\sigma_{3'}$ (MPa)	ε_{max}
1	−44.48	−142.16	−85.00	0.00501	−73.96	−174.26	−101.91	0.00270	−89.52	−181.38	−111.19	0.00196
2	−44.71	−128.38	−76.66	0.00523	−64.49	−137.95	−83.14	0.00305	−58.62	−116.89	−72.32	0.00194
3	−42.72	−108.77	−70.07	0.00433	−39.34	−69.62	−54.19	0.00185	−25.22	−50.52	−42.48	0.00083
4	−40.12	−87.33	−61.17	0.00323	−17.79	−38.71	−37.73	0.00080	−15.15	−29.09	−35.71	−0.00004
5	−32.66	−62.74	−49.07	0.00195	−17.15	−29.80	−36.13	−0.00007	−15.68	−26.52	−35.48	−0.00014
6	−23.86	−45.66	−40.42	0.00097	−16.17	−27.53	−35.79	−0.00006	−15.10	−24.65	−34.91	−0.00012
7	−20.75	−38.00	−37.25	0.00014	−15.95	−25.85	−35.30	−0.00008	−15.20	−23.19	−34.55	−0.00014
8	−19.36	−36.17	−37.26	0.00020	−16.14	−25.73	−35.25	−0.00006	−15.57	−23.52	−34.70	−0.00013
9	−19.50	−36.24	−37.52	0.00019	−16.71	−26.65	−35.51	−0.00006	−16.02	−24.40	−34.95	−0.00012
10	−19.25	−36.35	−37.60	0.00020	−16.87	−27.24	−35.63	−0.00004	−16.25	−25.08	−35.11	−0.00010

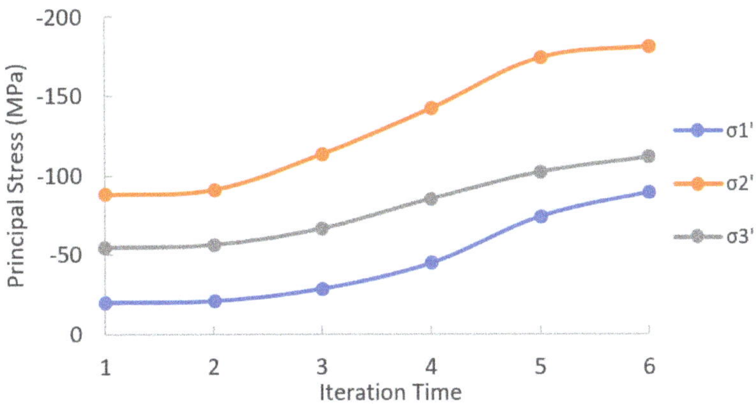

Figure 7. Principal stresses for the element in the breakout tip in the process of breakouts.

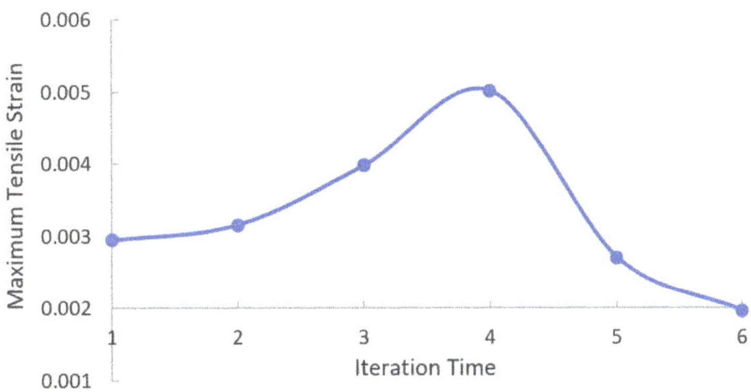

Figure 8. Maximum strain in the breakout tip in the process of breakouts.

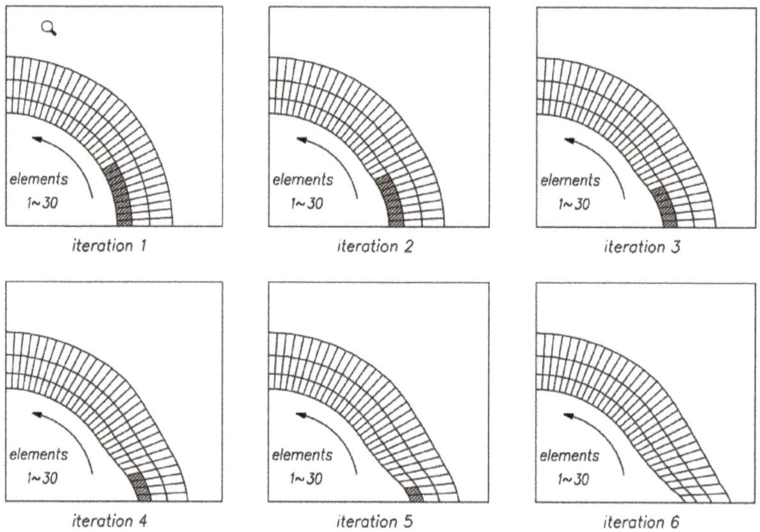

Figure 9. Changing of breakout region in the process of breakouts.

From Table 3 and Figures 7–9, the principal stresses of elements close to the tip of wellbore breakouts are increasing in the process of breakouts, but the maximum tensile strain changes to less than $\varepsilon_{breakout}$. As a breakout develops, the breakouts depth increases and the breakout width decreases, which means the depth of breakouts increases till a stable state, but the width of breakouts remains unchanged [3].

3.3. *Influence of Drilling Fluid Temperature*

For a vertical well shown in Figure 3, associated with horizontal in situ stresses σ_H = 85 MPa, σ_h = 72.5 MPa, and the temperature difference between drilling fluid and surrounding rock $\Delta T = -50 \sim 50$K, and other parameters are listed in Table 1, the breakout shape for different drilling fluid temperature is shown in Figure 10.

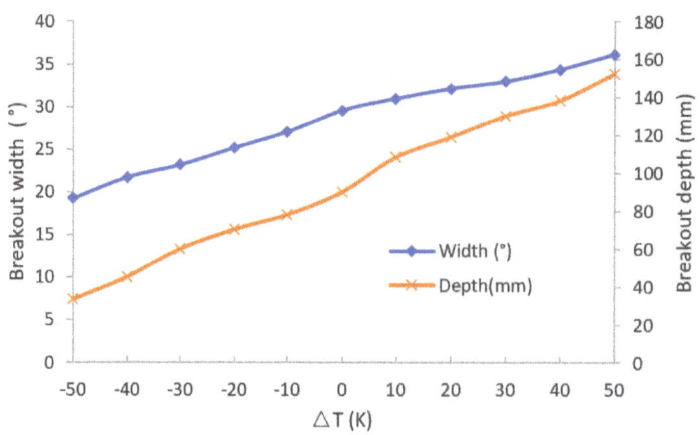

Figure 10. Breakout width and depth for different drilling fluid temperature.

Based on Figure 10, it can be seen that as temperature of drilling fluid increases, the breakout width and depth will increase.

3.4. Comparison between Thermoporoelastic and Thermoporoelastoplastic Modeling

For a vertical well, shown in Figure 3, which is subjected to horizontal in-situ stresses σ_H and σ_h, thermoporoelastic and thermoporoelastoplastic modeling are compared in this section, and parameters are listed in Table 1.

For the thermoporoelastoplastic model, the yield condition of rock is determined by Equation (4), and the failure (breakout) condition of rock is determined by Equation (5). For the thermoporoelastic model, the failure (breakout) condition of rock can be determined by Equation (4).

Figures 11 and 12 show different breakout depths and widths corresponding to different in situ stresses by thermoporoelastic and thermoporoelastoplastic modeling.

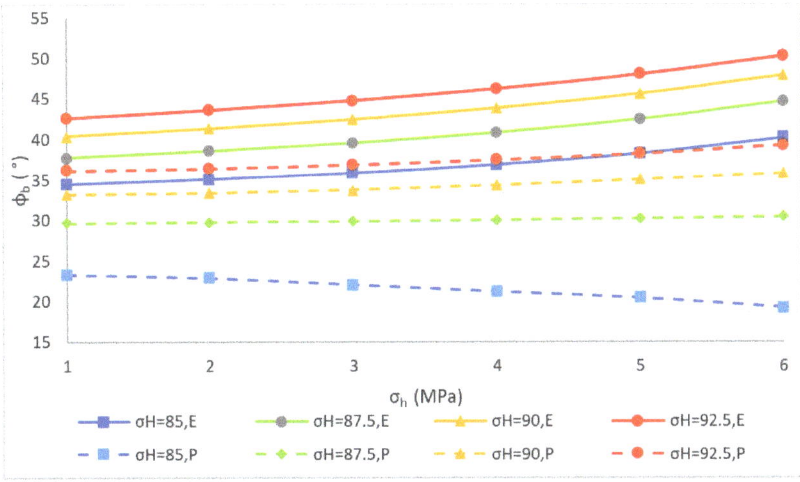

Figure 11. Comparison of breakout width between thermoporoelastic and thermoporoelastoplastic model.

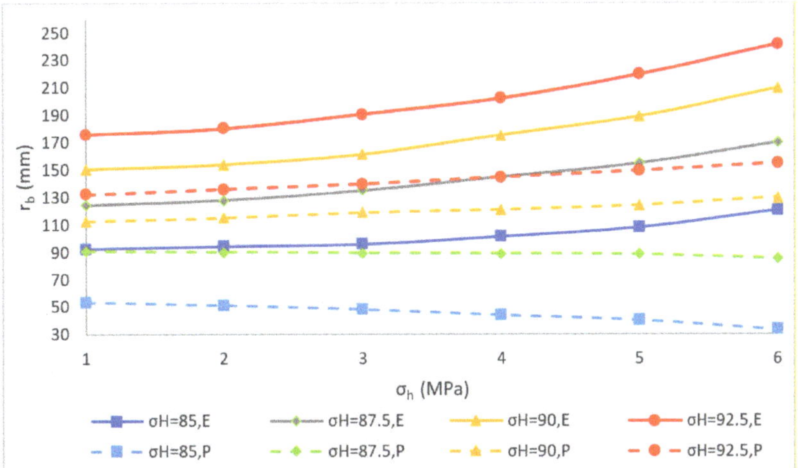

Figure 12. Comparison of breakout depth between thermoporoelastic and thermoporoelastoplastic model.

From Figures 11 and 12, it can be found that for different I -situ stresses, the width and depth of wellbore breakouts for thermoporoelastoplastic model are smaller than those for thermoporoelastic model. For the plastic model, the yield condition of rock is determined by Equation (4), and the failure (breakout) condition of rock is determined by Equation (5). For the elastic model, the failure (breakout) condition of rock can be determined by Equation (4). Therefore, in elastic model, when $F > 0$ in Equation (4), a breakout happens, but in plastic model, rocks just enter plastic state and no breakouts happened. Therefore, the breakout widths and depths are greater for the elastic model compared to the plastic model.

4. Discussion

In this paper, wellbore breakout process is simulated by the finite element method. Plasticity and temperature influence are considered, and results are compared with elastic model. There are some limits for this study. First, mud cake is assumed to be perfect, and there is no liquid transfer across the mud cake. Second, ideal elastictoplastic stress-strain relationship is used. Third, results are not verified by experimental data and field data. Future work is expected to address this in the next step.

5. Conclusions

In this paper, the finite element method is employed to simulate wellbore breakouts based on the thermoporoelastoplastic model. Numerical experiments on finite element modeling of wellbore breakouts show the contrasting tendency of breakouts shape with different in situ stress, and some conclusions can be made as follows:

1. The relationship between breakout shape and in situ stresses is nonlinear. If the σ_h is constant, as σ_H increases, the breakout width and depth become greater. If the σ_H is kept constant while the difference between σ_h and σ_H is relatively large, as σ_h increases, the breakout width and depth increase greater. However, if the σ_H is kept constant while the difference between σ_h and σ_H is relatively small, as σ_h increases, the breakout width and depth decrease instead.
2. In the process of wellbore breakouts, the breakout depth increases till a stable state, but the breakout width remains unchanged.
3. As temperature of drilling fluid increases, the breakout width and depth will increase.
4. For different in situ stresses, the width and depth of wellbore breakouts for thermoporoelastoplastic model are smaller than those for thermoporoelastic model.

Author Contributions: L.Z. and H.Z.: modeling; K.H., Z.C. and S.Y.: supervision. All authors have read and agreed to the published version of the manuscript.

Funding: This research received no external funding.

Data Availability Statement: Not applicable.

Conflicts of Interest: The authors declare no conflict of interest.

References

1. Bell, J.S. The stress regime of the scotian shelf offshore eastern Canada to 6 km depth and implications for rock mechanics and hydrocarbon migration. *Int. J. Rock Mech. Min. Sci. Geomech. Abstr.* **1991**, *28*, A88.
2. Zoback, M.D.; Moos, D.L.; Mastin, L.; Anderson, R.N. Wellbore breakout and insitu stress. *J. Geophys. Res.* **1985**, *90*, 5523–5538. [CrossRef]
3. Zheng, Z.; Kemeny, J.; Cook, N.G.W. Analysis of borehole breakouts. *J. Geophys. Res.* **1989**, *94*, 7171–7182. [CrossRef]
4. Ito, T.; Zoback, M.D.; Peska, P. Utilization of mud weights in excess of the least principal stress to stabilize wellbores: Theory and practical examples. *SPE Drill. Complet.* **2001**, *16*, 221–229. [CrossRef]
5. Haimson, B.C.; Herrick, C.G. In situ stress evaluation from borehole breakouts experimental studies. In Proceedings of the 26th US Rock Mechanics Symposium, Rapid City, South Dakota, 10 June 1985; Balkema: Rotterdam, The Netherlands, 1985; pp. 1207–1218.
6. Haimson, B.C.; Herrick, C.G. Borehole breakouts-a new tool for estimating in situ stress? In Proceedings of the First International Symposium on Rock Stress and Rock Stress Measurement, Stockholm, Sweden, 1–3 September 1986; Centek Publications: Lulea, Sweden, 1986; pp. 271–281.

7. Haimson, B.C. Micromechanisms of borehole instability leading to breakouts in rocks. *Int. J. Rock Mech. Min. Sci.* **2007**, *44*, 157–173. [CrossRef]
8. Papamichos, E. Borehole failure analysis in a sandstone under anisotropic stresses. *Int. J. Numer. Anal. Methods Geomech.* **2010**, *34*, 581–603. [CrossRef]
9. Yuan, S.C.; Harrison, J.P. Modeling breakout and near-well fluid flow of a borehole in an anisotropic stress field. In Proceedings of the 41st ARMA/USRMS, Golden, Colorado, 17–21 June 2006; pp. 6–1157.
10. Zhang, H.; Yin, S.; Aadnoy, B.S. Numerical investigation of the impacts of borehole breakouts on breakdown pressure. *Energies* **2019**, *12*, 888. [CrossRef]
11. Biot, A. General theory of three dimensional consolidations. *J. Appl. Phys.* **1941**, *12*, 155–164. [CrossRef]
12. Bratli, R.K.; Horsrud, P.; Risnes, R. Rock mechanics applied to the region near a wellbore. In Proceedings of the 5th ISRM Congress, International Society for Rock Mechanics, Melbourne, Australia, 10 April 1983.
13. Abousleiman, Y.; Roegiers, J.C.; Cui, L.; Cheng, A.H.D. Poroelastic solution of an inclined borehole in a transversely isotropic medium. In Proceedings of the 35th U.S. Symposium on Rock Mechanics (USRMS), American Rock Mechanics Association, Lake Tahoe, CA, USA, 1 January 1995.
14. Cui, L.; Cheng, A.H.D.; Abousleiman, Y. Poroelastic solution for an inclined borehole. *J. Appl. Mech.* **1997**, *64*, 32–38. [CrossRef]
15. Fakhimi, A.; Carvalho, F.; Ishida, T.; Labuz, J.F. Simulation of failure around a circular opening in rock. *Int. J. Rock Mech. Min. Sci.* **2002**, *39*, 507–515. [CrossRef]
16. Cook, B.K.; Lee, M.Y.; DiGiovanni, A.A.; Bronowski, D.R.; Perkins, E.D.; Williams, J.R. Discrete element modeling applied to laboratory simulation of near-wellbore mechanics. *Int. J. Geomech.* **2004**, *4*, 19–27. [CrossRef]
17. Rahmati, H. Micromechanical Study of Borehole Breakout Mechanism. Ph.D. Thesis, University of Alberta, Edmonton, AB, Canada, 2013.
18. Mostafa, G.; Goodarznia, I.; Shadizadeh, S.R. Transient thermo-poroelastic finite element analysis of borehole breakouts. *Int. J. Rock Mech. Min. Sci.* **2014**, *71*, 418–428.
19. Lee, H.; Moon, T.; Haimson, B.C. Borehole breakouts induced in Arkosic sandstones and a discrete element analysis. *Rock Mech. Rock Eng.* **2016**, *49*, 1369–1388. [CrossRef]
20. Jaeger, J.C.; Cook, N.G.W. *Fundamentals of Rock Mechanics*; Chapman & Hall: London, UK, 1979.
21. Drucker, D.C.; Prager, W. Soil mechanics and plastic analysis for limit design. *Q. Appl. Math.* **1952**, *10*, 157–165. [CrossRef]
22. Whittle, A.J.; Kavvadas, M. Formulation of the MIT-E3 constitutive model for overconsolidated clays. *J. Geotech. Eng.* **1994**, *120*, 173–198. [CrossRef]
23. Akl, S.A.; Whittle, A.J. Analysis of horizontal wellbore stability in clay shale. In Proceedings of the 46th US Rock Mechanics/Geomechanics Symposium, Chicago, IL, USA, 24–27 June 2012; ARMA12-559.
24. Zhang, H.; Yin, S. Poroelastoplastic borehole modeling by tangent stiffness matrix method. *Int. J. Geomech.* **2020**, *20*, 04020010. [CrossRef]
25. Li, X.; ELMohtar, C.S.; Gray, K.E. Numerical modeling of borehole breakout in ductile formation considering fluid seepage and damage-induced permeability change. In Proceedings of the 50th US Rock Mechanics/Geomechanics Symposium, Houston, TX, USA, 26–29 June 2016; ARMA16-244.
26. Mastin, L. The Development of Borehole Breakouts in Sandstone. Master's. Thesis, Stanford University, Stanford, CA, USA, 1984.
27. Yin, S.; Towler, B.F.; Dusseault, M.B.; Rothenburg, L. Fully coupled THMC modeling of wellbore stability with thermal and solute convection considered. *Transp. Porous Media* **2010**, *84*, 773–798. [CrossRef]
28. Lewis, R.W.; Majorana, C.E.; Schrefler, B.A. A coupled finite element model for the consolidation of non-isothermal elastoplastic porous media. *Transp. Porous Media* **1986**, *1*, 155–178. [CrossRef]
29. Aboustit, B.L.; Advani, S.H.; Lee, J.K. Variational principles and finite element simulations for thermo-elastic consolidation. *Int. J. Numer. Anal. Methods Geomech.* **1985**, *9*, 49–69. [CrossRef]
30. Vaziri, H.H.; Britto, A.M. Theory and application of a fully coupled thermalhydro-mechanical finite-element model. *Comput. Struct.* **1996**, *61*, 131–146.
31. Yin, S. Geomechanics-Reservoir Modeling by Displacement Discontinuity-Finite Element Method. Ph.D. Thesis, University of Waterloo, Waterloo, ON, Canada, 2008; pp. 22–23.
32. Zhu, B. *The Finite Element Method*; Tsinghua University Press: Beijing, China, 2018.
33. Neto, E.S.; Peric, D.; Owen, D. *Computational Methods for Plasticity*; Wiley: Chichester, UK, 2008.

mining

MDPI

Article

Cemented Paste Backfill (CPB) Material Properties for Undercut Analysis

Murray Grabinsky *, Mohammadamin Jafari and Andrew Pan

Department of Civil & Mineral Engineering, University of Toronto, Toronto, ON M5S 1A4, Canada; m.jafari@mail.utoronto.ca (M.J.); andrew.pan@mail.utoronto.ca (A.P.)

* Correspondence: murray.grabinsky@utoronto.ca

Abstract: A longstanding mine backfill design challenge is determining the strength required if the (partially) cured backfill is subsequently undercut. Mitchell (1991) called the undercut backfill a sill mat and proposed an analytical solution that is still often used, at least for preliminary design, and has motivated subsequent empirical design methods. However, fully employing the Mitchell sill mat solution requires knowledge of the backfill material's Unconfined Compressive Strength (UCS), tangent Young's modulus (E_t), tensile strength (σ_t), as well as estimates of stope wall closure. Conducting a high-quality UCS test poses challenges but relating the test result to the remaining material parameters is more difficult. Some new material testing data is presented and compared to available published results. Using the parameter $m_i = UCS/\sigma_t$ the range of available testing data is found to be $m_i = 3$ to 22, however, the most compelling data is obtained when the Mohr's failure circle in tension is tangential to the corresponding Mohr–Coulomb failure envelope determined from other strength tests. In these cases, the value $m_i = 4$ is found for the materials tested, which is much lower than the value $m_i = 10$ commonly assumed and implies a limiting UCS 60% lower compared to the conventional assumption. It is also found that the relationship between E_t and UCS is described by a power function that is close to linear, but the values for the constant and exponent in the power function depend on the material tested. However, for given tailings the power function is found to be independent of void ratio, binder type or concentration, curing time, and water salinity, within the ranges these parameters were investigated. Therefore, when E_t is used in the Mitchell sill mat solution it should be correlated with the UCS using the appropriate power function. These correlations are then used with the Mitchell sill mat solution and published measurements of backfill closure strains to estimate the Mitchell solution's range of applicability based on its underlying assumptions, and a similar analysis is extended to an "empirical design method" motivated by the Mitchell sill mat solution. It is demonstrated that these existing approaches have limited applicability, and more generally a full analysis in support of rational design will require numerical modeling that incorporates the effect of confining stress on the material's stiffness and mobilized strength.

Keywords: underhand cut and fill (UC&F) mining; cemented paste backfill (CPB); Mitchell sill mat solution; unconfined compressive strength (UCS); uniaxial tensile strength (σ_t); tangent Young's modulus (E_t)

1. Introduction

Optimizing mining methods and stope sequences can result in a decision to mine underneath previously placed backfill, as shown in Figure 1. One of the design questions to then address is: what backfill strength is required to keep the undercut span stable? Mitchell [1] considered the case shown in Figure 1a and called the lowest portion of backfill, adjacent to the fill barricade, the sill mat. However, this zone is also called a plug when the contained backfill has higher binder content than the overlying main backfill or is allowed to cure before starting to pour the main backfill. The cross-sectional stope dimensions in Figure 1a can be tens of meters in span and height in the case of long hole mining methods, or only a few meters span and up to 200 m high for Alimak mining methods. The elapsed time between pouring the plug (or sill mat) and the next mining stage when the backfill is undercut will depend on the mine's stope sequencing plans, but the backfill plug (or sill mat) must have sufficient strength by this time to maintain a self-supporting condition, including the stress imposed on it by the main backfill. The same considerations apply to Underhand Cut and Fill (UC&F) mining methods (Figure 1b), although the overall dimensions are modest by comparison and usually the hangingwall to footwall span is smaller than for long hole stopes. Additionally, in this case, there will be no overlying backfill stress if an air gap is left between successive pours.

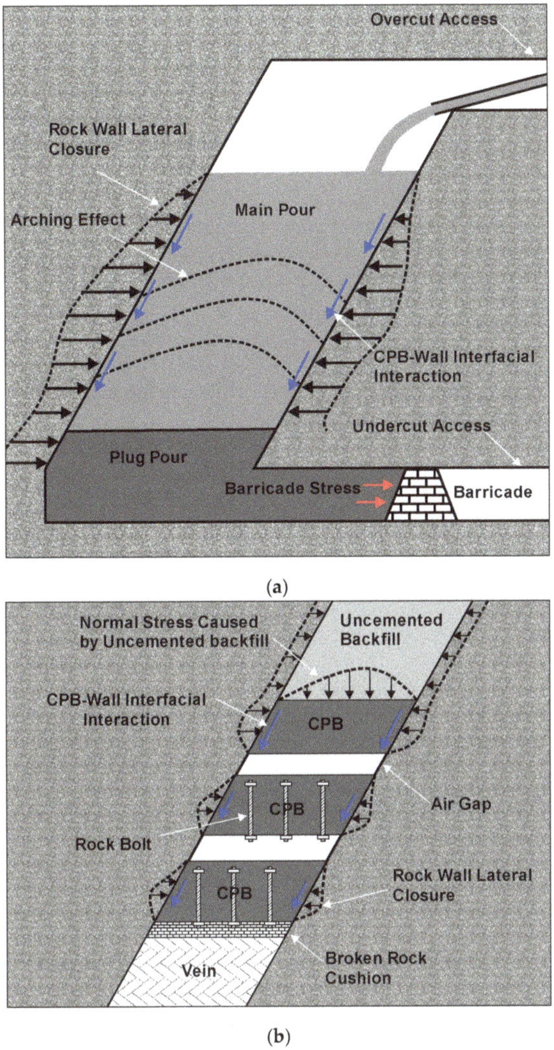

Figure 1. Mining scenarios involving undercutting previously placed backfill: (**a**) a "plug" or "sill mat" in a tall stope such as used in long hole mining; (**b**) a backfilled stope in Underhand Cut and Fill mining.

In 1991, Mitchell [1] proposed an analysis approach for the sill mat design problem based on four different failure mechanisms: flexural failure, tensile detachment (leading to caving), block sliding, and block rotation. The flexural failure mode is the one most often considered critical by mines reporting using Mitchell's approach for sub-vertical backfilled stopes [2,3]. Mitchell presented the limiting solution for this failure mode as Equation (1),

$$\left(\frac{L}{d}\right)^2 = 2(\sigma_t + \sigma_c)/w \tag{1}$$

where L is the undercut span, d is the sill mat height, σ_t is the backfill's tensile strength, σ_c is a "clamping stress" due to sidewall closure, and w is the vertical stress. Equation (1) will be referred to in the present work as the "Mitchell sill mat solution".

The Mitchell sill mat solution assumes a fully encastered (fixed-ended) beam-column, for which the critical stress concentrations are at the encastered ends. Assuming that the clamping stress is small or is ignored, and that the material strength in tension is less than in compression as is the case for cemented backfill, then the tensile failure mode dominates which is why the Mitchell sill mat solution is expressed in terms of tensile strength. The clamping stress term is modeled in the Mitchell solution as an average column compressive stress due to backfill compression caused by hangingwall to footwall closure, and this clamping stress reduces the tensile stress concentration. The clamping stress can be estimated by assuming the backfill behaves in unconfined compression, using Equation (2),

$$\sigma_c = E_t \varepsilon_c = E_t \frac{\Delta}{L} \tag{2}$$

where E_t is the backfill's tangent Young's modulus, ε_c is the sidewall-to-sidewall closure strain, and Δ is the sidewall-to-sidewall closure displacement.

There are two main challenges to using the Mitchell sill mat solution in practice, and both challenges arise from the fact that the UCS is the most commonly cited strength parameter used in parametric studies and for routine Quality Assurance and Quality Control (QA/QC) programs at mines. First, measuring the tensile strength of backfill presents testing challenges and few tensile strength test results are currently available. Second, in conducting individual unconfined compression tests, most studies and QA/QC programs focus on peak stress (the UCS) and not the elastic modulus. The first challenge is typically addressed by assuming the tensile strength is an order of magnitude less than the UCS (i.e., $\sigma_t = UCS/10$) [3,4], which is a heuristic arising from testing intact rock, even though it does not apply to all rock types. This assumption has important implications because backfill strength is approximately proportional to binder content in the strength ranges of interest to backfill design. Hence, using laboratory tests to validate that $\sigma_t > UCS/10$ has proportionate savings implications for binder cost. The second challenge is more problematic. The closure strains measured and reported from various mines range from several percent [5–13] up to 15% at Lucky Friday mine [2,14], and so closure can be significantly influential when using the Mitchell sill mat solution. In principle, a relationship should exist between E_t and UCS as appears in many concrete design codes (usually incorporating $\sqrt{UCS}$), but this has not been investigated thoroughly for mine backfills.

Given the practical challenges identified above, the current work has two objectives. First, to present recent backfill testing data and compare the results to similarly published data, thereby establishing correlations between σ_t and UCS, and between E_t and UCS. Second, to use these correlations to assess potential limitations of using Mitchell's sill mat solution for undercut backfill stability assessments, as well as other empirical design methods motivated by the original Mitchell sill mat solution.

The Mitchell sill mat solution has been used to design sill mats made of cemented rock fill, hydraulic fill, and Cemented Paste Backfill (CPB). The emphasis in this work is on CPB owing to its increasing popularity in underground mining. Because the UCS is critically important to applying the Mitchell sill mat solution, some issues with obtaining quality *UCS* test data are first identified.

2. Obtaining Quality *UCS* Test Data

A set of standardized laboratory test methods for CPB does not yet exist, but [15] review several equivalent standards meant for (cemented) soils, rock, and concrete, and [16] reviews how these should be used in backfill quality control (QC) programs at operating mines. The testing issues considered by [15] include specimen geometry (cylinders versus cubes, dimensions, and aspect ratios), curing conditions, edge preparation and tolerances, loading rates, results interpretation, and the number of tests and statistical evaluation of multiple test results. All the tests reviewed or reported in the current work fall within the scope of guidelines presented by [15].

It is now well established that reducing void space (as quantified by void ratio or porosity) for a given binder content results in stronger and stiffer backfill, as is the case for conventional concrete, because the hydration products generated by the available binder have smaller distances to span between solid particles (sand and silt) and form a denser network. This is shown at the microstructure level in Figure 2 where diamond-polished sections of a conventional mortar and a CPB are imaged using a Hitachi 2460N Scanning Electron Microscope fitted with a Backscattered Electron (BSE) imager and an Oxford Energy-dispersive X-Ray analyser (located in the Department of Civil & Mineral Engineering, University of Toronto, Toronto, ON, Canada): compared to the mortar, the CPB hydration products are diffuse and amorphous.

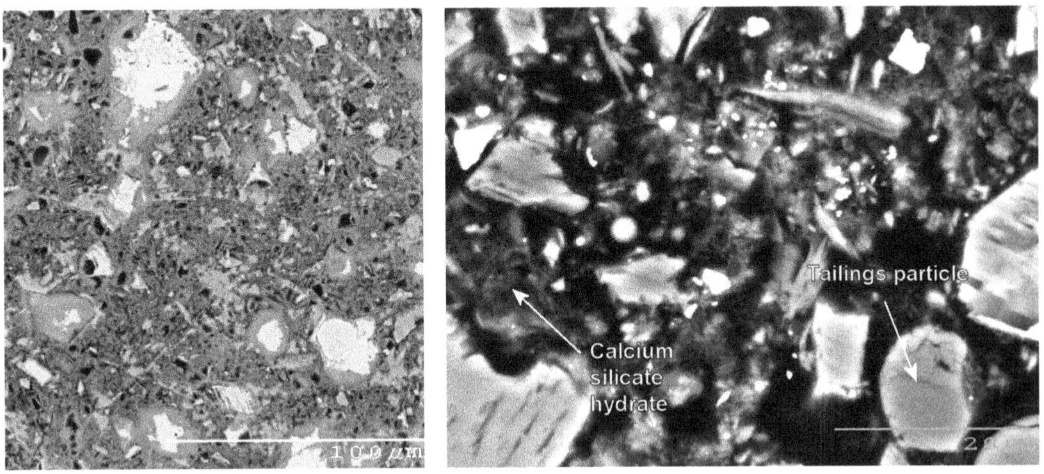

(a) (b)

Figure 2. Scanning electron microscope backscatter images of polished sections of (**a**) a conventional mortar with strong network of hydration products (note scalebar 0.1 mm) and (**b**) a CPB with 15% binder content (note scalebar 0.02 mm). (Imaged by Terry Ramlochan, University of Toronto).

It was also demonstrated in *UCS* tests by [17] and co-workers, where different samples were consolidated at different times and the resulting strengths were invariably greater than a control sample's strength. For example, initial consolidation using a 100 kPa surcharge increased the density from 1.64 to 1.77 t/m^3 and the resulting 7-day *UCS* approximately doubled. Importantly, they demonstrated that the applied stress had no intrinsic strengthening effect: two samples were rapidly consolidated just after mixing, with the load removed from one and remaining on the other (but with no further consolidation occurring) and the resulting strengths were essentially identical. The void-ratio strength dependency was also demonstrated in [18] where specimens were mixed at different water contents and comparing 78% with 74% solids content the 28-day *UCS* approximately doubled. Much more general load paths are considered by [19,20]. For example, ramping the vertical stress to 400 kPa over a 10 h period decreased the void ratio from about 0.9^+ to 0.7 and approximately doubled the 7-day *UCS*. It is therefore imperative in laboratory test work to carefully determine the bulk properties and achieve consistency in *UCS* results for multiple samples.

A testing issue not covered by the quoted standards is mitigating suction development during the dilation phase of the sample response. Figure 3a shows results for a relatively weak and soft CPB tested in two ways. First, three samples are tested underwater (submerged) at an axial strain rate of 0.03 %/min so that the samples can absorb water during their dilation phase to mitigate suction development. These three samples give consistent peak stress, $UCS \approx 30$ kPa. Second, three samples are tested in the usual way and at an axial strain rate of 1%/min to exaggerate the suction development effect. These samples exhibit a distinct bilinear response and achieve a peak stress at least double the true *UCS*. The second demonstration of this effect is shown in Figure 3b, with stronger and stiffer CPB. All tests are conducted following the second method but this time the sample indicated with open blue circle markers is sprayed with water from a misting bottle close to the artificial peak stress, and the suction within the sample is quickly suppressed and the sample fails under the applied load in the displacement-controlled load frame. The characteristic bilinear response can be seen in some of the published CPB literature, and those works were not included in the present analysis if it was deemed that their results could be affected by suction development during material testing.

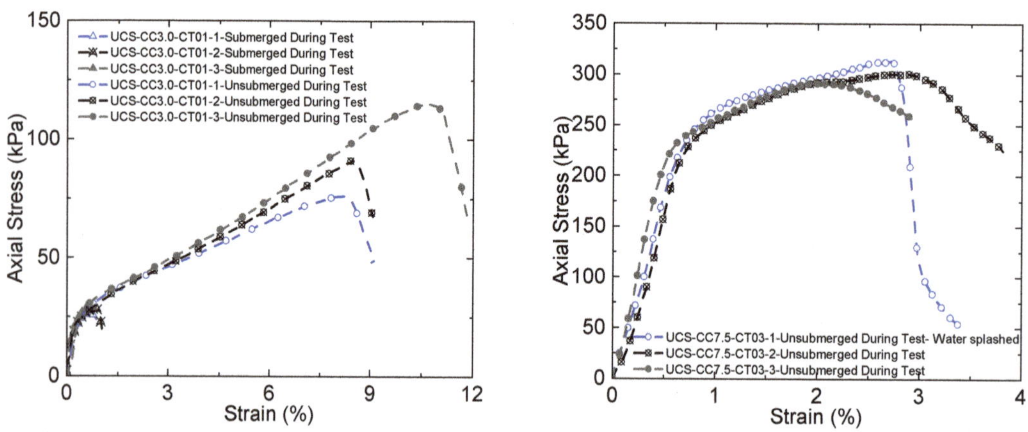

(a) (b)

Figure 3. (**a**) Comparison of peak stresses obtained from three samples tested submerged, and three samples tested in air; (**b**) comparison of three samples tested in air, with one of these samples sprayed with water near its peak stress leading to internal suction loss and sample failure.

In contrast to obtaining the peak strength (*UCS*), assessing the elastic response is an even greater challenge. It is now established that determining the axial strain based on end-to-end platen displacement is inappropriate at small strain levels [21], however, most of the tests reviewed or reported in the current work rely on determining the tangential Young's modulus, E_t, from strains at about $\frac{1}{2}$ *UCS* calculated based on end-to-end platen displacement. Only [14] compare a variety of strain determination methods, including strain gauges glued on the sample at mid-height, but they do not provide strains calculated in the usual way for comparison. The use of Particle Velocity Imaging, PVI, to determine radial strain and therefore estimate Poisson's ratio is described in [22], but those results were assessed as unreliable, and more research and development is required before this method might become generally practical and reliable.

Finally, the tangential modulus E_t is an apparent instantaneous quantity and even if it is linear it cannot necessarily be taken as elastic. This is demonstrated in Figure 4 using load–unload–reload cycles on Macassa backfill (considered in detail subsequently). Similar results are shown in [3] for Stillwater CPB. Nevertheless, it is arguably E_t that should be used in the Mitchell sill mat solution if the backfill's closure response is believed to be essentially one-dimensional and unconfined. The issue is revisited at the end of this work. Another complicating factor in determining a suitable modulus to be used in the Mitchell sill mat solution is that, similar to other geomaterials, a CPB's Young's modulus is pressure-dependent and increasing confining stress increases the elastic modulus. This effect is quantified using consolidated drained triaxial test data in [23] where increasing the confining stress from 50 to 200 kPa increased the CPB's E_t from about 25% for 5% binder content at 28-day cure time, to 100% for 2% binder content at 3-day cure time. Similar effects are shown in the consolidated drained triaxial test results for Williams CPB in [24].

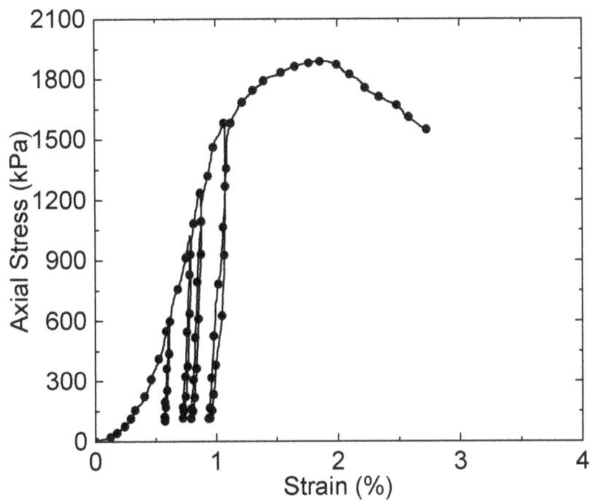

Figure 4. Load–unload–reload cycles applied to Macassa CPB (19% water content, 5.0% binder content, 28-day cure time).

3. Correlating σ_t and UCS

A meta-analysis of tensile and other strength testing data for intact rock is provided by [25]. The test methods covered include direct tensile tests (DTSs) using four different arrangements (split grips, glued end caps, biaxial extension in a Hoek cell, and compression-to-tension load conversion), and indirect tests based on 3-point and 4-point beam bending, Brazilian Tensile Strength (BTS), Ring Test, Disc Bending, and hydraulic fracture. They conclude that the DTS is the truest measure of tensile strength, however, it is impractical for large-scale test programs and therefore corrected BTS test results are suggested, which can then be correlated to UCS. A similar meta-analysis will be carried out next for CPB materials, however, there is significantly less data available than is the case for rocks. For the current analysis it will be convenient to use the parameter m_i from the Hoek–Brown failure criterion, where $m_i = UCS/\sigma_t$. In [26], the author gives tabulated values of m_i by rock type, showing trends of decreasing m_i going from igneous to metamorphic to sedimentary rocks, and decreasing m_i going from coarse to very fine-grained texture. The common assumption for CPB, $m_i = 10$ is representative of fine-grained clastic materials and is similar to the value assumed for normal strength concrete in North American reinforced concrete design codes. However, the authors of [26] indicate a value as low as $m_i = 4 \pm 2$ is appropriate for very fine-grained clastic rocks (e.g., claystone), and $m_i = 7 \pm 2$ is appropriate for fine-grained clastic rocks (e.g., siltstone), therefore, it is possible that values of m_i as low as 3 or 4 might be found for CPB. Given the significance this parameter has on the assessed required UCS when determining σ_t using the Mitchell sill mat solution, direct measurements using the mine's CPB are clearly desirable. Table 1 summarizes CPB tensile test data found in the literature, and the details of these are described next.

Table 1. Summary of $m_i = UCS/\sigma_t$ for various mines. n is the number of datapoints in the dataset.

Mine	n	Test Type	m_i
Lucky Friday	14 5 3 14	Brazilian Splitting Direct Tension In Situ Direct Tension	8.8 avg 7.3 avg 16 avg 22 avg
Cannington	8 *	Direct Tension	7–10
Stillwater	4	Brazilian	4.5–7
Raleigh	2 3	3-Point Bending Brazilian	3 8
Brunswick	4	3-Point Bending	11
Williams	18	Direct Tension	4
Macassa (new)	6	Direct Tension	3–5
Uncemented	32	4-Point Bending	4

* 7- and 14-day subsets.

Lucky Friday CPB (i.e., the CPB produced at Lucky Friday mine) tensile strength was tested by [27] using Brazilian tests, splitting tests (loading along the long axis of cylindrical core samples), direct tension with glued end caps, and pull tests on samples partially cored (i.e., not end detached) from large block samples (which the authors call an "in situ test"). These resulted in m_i values in the range of about 7–22, but the photograph of a failed sample from a direct tension test showed a failure plane inclined to the tensile loading axis, suggesting that these test results may not be reliable.

Cannington CPB tensile strength was tested by [22] using direct tension with glued end caps, with reported values equivalent to m_i of about 7–10, however the tensile strength was relatively invariant between 7 and 56 days which does not seem reasonable given that compressive strength continued to increase after 28 days. It was noted that samples dried out (indicated by decreasing degrees of saturation) during the test period, but it is unclear if the samples were re-saturated prior to tensile testing.

Stillwater CPB tensile strength was tested by [3] using Brazilian tests conducted on samples obtained from mass pours at the mine, and although only four tests were conducted the range of corresponding m_i was 4.5–7.

Raleigh CPB tensile strength test results are reported in [28] using three Brazilian tests and two 3-point beam bending tests, and these are compared with Cannington direct results [22], and Brunswick CPB using four 3-point beam bending test results [29]. Significant scatter is observed, with equivalent values for m_i ranging from about 3 (Rayleigh 3-point beam bending) to 11 (Brunswick 3-point beam bending), but the authors conclude that the most credible results are from Brazilian tests with an upper bound m_i about 8.

More recently, Pan et al. [30,31] developed a direct tension test based on a compression-to-tension load conversion method, and compared the measured Williams CPB tensile strengths to the UCS and also to direct shear test results from samples with corresponding mix designs. Importantly, all tested samples were cured in virtually saturated conditions and the UCS and direct shear tests were carried out with the samples submerged to prevent suction development associated with sample dilation. The tensile tests could not be carried out submerged, but water was sprayed on the sample during testing as discussed for the test results shown in Figure 3. An example of the combined results is shown in Figure 5, which shows the Mohr's failure circles for three UCS tests, three direct tension tests, and seven direct shear tests at varying normal stress from 0 to about $\frac{1}{4}UCS$. The consistency between the three independent test methods adds confidence to the reliability of the combined results, and the conclusion is that $m_i = K_p$, the passive earth pressure coefficient which for the material tested is about 4. This combination of tests is now recommended for high-quality CPB strength determination at low confining stress levels. For comparison, Williams tensile strengths were also determined using the Brazilian test method [32], with resulting strengths much lower than from direct tension tests. This is contrary to the trends for rock reported in [25] where the Brazilian strength results are typically higher than direct strength results. The differences may be due to the relatively soft nature of CPB as opposed to intact rock.

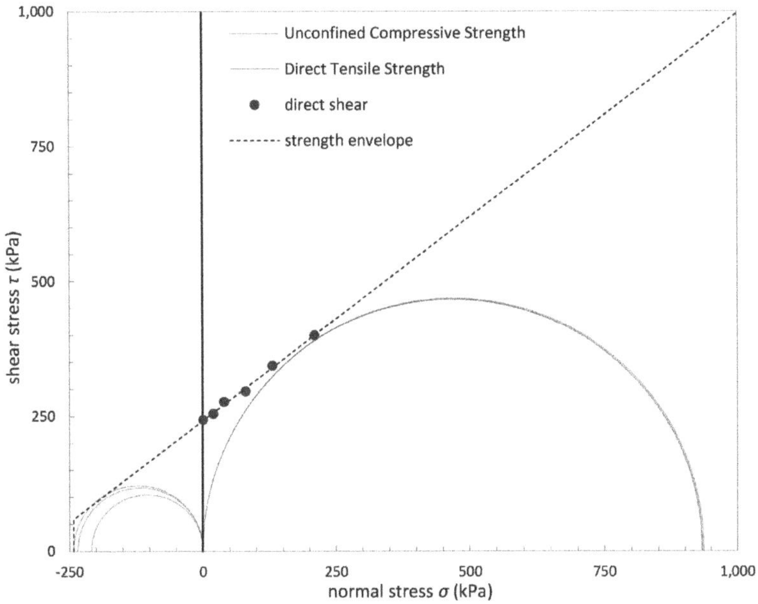

Figure 5. Mohr–Coulomb failure envelope from direct shear tests (dot markers) compared with Mohr's failure circles for direct tension (3 tests) and unconfined compression (3 tests).

Macassa CPB was tested using the same direct tension device as [30,31]. For 10% binder content and testing at 3, 7, and 14 days the strength ratios varied from $m_i = 5$ to 3. However, it was found that a maximum tensile strength of about 1 MPa could be reliably determined, beyond which the samples did not fail in the expected fashion (meaning, a fracture surface normal to the tension loading axis and in the middle third specimen length) and so the test frame and specimen casting mold are being modified to overcome this limitation.

An interesting contrast to cohesion created by hydraulic binders is presented in [33] where matric suction in uncemented samples creates an apparent cohesion, and the unsaturated tensile strength was determined using carefully measured 4-point beam bending tests. The unsaturated backfill samples were assumed to retain the effective stress friction angle (35°) determined from triaxial test results on the same backfill, and the measured tensile strength and failure envelope results were consistent with the findings of [30,31] shown in Figure 5.

Based on the reviewed and presented new results, it is recommended that wider consideration is given to carrying out the direct tension, direct shear, and UCS test methods following the procedures discussed in [30,31] and shown in Figure 5. For the Williams and

Macassa CPB, it was reliably determined that the tensile strength is much higher than the conventional assumption $\sigma_t = UCS/10$, as may well be the case for many other CPBs.

4. Correlating E_t and UCS

The Williams and Macassa datasets mentioned previously contain a relatively large number of UCS tests (25 and 45, respectively) which are useful for evaluating the functional relationship between the tangential Young's modulus, E_t, and the UCS. The Williams dataset is in two test series: the first considers binder contents of 4.2%, 6.9%, and 9.7% tested at 3, 7, 14, and 28 days (12 tests total), all mixed at the same water content resulting in a void ratio $e \approx 1.0$; and the second considers binder contents 3%, 5%, 7%, and 10% all tested at 56 days but mixed at lower water contents resulting in void ratios as low as $e \approx 0.7$. Recall that mixing at lower water contents to achieve lower void ratios is like the approach used in [18]. The Macassa dataset uses binder contents of 5%, 7%, and 10% tested at 1, 3, 7, 14, and 28 days, and at water contents of 19.0%, 20.5%, and 22.0%. The Macassa CPB contains sand, and its void ratio is lower than the Williams samples, generally $e \approx 0.65$. The relationship between E_t and UCS for the combined Williams datasets and the Macassa dataset are shown in Figure 6.

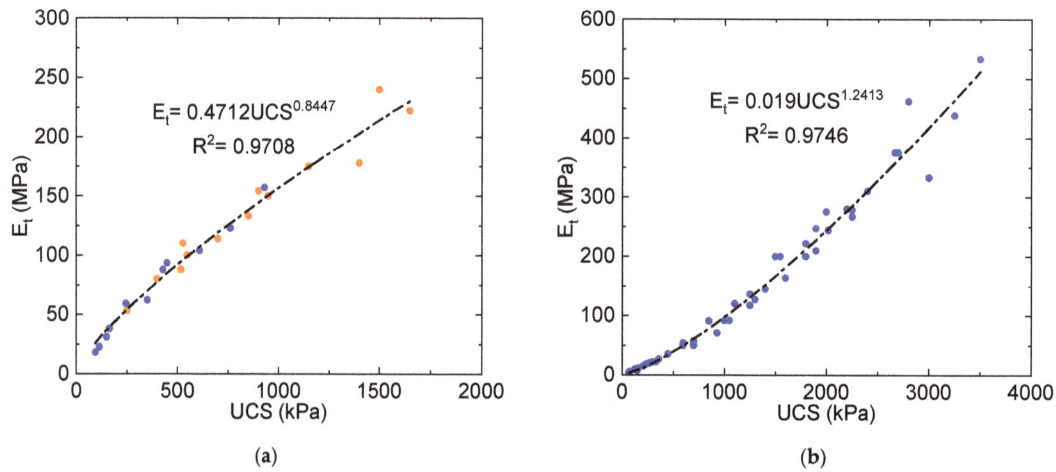

Figure 6. Relationship between E_t and UCS for (**a**) Williams showing standard water contents (blue markers) and reduced water contents (orange markers), and (**b**) Macassa.

Both datasets are fit with power functions and show good statistical significance (i.e., coefficients of determination close to unity). For these datasets, it is important to note that the trends are independent of cure time and void ratio. For example, a sample prepared at 4.2% binder content and tested at 28-day cure time has about the same strength as a sample prepared at 6.9% binder content and tested at 3-day cure time (although this is not shown), and correspondingly similar tangent Young's moduli. Similarly, for the Williams results the combined datasets are shown to overlap in the range 250–935 kPa UCS, although the void ratios are significantly different between the two preparation methods. However, the statistical fits for the Williams and Macassa datasets are demonstrably different, with the exponent in the power function being < 1 for Williams and > 1 for Macassa. Given that consistent preparation and testing procedures were used, it must be concluded that the relationship between E_t and UCS is material dependent.

Another high-quality dataset is provided for Garpenberg by [19], where two types of binder were trialed: unblended Portland cement, and a blend of 20% Portland cement to 80% ground blast furnace slag. In each case, 3, 5, 7, and 10% binder contents were tested, and a range of water contents and corresponding void ratios were achieved through different rates of load application resulting in $e \approx 0.7 - 0.9$ for the Portland-only series and $e \approx 0.6 - 0.9$ for the Portland/slag blend series. The corresponding 28-day strength results are shown in Figure 7. Although the Portland/slag blend is more effective in achieving higher strengths for equivalent mix design and cure time, the trend between E_t and UCS appears to be consistent suggesting the different binder type is not significant to this trend for the range of parameters considered. Additionally, note that the exponent in the power function is similar to the Williams CPB, however, the constant is about 2.7 times higher.

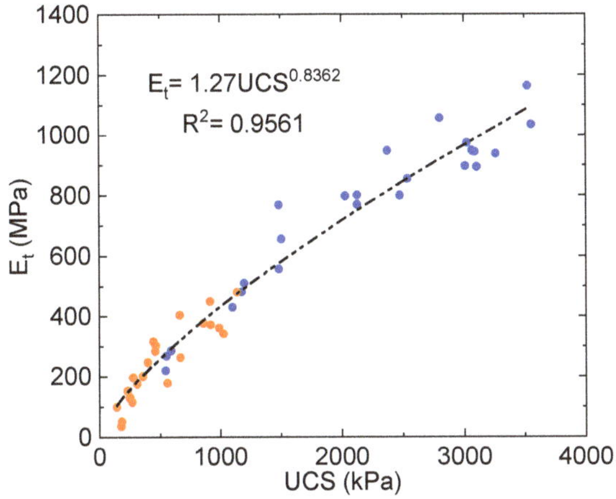

Figure 7. Relationship between E_t and UCS at 28-day cure time for Garpenberg with Portland (orange markers) and Portland/slag blended (blue markers) binders (data from [19]).

A similar dataset is generated by the same authors for an unidentified mine in Quebec [20], this time using only the Portland/slag blend but with a range of curing times between 7 and 28 days. The best-fit power function to this dataset has a constant of 0.4328 and exponent of 0.9526, with a coefficient of determination of 0.9267. Presumably the authors' testing procedures were consistent for the Garpenberg and Quebec datasets, indicating that the same binder was less effective at strength and stiffness gains for the Quebec mine. This also supports the hypothesis that the E_t versus UCS relationship is unique to each CPB.

A further previously unpublished dataset comes from Kidd mine. Kidd uses blended aggregates from an esker deposit, which can make sample preparation and testing difficult especially if the samples come from field coring (Figure 8b). The results shown in Figure 8a show considerable scatter (and a correspondingly low coefficient of determination) but its significance is that the data is combined from three different organizations (the University of Toronto and two industry laboratories) tested at three different times, yet similar results were obtained in terms of both trend and variability.

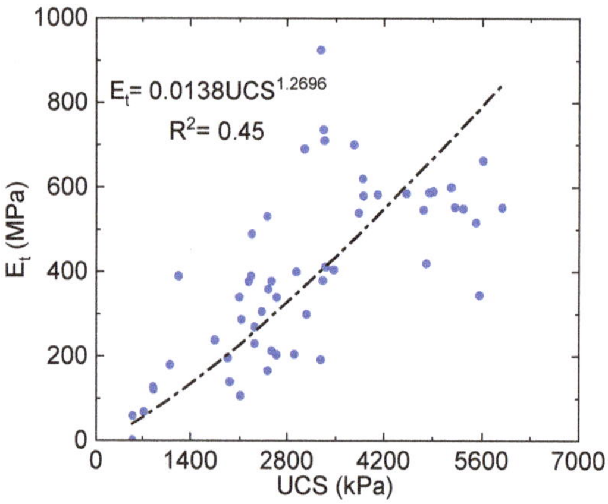

(**a**)

(**b**)

Figure 8. (**a**) Relationship between E_t and UCS for Kidd, and (**b**) example of NQ diamond drill core from field tests.

The effect of the mixing water's salinity on hydration at cold temperatures (−6 °C) was considered in [34] using a synthetic tailing (ground silica), 4.5% Portland cement, water:cement ratio 7.35, and water salinities of 0, 5, 35, and 100 g/L. Results are shown in Figure 9 for testing times of 7, 28, and 90 days, and indicate that the range of studied salinities profoundly influences hydration effectiveness in the studied timeframes, however, the salinity concentration does not seem to influence the functional relationship between E_t and UCS in any statistically significant way.

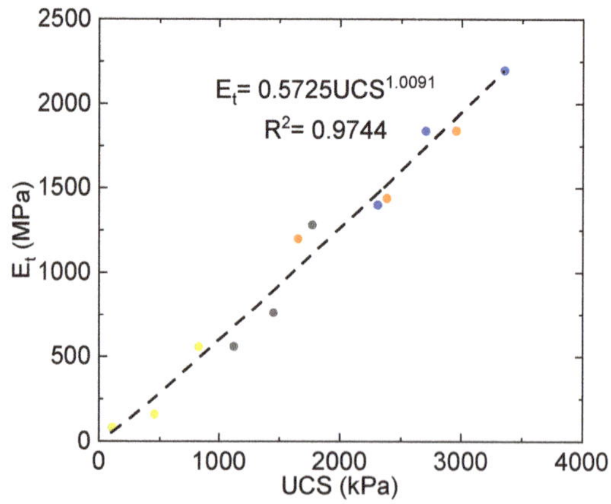

Figure 9. Relationship between E_t and UCS for salinity study with samples tested at 7, 28, and 90 days: marker colors are 0 g/L (blue), 5 g/L (orange), 35 g/L (grey), and 100 g/L (yellow). Data from [34].

Additional data was found in the literature and is summarized in Table 2 and plotted in Figures 10 and 11. The entries in Table 2 are approximately ordered so that the strongest datasets (those with the most datapoints, n, and strongest statistical fit, R^2) are first. Some of these datasets have been parsed or combined, indicated by the highlighted cells. Fitting parameters are not provided for datasets that have low numbers of datapoints and/or very poor statistical fits, but these data are used in the figures.

All datapoints from the literature review and the new data are plotted in Figure 10 to provide an indication of the variability within individual datasets and the ranges of E_t, UCS, and the ratio of these for all available data. Note that the ratio of upper-bound to lower-bound E_t values for a given UCS is approximately an order of magnitude. Where a given organization produces multiple datasets (e.g., the Williams vs. Macassa datasets, and the Garpenberg vs. Quebec datasets) the differences between those datasets should be considered "statistically significant". However, the Kidd dataset notwithstanding, it is plausible that differences in testing procedures between organizations (e.g., sample preparation, machine stiffness, loading rates, transducer types and locations of testing equipment, data reduction and interpretation of raw test results, etc.) may be responsible for some of the variability observed. Regardless, it is useful to understand the range of results obtained by different organizations testing different CPB materials, so that new test results can be interpreted in the context of this combined data. Furthermore, the wealth of available evidence compellingly indicates that the $\sqrt{UCS}$ type functional relationships commonly accepted for normal strength concrete are inappropriate for CPBs.

Taking the data in Table 2 at face value, it is not clear that the functional relationship between E_t and UCS can be related to any of the indicated parameters, such as % sand content, void ratio (or porosity), binder type, or specific gravity of the solids. Therefore, when a new CPB is being considered, high-quality laboratory test results will remain necessary to establish that material's E_t versus UCS functional relationship.

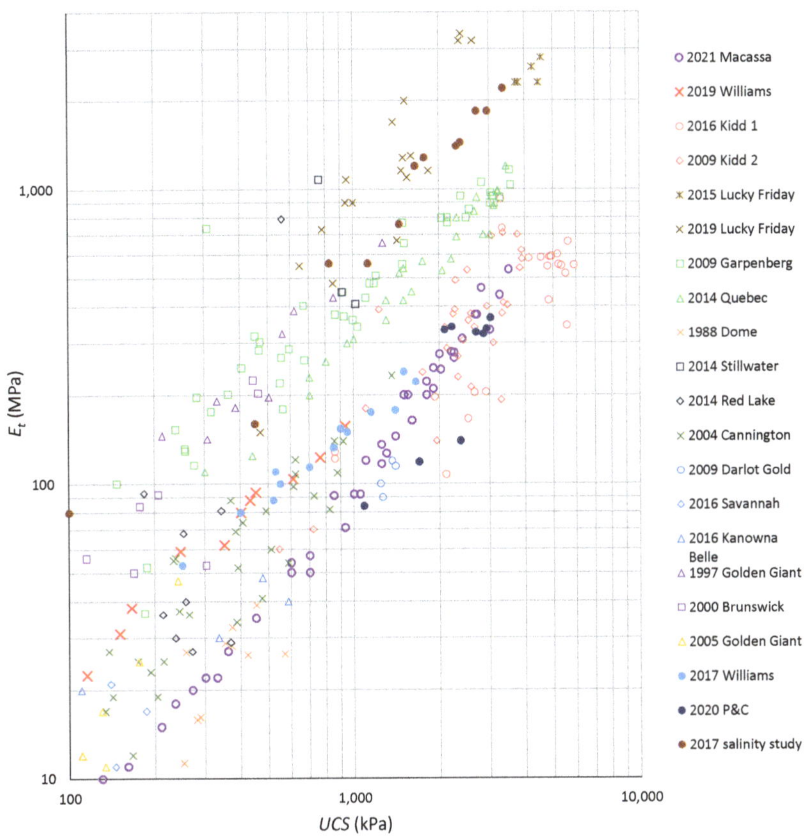

Figure 10. Relationship between E_t and UCS: combined results (the salinity study markers may be an order of magnitude too high, see footnote of Table 2).

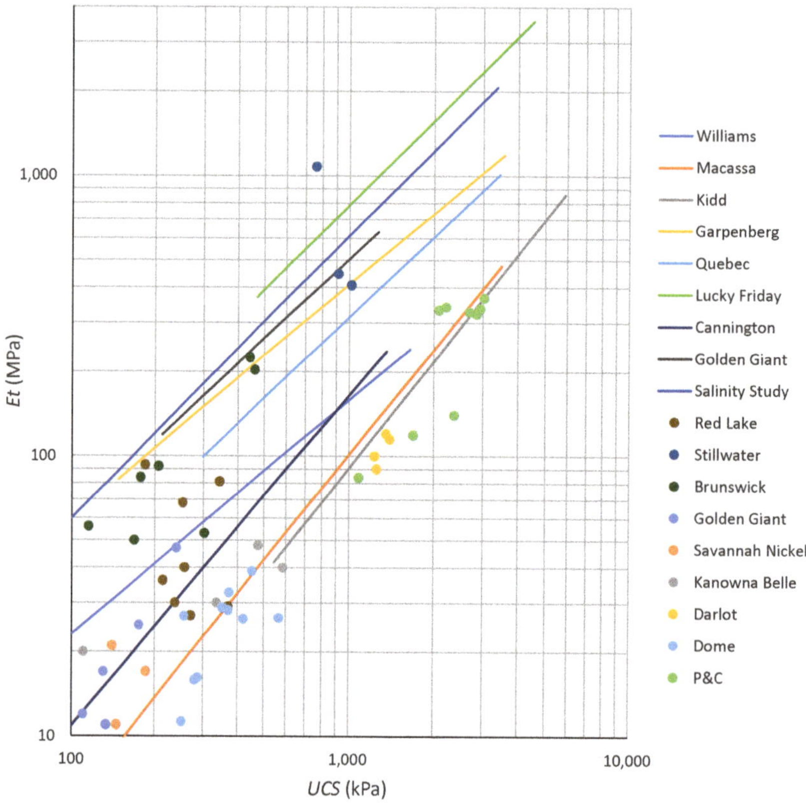

Figure 11. Relationship between E_t and UCS: summarized results (the salinity study markers may be an order of magnitude too high, see footnote of Table 2).

Table 2. Summary of E_t and UCS datasets, and fitting parameters for $E_t = \text{const } UCS^{\text{exp}}$ where the modulus is in MPa and the strength is in kPa; n is the number of datapoints in the dataset; the strength range is UCS_{min} to UCS_{max}; R^2 is the coefficient of determination; e is void ratio; G_s is specific gravity of tailings; % sand is the fraction of tailings (and any blended aggregates) greater than 0.075 mm; values may be approximated from the original publications and the reader should refer to the original if detailed information is required. Regression analysis is performed using MS Excel for power functions. The abbreviations used for binder type are Portland Cement (PC), fly ash (FA), and ground granulated blast furnace slag (Slag) and their relative proportions when blended. The regression analysis parameters and other mix design information are not reported for small datasets nor datasets with poor coefficients of determination. For the remaining datasets, mix design parameters unavailable from the original article are indicated using n/a.

Mine	n	UCS_{min} (kPa)	UCS_{max} (kPa)	const	exp	R^2	e	Binder Type	Binder %'s	Curing Days	G_s	% Sand
Williams(a) [31]	12	95	935	0.427	0.858	0.995	1.0	PC	4, 7, 10	3, 7, 14, 28	2.85	10
Williams(b) (new)	13	250	1650	0.838	0.756	0.971	>0.7	PC	3, 5, 7,10	56	2.85	10
combined (new)	25	95	1650	0.471	0.845	0.971						
Macassa (new)	45	70	3500	0.019	1.241	0.975	0.65	10PC:90Slag	5, 7, 10	1, 3, 7, 14, 28	2.76	67

Table 2. *Cont.*

Mine	n	UCS_{min} (kPa)	UCS_{max} (kPa)	const	exp	R^2	e	Binder Type	Binder %'s	Curing Days	G_s	% Sand
Garpenberg(a) [19]	23	147	1140	0.687	0.939	0.760	0.7–0.9	PC	3, 5, 7, 10	28	2.96	49
Garpenberg(b) [19]	24	549	3558	2.154	0.765	0.949	0.6–0.9	20PC:80Slag	3, 5, 7, 10	28	2.96	49
combined	48	147	3558	1.270	0.836	0.956						
Quebec [20]	27	300	3425	0.433	0.953	0.927	0.6–0.9	20PC:80Slag	3, 4.5, 7	7, 14, 28	3.70	20
Lucky Friday(a) [27]	5	3720	4540									
Lucky Friday(b) [2]	18	470	2625	0.032	1.458	0.808	0.55–0.67	25PC:75Slag	8–10	>28	n/a	29
combined	23	470	4540	0.784	1.000	0.741						
Cannington [22]	36	58	1361	0.048	1.178	0.818	4 wc	PC	2, 4, 6	7, 14, 28	3.20	35
Kidd (new)	54	544	5880	0.014	1.270	0.450	0.7	10PC:90Slag	2–3, 4.5	>28	2.74	55
Salinity study * [34]	12	100	3550	0.573	1.009	0.974	n/a	PC	4.5	7, 28, 90	2.7	n/a
Golden Giant [35]	9	213	1267	0.801	0.933	0.905	1.0	50PC:50FA	3, 5, 7	28, 56, 112	2.86	40
Red Lake [3]	13	54	562									
Dome [36]	12	246	566									
P&C [37]	9	1085	3025									
Brunswick [28,29]	7	115	461									
Golden Giant [38]	6	90	240									
Darlot [17]	4	1240	1400									
Kenowna Bell [39]	4	110	585									
Savannah Nickel [39]	3	140	186									
Stillwater [3]	3	760	1018									

* There appears to be an inconsistency between the reported E_t values and the corresponding stress–strain curves in the original publication, and it is possible that the E_t values used here are an order of magnitude too large. The authors [34] were asked to clarify this issue, but a response was not obtained by the publication date.

A clearer comparison between the different datasets is provided in Figure 11 where those datasets with fitted power functions are plotted as straight lines in log-log space, and the remaining datasets are shown as individually plotted datapoints. For the sake of comparison with other geomaterials, it is useful to define the Modulus Ratio $MR = E_t/UCS$. From Figures 10 and 11, reasonable limits for the reviewed CPB materials are $100 < MR < 1000$. The upper bound is like normal strength concrete, while the lower bound compares well with very-fine grained clastics, and the mid-range is consistent with many of the rock types reported in [26].

Finally, direct tension tests of the type described by [30,31] allow comparing E_t in uniaxial tension and compression. For the nine cases considered by [30,31], with 6.9% and 9.7% binder content tested at 7, 14, and 28 days, the modulus in tension was approximately half of that in compression. However, this is a limited dataset using only one material, and much more testing using other materials is required to determine if the heuristic is generally applicable to other CPBs.

5. Implications for Mine Backfill Design Using Mitchell's Sill Mat Solution

The Mitchell sill mat solution assumes a fully fixed-ended (encastered) ideal beam for which the critical tensile stress concentrations are located at the top corners, which for sill mats (or plugs) correspond to the contacts at the sidewalls. If the side wall closure effect is ignored (i.e., $\sigma_c = 0$) then the Mitchell sill mat solution can be rewritten to determine the limiting UCS (strength factor = 1) as Equation (3),

$$UCS = m_i \frac{1}{2} w \left(\frac{L}{d} \right)^2 \tag{3}$$

making it immediately apparent that the required strength is directly proportional to m_i, commonly assumed to be 10. However, the values summarized in Table 1 show many CPBs having m_i values less than $m_i = 10$. More important, the results obtained by [30,31] which are substantiated by consistency with direct shear and UCS test results indicate $m_i = K_p$, or about 4 for the material tested. This result is consistent with the observations of suction-induced tensile strengths in unsaturated beam tests conducted by [33]. It should be noted that $K_p = 4$ corresponds to an effective stress friction angle $\phi' = 37°$ which is a commonly reported value for CPBs composed of non-plastic silt (ML) tailings. Furthermore, $m_i = 4$ is consistent with the Hoek–Brown recommendations for very fine-grained clastic materials. It is therefore plausible that many CPBs have greater tensile strength than is currently thought to be the case, and future direct tensile strength testing of the type described in [30,31] is therefore encouraged.

If the fixed-ended beam analogy is strictly accepted, then it is necessary to question if the backfill tensile strength is relevant or if it is instead the backfill-to-host rock contact tensile strength that should be used, which will likely be generally much smaller and perhaps negligible. If the contact tensile strength is taken as zero, then the assumed failure mechanism is no longer relevant and the underlying mechanism must account for the progressive failure of a fixed-ended beam. However, this additional complexity is not warranted if sidewall closure is considered, described next.

If the Mitchell sill mat solution is used and reasonable values of E_t are determined based on laboratory studies (or the trends shown in Figure 11, if considered in a parametric study) then one finds the results are very sensitive to the assumed closure displacements and therefore closure strain, $\varepsilon_c = E_t \frac{\Delta}{L}$. This is because the UCS is typically reached at less than 2% axial strain for the lower-bound trends shown in Figures 10 and 11 and milli-strain for the upper-bound trends, and proportionately less for the tensile strength. The literature reviewed in Section 1 indicated that many mines have measured several percent closure strains, and up to 15% closure strain at Lucky Friday mine. Even if the effects of confinement on the elastic modulus and mobilized strength are accounted for, such strains must drive the CPB into its post-peak response. For example, drained triaxial tests were conducted by [23] with confinement 50–200 kPa on Shandong CPB, and by [24] with confinement 25–350 kPa on Williams CPB, and for the strongest materials tested by each, the axial strains to failure ranged from about 4% at the lowest confinement levels to 12% at the highest. Similarly, the authors of [40] interpreted 1-dimensional consolidation results in the context of ground reaction curves between CPB and the host rock during stope closure and concluded that the strains can essentially be ignored until the induced stress in the CPB reaches $2UCS$, beyond which the stiffness increases exponentially and so too does the reaction. Considering the likelihood that many undercut CPB sill mats or plugs will be forced into the post-peak regime of their stress–strain response, design based on elastic beam analysis is tenuous. This interpretation is consistent with recent numerical modeling simulations considered in [41–44], where a reasonable expanse of rock mass was included surrounding the modeled stope to better capture the closure strains induced by excavation. In [43] a parametric study considered sill mats in (sub-) vertical orebodies at depths from 300 to 800 m, corresponding to horizontal stresses of 16–43 MPa. The CPB's cohesion and Young's modulus were not varied in a consistent way, but instead the modulus was varied from 0.6 to 1.8 GPa and the limiting cohesion for stability was determined. For Young's modulus of 0.6 GPa the corresponding UCS based on the assessed cohesion was about 1–2 MPa; and for Young's modulus 1.8 GPa it was 3–7 MPa. Note that these ranges lie within the composite data presented in Figure 11 and are therefore plausible, but do not consistently represent a single CPB. (The authors acknowledge that strength and stiffness should increase in some systematic fashion.) They found that the hypothesized fixed-ended beam type response failed to materialize, and the resulting closure displacements drove the backfill to crushing failure for undercut spans up to 7 m. This finding is only applicable to the range of parameters studied but does indicate that the Mitchell sill mat solution cannot be applied indiscriminately.

6. Implications for Empirical Mine Design Motivated by Mitchell's Sill Mat Solution

Simplified design charts "based on" or "motivated by" Mitchell's sill mat solution have been proposed [45–48], but their development has never been fully explained. Pakalnis's "empirical" design chart [46,48] is particularly interesting because it compares the design chart strength recommendations with case histories. The presentation of the chart does not, however, clarify if each plotted case study obtains the intended Factor of Safety FS = 2, or if the design has a higher or lower FS. It can be shown that the design chart has the underlying formula, for strength in MPa and lengths in meters, given in Equation (4),

$$UCS_{FS=2} = 0.15L\left(\frac{L}{d}\right)^{1.5} \tag{4}$$

The motivation for Equation (4) as compared to the modified Mitchell sill mat solution (Equation (3)) can be assessed as follows: the factor $0.15L$ in Equation (4) can be explained using reasonable assumptions (e.g., $w = \sigma_{v,max} = \frac{\gamma L}{2K \tan\phi'}$, unit weight $\gamma = 0.02\ \mathrm{MN/m^3}$, stress ratio $K = 1.0$, $\tan\phi' = \frac{2}{3}$, $m_i = 10$, ignore clamping stress); but the reduction of the exponent from 2 to 1.5 must be considered empirical, as indicated by the authors. If the results are instead presented in terms of required strength for a given span and depth of sill mat or plug, then the strength factor for each case history can be computed and compared to the modified Pakalnis empirical design chart shown in Figure 12. Many of the case histories included in the database use strengths considerably larger than recommended by the design chart, especially the cases with narrow spans. This may be because the underlying basis for the design chart is not too dissimilar from Mitchell's fixed beam analogy, and so also does not account for backfill stresses induced by large (several percent) closure strains.

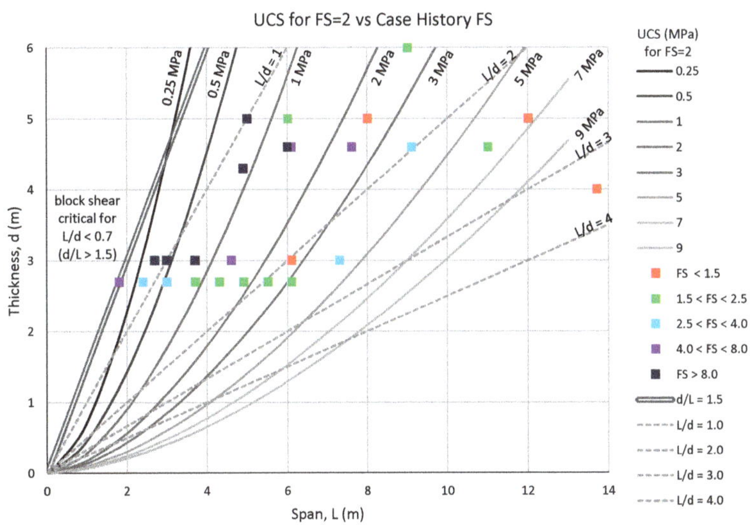

Figure 12. Modified Pakalnis design chart showing actual case history strength factors. (The "block shear critical" criterion for $\frac{d}{L} > 1.5$ assumes the mobilized sidewall cohesion is $1/4 UCS$ and ignores frictional resistance that might arise from clamping stress.)

7. Conclusions

Two fundamental contributions are arising from this work. First, understanding backfill material behavior under the range of loading conditions relevant to undercut CPB requires careful laboratory testing using a variety of test methods. The advantage of the direct tensile test is that the failure geometry can be observed to confirm whether the failure mode is consistent with expectations of a brittle fracture surface normal to the tensile loading direction; and the pre-fracture response can be assessed to determine a tangent Young's modulus in tension. More importantly, the tensile strength test results can be directly compared with direct shear test results and UCS test results to confirm a consistent interpretation of the CPB's strength envelope. The tangent Young's modulus, E_t, can be correlated to UCS using a power function, but this functional relationship appears to be unique for each CPB. The E_t to UCS relationship was found to be independent of binder type (Portland cement only versus a Portland/Slag blend) for Garpenberg CPB; and independent of void ratio for Garpenberg, Quebec, Williams, and Macassa CPBs; and independent of water salinity (for synthetic tailings) over the range of parameters studied in each case, however, these findings should not be automatically assumed for all CPBs. While the combination of direct tension, direct shear, and UCS tests is useful to characterize the material strength and stiffness under low confining stresses, undercut CPB analysis is generally more complicated and needs to consider material response under higher ranges of confinement and strain. This may require drained triaxial testing, one-dimensional (oedometer) consolidation testing, and isotropic consolidation testing for massive orebodies where the deformation may occur in multiple directions.

The second contribution is to a better understanding of the design scenarios to which the original Mitchell sill mat solution and the Pakalnis empirical design chart may apply. If the rock mass is very stiff and field stresses are low, the closure displacements and strains may be low enough (i.e., less than 1% for the lower-bound trends reviewed) to prevent crushing stresses in the CPB. In this case, the Mitchell sill mat solution can be used but care must be taken when evaluating the clamping stress. Similarly, the trends in the (modified) Pakalnis empirical design chart, which presumably ignores clamping stress, may be useful for parametric studies during preliminary design. In either case, laboratory test results are recommended in support of final design. Where larger closure displacements and strains can be anticipated, such as deep and high stress mining, additional consideration must be given to the CPB's strength and stiffness dependency on confining stress levels, in which case drained triaxial and consolidation test results are required, and numerical modeling is generally needed to fully understand the mechanisms by which the CPB remains stable and transmits stress across the undercut span.

Finally, it is imperative to remember that any analysis of undercut stability inherently assumes that the CPB's material properties in the sill mat or plug are homogeneous and isotropic. In contrast, the authors of [27] described an incident at Lucky Friday mine where a large backfill roof failure occurred which was attributed to, in part, the presence of flat-lying cold joints. Therefore, quality assurance and control procedures are needed to ensure the as-placed CPB properties are consistent with design expectations, and in particular any incidents of inconstant or irregular backfill pouring are noted and accounted for.

Author Contributions: The second and third authors completed graduate degrees under the supervision of the first author and their previous research contributions are indicated in the cited papers. This background motivated the authors to examine existing undercut analysis and design procedures and conceptualize the analysis presented herein. M.G. wrote the first draft and M.J. and A.P. provided review and feedback. All authors have read and agreed to the published version of the manuscript.

Funding: This research extends previously funded work at Kirkland Lake Gold Inc.'s Macassa mine, Xstrata PLC (now Glencore Inc.) Kidd mine, and Barrick Gold Corp.'s Williams mine. Ongoing financial support is being provided by Kirkland Lake Gold and Mitacs (previously known as The Mathematics of Information Technology and Complex Systems, MITACS NCE).

Data Availability Statement: Not applicable.

Acknowledgments: The authors thank Glencore and Kirkand Lake Gold for permission to use the previously unpublished data from Kidd and Macassa mines, respectively.

Conflicts of Interest: The authors declare no conflict of interest. The funders had no role in the design of the study; in the collection, analyses, or interpretation of data; in the writing of the manuscript, or in the decision to publish the results.

References

1. Mitchell, R. Sill mat evaluation using centrifuge models. *Min. Sci. Technol.* **1991**, *13*, 301–313. [CrossRef]
2. Raffaldi, M.; Seymour, J.; Richardson, J.; Zahl, E.; Board, M. Cemented Paste Backfill Geomechanics at a Narrow-Vein Underhand Cut-and-Fill Mine. *Rock Mech. Rock Eng.* **2019**, *52*, 4925–4940. [CrossRef] [PubMed]
3. Hughes, P. Design Guidelines: Underhand Cut and Fill Cemented Paste Backfill Sill Beams. Ph.D. Thesis, University of British Columbia, Vancouver, BC, Canada, 2014. Available online: open.library.ubc.ca/collections/ubctheses/24/items/1.0167519?o=1 (accessed on 5 February 2022).
4. Tesarik, D.R.; Seymour, J.B.; Martin, L.A.; Jones, F.M. Numeric model of a cemented rockfill span test at the Turquoise Ridge Mine, Golconda, Nevada, USA. In Proceedings of the Minefill 2007, Montreal, QC, Canada, 29 April–3 May 2007; p. 2512.
5. Corson, D.R. *Field Evaluation of Hydraulic Backfill Compaction at the Lucky Friday Mine, Mullan, Idaho*; Report of investigation no. 7546; US Department of Interior, Bureau of Mines: Washington, DC, USA, 1971.
6. Corson, D.R.; Wayment, W.R. *Load-Displacement Measurement in a Backfilled Stope of a Deep Vein Mine*; Report of investigation no. 7038; US Department of the Interior, Bureau of Mines: Washington, DC, USA, 1967.
7. Gay, N.C.; Jager, A.J.; Piper, P.S. Quantitative evaluation of fill performance in South African gold mines, Backfill in South African Mines. In *Symposium on Backfill in South Africa Mines*; South African Institute of Mining and Metallurgy: Johannesburg, South Africa, 1988; pp. 167–202.
8. Bruce, M.F.G.; Klokow, J.W. A follow-up paper on the development of the West Driefontein tailings backfill project. In *Proceedings of the Symposium on Backfill in South Africa Mines*; South African Institute of Mining and Metallurgy: Johannesburg, South Africa, 1988; pp. 473–490.
9. Hassani, F.; Ouellet, J.; Servant, S. In situ measurements in a paste backfill: Backfill and rock mass response in the context of rockburst. In Proceedings of the 17th International Mining Congress and Exhibition of Turkey—IMCET2001, Ankara, Turkey, 19–22 June 2001; pp. 165–175. Available online: www.maden.org.tr/resimler/ekler/d38dd921e155207_ek.pdf (accessed on 5 February 2022).
10. McNay, L.M.; Corson, D.R. *Hydraulic Sandfill in Deep Metal Mines*; US Bureau of Mines: Washington, DC, USA, 1975.
11. Clark, I.H. The Strength and Deformation Behaviour of Backfill in Tabular Deep-Level Mining Excavations. PhD Thesis, University of the Witwatersrand, Johannesburg, South Africa, 1988.
12. Clark, I.H.; Gurtunca, R.G.; Piper, P.S. Predicting and monitoring stress and deformation behaviour of backfill in deep-level mining excavations. In Proceedings of the 5th Australia-New Zealand Conference on Geomechanics: Prediction Versus Performance; Preprints of papers, Barton, ACT, Institution of Engineers, Sydney, Australia, 22–23 August 1988; pp. 214–218.
13. Thibodeau, D. In situ determination of high density alluvial sand fill. In Proceedings of the Innovations in Mining Backfill Technology: Fourth International Symposium, Montreal, QC, Canada, 2–5 May 1898; Balkema, A.A., Ed.; Engineering & Technology: London, UK; pp. 267–273.
14. Seymour, J.B.; Rafaldi, M.J.; Abraham, H.; Johnson, J.C.; Zahl, E.G. Monitoring the in situ performance of cemented paste backfill at the Lucky Friday Mine. In Proceedings of the The 12th International Symposium on Mining with Backfill, Denver, CO, USA, 19–22 February 2017; pp. 19–22. Available online: www.researchgate.net/profile/Michael-Raffaldi/publication/326234977_Monitoring_the_In_Situ_Performance_of_Cemented_Paste_Backfill_at_the_Lucky_Friday_Mine/links/5b3fd6f8458515f71cad008f/Monitoring-the-In-Situ-Performance-of-Cemented-Paste-Backfill-at-the-Lucky-Friday-Mine.pdf (accessed on 5 February 2022).
15. Snyman, B.; van der Spuy, B.; Correia, L. A critical look at uniaxial test procedures applied in the backfill industry. In *Mine Fill 2014: Proceedings of the Eleventh International Symposium on Mining with Backfill*; Australian Centre for Geomechanics: Crawley, Australia, 2014. [CrossRef]
16. Stone, D. *Paste Quality Control Benchmarks. Minefill 2020–2021*; Taylor & Francis Group: London, UK, 2021; ISBN 9781032072036. Available online: library.oapen.org/bitstream/handle/20.500.12657/48839/9781000454215.pdf?sequence=1#page=52 (accessed on 5 February 2022).
17. Fahey, M.; Helinski, M.; Fourie, A. Development of specimen curing procedures that account for the influence of effective stress during curing on the strength of cemented mine backfill. *Geotech. Geol. Eng.* **2011**, *29*, 709–723. [CrossRef]
18. Rankine, R.; Sivakugan, N. Geotechnical properties of cemented paste backfill from Cannington Mine, Australia. *J. Geotech. Geol. Eng.* **2007**, *24*, 383–393. [CrossRef]
19. Yilmaz, E.; Benzaazoua, M.; Belem, T.; Bussière, B. Effect of curing under pressure on compressive strength development of cemented paste backfill. *Miner. Eng.* **2009**, *22*, 772–785. [CrossRef]

20. Yilmaz, E.; Belem, T.; Benzaazoua, M. Effects of curing and stress conditions on hydromechanical, geotechnical and geochemical properties of cemented paste backfill. *Eng. Geol.* **2014**, *168*, 23–37. [CrossRef]
21. Jardine, R.J.; Symes, M.J.; Burland, J.B. The measurement of soil stiffness in the triaxial apparatus. *Geotechnique* **1984**, *34*, 323–340. Available online: www.icevirtuallibrary.com/doi/pdf/10.1680/geot.1984.34.3.323 (accessed on 5 February 2022).
22. Rankine, R. The Geotechnical Characterisation and Stability Analysis of BHP Billiton's Cannington Mine paste Fill. Ph.D Thesis, James Cook University, Douglas, Australia, 2004. Available online: .http://researchonline.jcu.edu.au/1260/ (accessed on 5 February 2022).
23. Xu, W.; Liu, B.; Wu, W. Strength and deformation behaviors of cemented tailings backfill under triaxial compression. *J. Cent. South Univ.* **2020**, *27*, 3531–3543. [CrossRef]
24. Jafari, M.; Shahsavari, M.; Grabinsky, M. Drained triaxial compressive shear response of Cemented Paste Backfill. *Rock Mech. Rock Eng.* **2021**, *54*, 3309–3325. [CrossRef]
25. Perras, M.A.; Diederichs, M.S. A review of the tensile strength of rocks: Concepts and testing. *Geotechnol. Geol. Eng.* **2014**, *32*, 525–546. [CrossRef]
26. Hoek, E. Practical Rock Engineering: Chapter 11, Rock Mass Properties. Available online: https://www.rocscience.com/assets/resources/learning/hoek/Practical-Rock-Engineering-Chapter-11-Rock-Mass-Properties.pdf (accessed on 18 January 2022).
27. Johnson, J.; Seymour, J.B.; Martin, L.; Stepan, M.; Arkoosh, A.; Emery, T. Strength and Elastic Properties of Paste Backfill at the Lucky Friday Mine, Mullan, Idaho. In Proceedings of the 49th US Rock Mechanics/Geomechanics Symposium, San Francisco, CA, USA, 29 June–1 July 2015; American Rock Mechanics Association: San Francisco, CA, USA; Volume 3, pp. 2321–2332. Available online: www.cdc.gov/niosh/mining/UserFiles/works/pdfs/Strength%20and%20Elastic%20Properties%20of%20Paste%20Backfill%20at%20the%20Lucky%20Friday%20Mine%20Mullan%20Idaho.pdf (accessed on 5 February 2022).
28. Sainsbury, D.; Uri, R. *Stability Analysis of Horizontal and Vertical Paste Fill Exposures at the Raleigh Miine*; Canadian Institute of Mining, Minefill: Montreal, QC, Canada, 2007.
29. Pierce, M. Stability of paste fill exposures at Brunswick mine. In *FLAC and Numerical Modelling in Geomechanics*; Balkema: Rotterdam, The Netherlands, 2000. Available online: https://www.researchgate.net/profile/Salah-Badr-5/publication/281267830_Three-dimensional_strain_softening_modeling_of_deep_longwall_coal_mine_layouts/links/5fee27b6a6fdccdcb81e9100/Three-dimensional-strain-softening-modeling-of-deep-longwall-coal-mine-layouts.pdf (accessed on 5 February 2022).
30. Pan, A.; Grabinsky, M. Tensile strength of Cemented Paste Backfill. *Geotechnol. Test. J.* **2021**, *44*, 1886–1897. [CrossRef]
31. Pan, A. Mechanical Properties of Cemented Paste Backfill under Low Confining Stress. Master's Thesis, University of Toronto, Toronto, ON, Canada, 2019. Available online: http://hdl.handle.net/1807/98310 (accessed on 5 February 2022).
32. Jafari, M.; Shahsavari, M.; Grabinsky, M. Effect of hydration on failure surface evolution of a low Sulfide content Cemented Paste Backfill. *Int. J. Rock Mech. Min. Sci.* **2021**, *144*, 104749. [CrossRef]
33. Narvaez, B.; Aubertin, M.; Saleh-Mbemba, F. Determination of the tensile strength of unsaturated tailings using bending tests. *Can. Geotech. J.* **2015**, *52*, 1874–1885. [CrossRef]
34. Jiang, H.; Fall, M. Yield stress and strength of saline cemented tailings in sub-zero environments: Portland cement paste backfill. *Int. J. Miner. Processing* **2017**, *160*, 68–75. [CrossRef]
35. Pierce, M. Laboratory and Numerical Analysis of the Strength and Deformation Behaviour of Paste Backfill. Master's Thesis, Queen's University, Kingston, ON, Canada, 1999. Available online: www.collectionscanada.gc.ca/obj/s4/f2/dsk2/ftp01/MQ28246.pdf (accessed on 5 February 2022).
36. Aref, K. A study of the geotechnical characteristics and liquefaction potential of paste backfill. Ph.D Thesis, University of Montreal, Montreal, QC, Canada, 1988. Available online: escholarship.mcgill.ca/concern/theses/b5644s191 (accessed on 5 February 2022).
37. De Villiers Wickens, J.; Wilson, S. Paste Fill Stiffness Investigation. Minefill 2020–2021. Available online: oa-edit/10.1201/9781003205906-2 (accessed on 5 February 2022).
38. le Roux, K. In Situ Properties and Liquefaction Potential of Cemented Paste Backfill. Ph.D Thesis, University of Toronto, Toronto, ON, USA, 2004.
39. Walske, M.; McWilliam, H.; Doherty, J.; Fourie, A. Influence of curing temperature and stress conditions on mechanical properties of cementing paste backfill. *Can. Geotech. J.* **2016**, *53*, 148–161. [CrossRef]
40. Jafari, M.; Shahsavari, M.; Grabinsky, M. Cemented Paste Backfill 1-D consolidation results interpreted in the context of ground reaction curves. *Rock Mech. Rock Eng.* **2020**, *53*, 4299–4308. [CrossRef]
41. Sobhi, M.; Li, L. Numerical investigation of the stresses in backfilled stopes overlying a sill mat. *J. Rock Mech. Geotech. Eng.* **2017**, *9*, 490–501. [CrossRef]
42. Pagé, P.; Li, L.; Yang, P.; Simon, R. Numerical investigation of the stability of a base-exposed sill mat made of cemented backfill. *Int. J. Rock Mech. Min. Sci.* **2019**, *114*, 195–207. [CrossRef]
43. Keita, A.; Jahanbakhshzadeh, A.; Li, L. Numerical analysis of the failure mechanisms of sill mats made of cemented backfill. *Int. J. Geotech. Eng.* **2021**. [CrossRef]
44. Keita, A.; Jahanbakhshzadeh, A.; Li, L. Numerical analysis of the stability of arched sill mats made of cemented backfill. *Int. J. Rock Mech. Min. Sci.* **2021**, *140*, 104667. [CrossRef]
45. Stone, D. *The Optimization of Mix Designs for Cemented Rockfill. Minefill 93*; SAIMM: Johannesburg, South Africa, 1993; pp. 249–253.

46. Pakalnis, R.; Caceres, C.; Clapp, K.; Morin, M.; Brady, T.; Williams, T.; Blake, W.; MacLaughlin, M. Design Spans-Underhand Cut and Fill Mining. In Proceedings of the 107th CIM-AGM, Toronto, ON, Canada, April 2005; pp. 1–9. Available online: www.cdc.gov/niosh/mining/userfiles/works/pdfs/dsuca.pdf (accessed on 5 February 2022).
47. Pakalnis, R. Empirical design methods in practice. In Proceedings of the International Seminar on Design Methods in Underground Mining, Australian Center for Geomechanics, Perth, Australia, 17–19 November 2015; pp. 37–56. [CrossRef]
48. Caceres, C.; Moffat, R.; Pakalnis, R. Evaluation of flexural failure of sill mats using classical beam theory and numerical models. *Int. J. Rock Mech. Min. Sci.* **2017**, *99*, 21–27. [CrossRef]

mining

MDPI

Article

Hyperbolic Representation of Lateral Force–Displacement Relationship for Underground Installed Pipe

Richard D. Yovichin III [1], Jai K. Jung [2,*] and Wei-Ju Lee [3]

1 McMillen Jacobs Associates, Mayfield Heights, OH 44124, USA; yovichin@mcmjac.com
2 Department of Civil & Environmental Engineering, Virginia Military Institute, Lexington, VA 24450, USA
3 The Republic of China Marine Corps, Zuoying, Kaohsiung 813779, Taiwan; wjlmarinecorps@gmail.com
* Correspondence: jungjk@vmi.edu

Abstract: Extreme natural hazards such as earthquakes, landslides, and liquefaction create permanent ground deformation (PGD). With the recognition that PGD often causes the most serious local damage in underground structures such as buried pipelines and mining facilities, research and engineering practices for underground structures have focused on soil–structure interaction under PGD. In this study, an underground pipeline was investigated due to its simple geometry. Geotechnical data collection and analysis were used as a study method. Of key importance is the soil–pipe interaction with respect to PGD below the subsurface. This response is typically highlighted by a force vs. displacement relationship and is primarily a function of soil unit weight, depth from surface to the pipe centerline, and the pipe diameter. The non-linear force vs. displacement relationship for transverse horizontal force on a pipe subjected to lateral ground movement, can be represented by a hyperbola. The nonlinear hyperbola can then be turned into a linear line by transforming the axis. This paper investigates a wide range of soil characteristics and summarizes representative hyperbolic parameters for dry medium, dense, and very dense sand for lateral ground movement. The approach is convenient for modeling the soil–pipe interaction and is critical for addressing the complexities of soil and pipe performance, consistent with real-world soil–pipe behavior. The ideas and data analysis techniques presented in this study may be fine-tuned and applied to more complex problems including mining and could ultimately contribute to the management of geotechnical risks.

Keywords: soil–pipe interaction; hyperbolic response; lateral pipe movement

1. Introduction

Earthquake-induced permanent ground deformation (PGD) often involves large, irrecoverable soil distortion with geometric soil mass changes and large plastic underground structure deformation, involving both material and geometric nonlinearities [1,2]. The effects of PGD not only apply to earthquakes, but also occur in response to floods, landslides, tunneling, deep excavations, and subsidence caused by dewatering or the withdrawal of minerals and fluids during mining and oil production. Such loaded conditions are becoming more important as the concern for public safety increases regarding natural hazards, human threats, and construction in a congested urban environment. These behaviors impose significant demands on the modeling of the soil–structure interaction. Research and engineering practices for an earthquake response for underground structures have focused on PGD and transient ground deformation effects, with the recognition that PGD often causes the most serious local damage in buried structures [3–6]. In this study, an underground pipeline under PGD was investigated due to its simple geometry. However, the ideas and geotechnical data collection/analysis techniques used in this study may help others investigate other soil–structure applications such as mining and limit the potential risk of geotechnical failures.

The soil–pipe interaction under PGD is often performed with one-dimensional finite element models to represent the pipeline and soil force vs. displacement relationships that are mobilized by various types of ground movement. As described by several design guidelines and previous researchers [7–11], soil–pipeline interaction is represented by components in the axial, transverse, and vertical bearing directions, as represented by the soil springs [1,12,13]. This approach benefits from ease of application and its incorporation into available finite element codes [8] but suffers from the uncoupled representation of soil as a series of spring-slider reactions [14–16]. Many researchers have developed nonlinear soil–pipe force vs. displacement relationships more realistically, using advanced numerical analysis [5,17–21], but they require significant computing power and an in-depth understanding of numerical modeling. The nonlinear behavior of force vs. displacement relationships can be represented by a hyperbola [8,11,14,22]. Such hyperbolas can then be transformed into a linear representation, which makes an analysis of soil–pipe interaction much easier as illustrated in Figure 1. In the figure, F is the measured lateral pipe force, γ_d is the dry unit weight of the soil, H_c is the depth from the top of the soil to the center of the pipe, D is the external diameter of the pipe, L is the length of the pipe, Y is a relative displacement of the pipe, and A and B are the hyperbolic parameters. Y' and Y'' in Figure 1b are defined in Equations (3) and (7), respectively, and F' and F'' are defined in Equations (4) and (8), respectively.

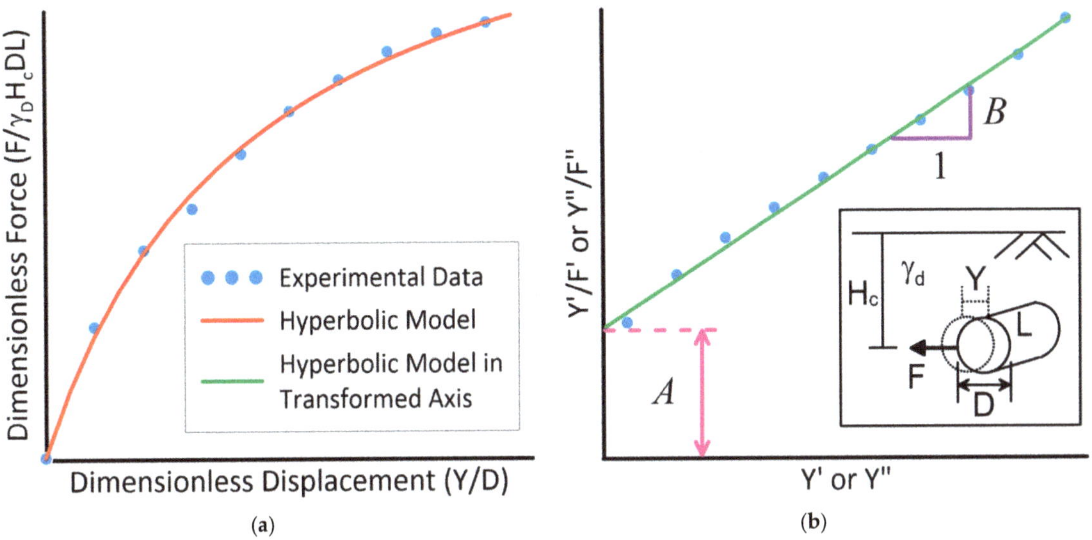

Figure 1. Diagram of transformation of (**a**) Nonlinear dimensionless force vs. displacement to a (**b**) Linear representation emphasizing A and B parameters.

The overall objective of this paper is to represent the nonlinear soil–pipe interaction using the hyperbolic parameters *A* and *B* as shown in Figure 1b. As described by Wong and Duncan [23], the *A* and *B* parameters are related to the line of best fit when describing the hyperbolic curve. This will be done by transforming large-scale experimental data into a dimensionless form and using a normalization process plotted on a transformed axis. Providing a simplified hyperbolic method to represent soil–pipe behavior can allow for a user-friendly and relatively quick process based on numerous experimental data as an alternative to advanced numerical analysis.

2. Experimental Literature Review

In this study, lateral pipe movement in dry sand is considered and the soil is categorized as medium ($16 \le \gamma_d \le 16.7$ kN/m^3), dense ($16.7 < \gamma_d \le 17.5$ kN/m^3), and very dense sand ($\gamma_d > 17.5$ kN/m^3). In addition, the scope of this research is on the force exerted on the pipe due to the relative displacement between the soil and the pipe. Therefore, the pipes used in the experiments are very rigid compared to the adjacent soil. Trautmann and O'Rourke [7,9] conducted tests on lateral loaded pipes with ranging overburden ratios from 2 to 11 and pipe diameters between 102 mm and 324 mm. The scope of their research was to investigate how the depth and unit weight of soils effect a force vs. displacement curve for a laterally loaded pipe. The results from their experimentation for medium and dense sand were compared with an analytical model developed by Rowe and Davis [24] and, more recently, by Jung et al. [15] with very good agreement. Based on the test results, Trautmann and O'Rourke [7] derived hyperbolic curves for medium and very dense sand.

Hsu [25] performed a series of tests on two soil types with overburden ratios ranging from 1.5 to 20.5 and pipe diameters between 38.1 mm and 76.2 mm. Hsu's [25] findings for the maximum soil loads were between the results found by Trautmann and O'Rourke [7] and Audibert and Nymann [26]. Hsu [25] investigated the relationship between strain rate and maximum soil loads on the pipe. This allowed Hsu [25] to develop a power–law relationship between the two. By using different strain rates found from the soil loads and pullout rates, Hsu [25] developed a series of rectangular hyperbolic relationships.

Turner [27] performed experiments using glacio-fluvial well-graded sand (referred to as RMS graded sand) for dry and partially-saturated conditions. The density of the dry soil ranged from 16.9 kN/m^3 to 17.2 kN/m^3, which was checked by a nuclear density gage. The results from testing were slightly higher than Trautmann and O'Rourke [7] and provided modifications to the charts presented by Trautmann and O'Rourke [7]. In addition, Turner [27] provided force vs. displacement curves for soils with 4~10% moisture content and concluded that moist soil loads can be as high as double when compared to dry sand.

Karimian [28] developed the instrumentation capable of recording normal stresses, pullout resistance, displacement of geosynthetics, pipes, and sand grains during testing, and deformation occurring at the surface. Axial and lateral pullout tests were performed on trenched backfill and geotextile-lined trenches, while developing a model for comparison. Karimian [28] found that geosynthetic-wrapped pipes decreased axial soil loads and mentioned that the lateral loading tests were consistent with the rectangular hyperbola reported by Das and Seeley [29] and described by Trautmann and O'Rourke [7]. Karimian [28] suggested that the findings described by Trautmann and O'Rourke [7] and Turner [27] were slightly higher when predicting soil loads on pipes. The developed model was created using a modified hyperbolic model for transverse ground movement. The trenched geotextiles analytical model soil loads were slightly higher than the tests performed, and Karimian [28] stated that this occurrence could be due to localized shear failure.

Olson [30] performed several large-scale lateral pipe movement tests. In general, the dimension of the test box greater than 10 times the pipe diameter is recommended to eliminate the boundary effect [30] and is considered a large-scale test. The purpose of his research was to understand the factors influencing soil performance in soil–pipe interaction, and how to improve accuracy of large-scale testing. The nuclear gage and density scope were used to measure a unit weight of dry and partially saturated RMS graded sand. The comparison of the unit weight, using two different methods, agrees. Olson [30], however, suggested using the nuclear gage for large-scale tests because it is easy to use and a relatively quick process. Using tactile pressure sensors, Olson [30] accurately measured normal stress on the pipe and reported several force vs. displacement curves.

Robert et al. [31] performed two separate tests, under both dry and unsaturated conditions, with two types of soils. Fine Chiba sand [32–34] and coarser RMS graded sands [3,15] were used in these tests. The results from the tests were compared with the numerical simulation. The peak load for unsaturated Chiba sand experiments was greater than the load for dry sand. The RMS graded sand contained nearly the same peak load for unsaturated as for the dry sand conditions; however, the peak load for the unsaturated condition was 10% higher. The unsaturated soil model found similar results for both cases where the pre-failure stiffness of the unsaturated condition was greater than the dry condition, which agrees with Jung et al. [15]. The types of soils used were found to affect the mechanical behavior and were more apparent in the finer Chiba sand.

Minh and Zhang [18] performed five tests under the monotonic and cyclic forces. The British Standard 137 BSI [35] was used to prepare the very dense sand with H_c/D ratio of 4.77. The diameter and the length of the pipe were 0.351 m and 1.2 m, respectively. The test results show that the soil–pipe behavior basically follows the monotonic rule. They also found that the ultimate soil resistance captured from a monotonic load may differ from that captured from a cyclic load. Of the five tests, only the test performed with the monotonic load was used in this study.

3. Experimental Data

3.1. Medium Sand Experimental Data

The dry unit weight (γ_d) of medium sand was categorized as ranging from 16 kN/m^3 to 16.7 kN/m^3. The tests performed by Trautmann and O'Rourke [7], Karimian [28], Li [36], and Robert [5] fall in this category. The raw experimental data were acquired from Trautmann and O'Rourke [7] and Li [36]. However, the original experimental data were not available from Karimian [28] and Robert [5], and therefore, their data were digitized from their published work. Data shown in Figure 2 encompass these researchers' findings, totaling nineteen experiments. Table 1 describes the specific parameters from each researcher used in the construction of Figure 2.

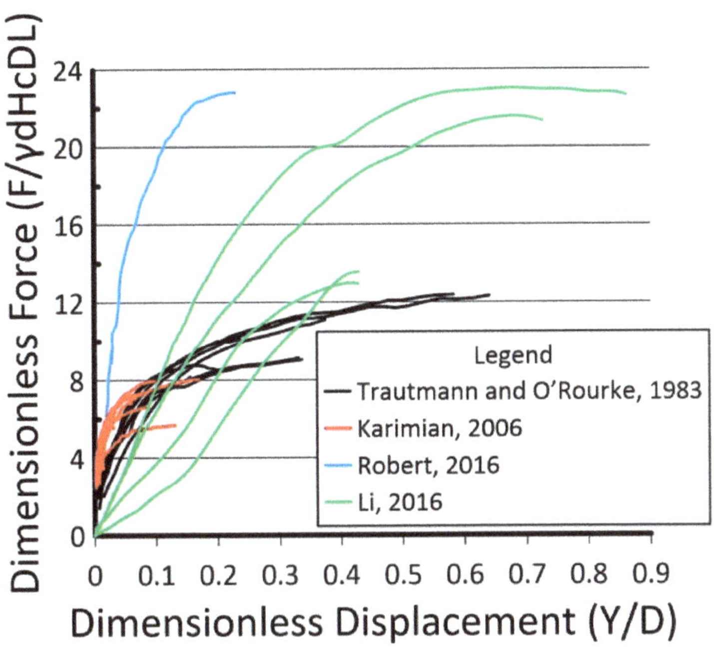

Figure 2. Dimensionless force vs. displacement test data for medium sand (16 kN/m^3 < $\gamma_d \leq$ 16.7 kN/m^3). Experiments performed by [5,7,28,36].

3.2. Dense Sand Experimental Data

The dense sand is defined as the soil dry unit weight from 16.7 kN/m^3 to 17.5 kN/m^3. Figure 3 shows dimensionless force vs. displacement relationship for lateral pipe movement and is composed of data from four research groups: Hsu [25], Turner [27], Olson [30], and Li [36]. This graph was created using eighteen experiments from the various researchers mentioned above. The original test data were not available for Hsu [25], and the data were digitized from the experimental data presented by Hsu [25], which were overlapping at various points. This issue caused some difficulty in the data collection process. For Turner [27], Olson [30], and Li [36], raw test data were available and included in the figure. The majority of the data presented in Figure 3 follow a general trend, whereas Li's data [36] have a higher maximum dimensionless force and soft pre-peak behavior when compared to the other researchers' data. Li [36] stated that the compaction while preparing the test bed was not adequately performed and may have affected the test results. Detailed test parameters for dense sand are summarized in Table 2.

Table 1. The parameters used for each medium sand test with the corresponding researcher.

γ_d (kN/m^3)	Length (mm)	Diameter (mm)	H_c/D	Experiment	Note
16	2400	457	1.92	Karimian [28]	2 tests
16	2400	324	1.92	Karimian [28]	3 tests
16	2400	324	2.75	Karimian [28]	
16	1000	60	3	Li [36]	
16	1000	60	5	Li [36]	
16	1000	60	8	Li [36]	
16	1000	60	10	Li [36]	
16.4	1200	102	3.5	T & O [1] [7]	3 tests
16.4	1200	102	3.5	T & O [7]	
16.4	1200	102	5.5	T & O [7]	
16.4	1200	102	8	T & O [7]	
16.4	1200	102	11	T & O [7]	2 tests
16.6	2400	114.6	6	Robert [5]	

[1] Trautmann and O'Rourke.

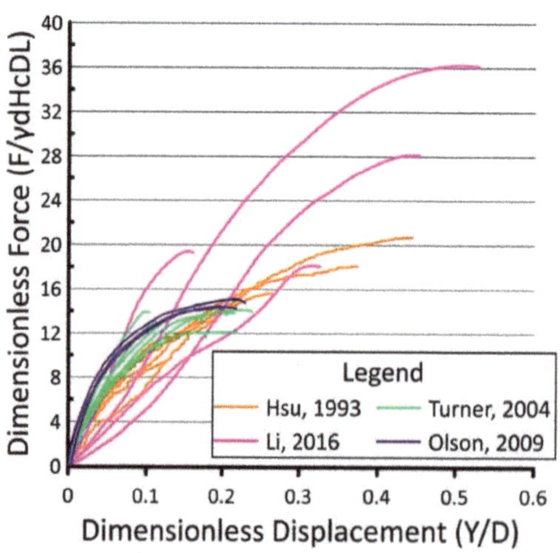

Figure 3. Dimensionless force vs. displacement test data for dense sand ($16.7\ kN/m^3 < \gamma_d \leq 17.5\ kN/m^3$). Experiments performed by [25,27,30,36].

3.3. *Very Dense Sand Experimental Data*

Very dense sands are described as having a $\gamma_d > 17.5\ kN/m^3$. Trautmann and O'Rourke [7], and Minh and Zhang [18] performed lateral pipe movement experiments on very dense sand. The test results are displayed in Figure 4. Please note that the data shown for Trautmann and O'Rourke [7] in Figure 4 does not represent the entire test recorded by the original researchers. For dense to very dense sands, the hyperbola characterizes the actual curve for displacement less than Y'_{max}. A limiting force of F'_{max} is used for displacement greater than Y'_{max}. Therefore, the data points exceeding the maximum dimensionless force are not shown in the figure. The test data from Minh and Zhang [18] as well as $H_c/D = 1.5$ from Trautmann and O'Rourke [7] were digitized. Table 3 shows the different variables for each test.

Table 2. The parameters used for each dense sand test with the corresponding researcher.

γ_d (kN/m^3)	Length (mm)	Diameter (mm)	H_c/D	Experiment	Note
16.9	2440	120	5.47	Olson [30]	
16.9	1200	120	5.5	Turner [27]	
17	1000	60	3	Li [36]	
17	1000	60	5	Li [36]	
17	1000	60	8	Li [36]	
17	1000	60	10	Li [36]	
17	1200	120	5.5	Turner [27]	
17.1	2440	124	5.29	Olson [30]	
17.1	1210	120	5.5	Turner [27]	
17.2	1200	76.2	2.5	Hsu [25]	
17.2	1200	76.2	4.5	Hsu [25]	
17.2	1200	76.2	6.5	Hsu [25])	
17.2	1200	76.2	8.5	Hsu [25]	
17.2	1200	76.2	10.5	Hsu [25]	
17.2	2440	124	5.29	Olson [30]	
17.2	1200	120	5.5	Turner [27]	3 tests

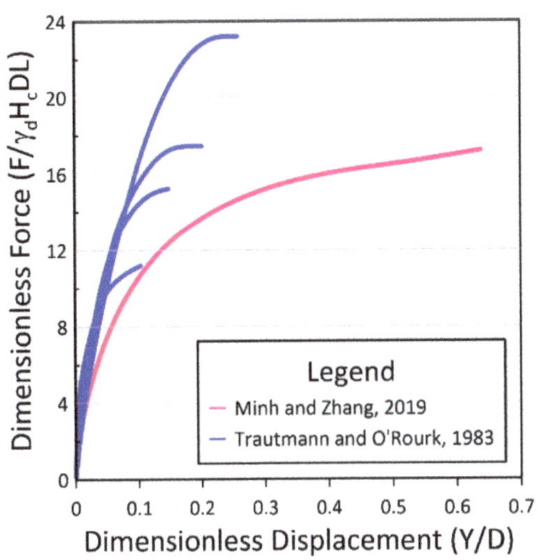

Figure 4. Dimensionless force vs. displacement test data for dense sand (16.7 kN/m^3 < $\gamma_d \leq$ 17.5 kN/m^3). Experiments performed by [7,18].

Table 3. The parameters used for each very dense sand test with the corresponding researcher.

γ_d (kN/m^3)	Length (mm)	Diameter (mm)	H_c/D	Experiment
17.7	1200	120	3.5	T & O [1] [7]
17.7	1200	120	5.5	T & O [7]
17.7	1200	120	8	T & O [7]
17.7	1200	120	11	T & O [7]
20.9	1200	351	4.8	Minh and Zhang [18]

[1] Trautmann and O'Rourke.

4. Data Analysis

To begin performing the data analysis, an intermediate step is needed. This step involves finding the maximum force (F_{max}) with the corresponding maximum displacement (Y_{max}), as described by Jung et al. [1]. The maximum dimensionless force is selected where a clear peak can be seen in the force vs. displacement graph or in a dimensionless graph as shown in Figure 5b. However, in some cases, especially for medium sand, a definitive peak is not visible. In such cases, a methodology as illustrated in Figure 5a is used to define F_{max}. In the figure, $(F)_{ult}$ is the asymptotic value of the principal force, which the force vs. displacement curve approaches at the infinite displacement. As shown in the figure, the hyperbola remains below $(F)_{ult}$ within all finite values of displacement. The force at the maximum lateral pipe force, F_{max}, is expressed as:

$$F_{max} = R_f \times (F)_{ult} \tag{1}$$

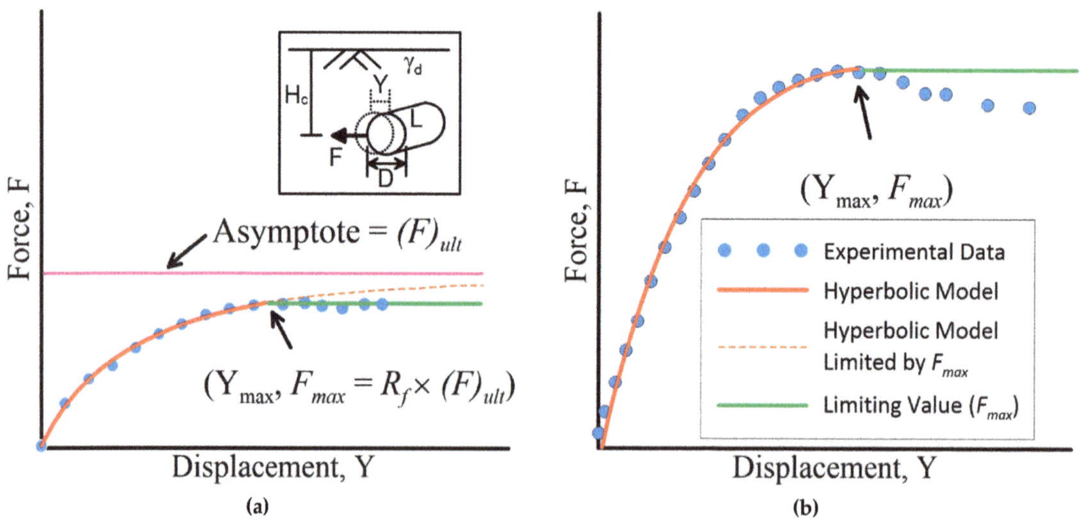

Figure 5. Determining F_{max} and Y_{max} from experimental data for (**a**) when a clear peak is not visible and (**b**) when a clear peak is visible (Reprinted with permission from Ref. [7]. Copyright 1983 ProQuest).

Duncan [37] explains how the over estimation of maximum force is typically around 11% on average, therefore, a reduction factor (R_f) is needed to correct the over estimation. Wang and Duncan [23] reported that R_f typically ranged between 0.5 and 0.9 for most soils. They also reported that the compressive strength of the soils is always less than $(F)_{ult}$. In the analysis, F_{max} is defined as $F_{max} = R_f\,(F)_{ult}$, where $R_f = 0.9$, which is in agreement with the information used by Trautmann and O'Rourke [7,9] and falls in the range of R_f

reported by Wang and Duncan [23]. This approach is also used by Yimsiri et al. [17] and Jung et al. [15,22], which was developed from the method discussed and used by several researchers [7,37–39]. When a clear peak was not visible, a hyperbolic curve was fitted from the obtained force vs. displacement data by means of maximum force extrapolation, using a similar approach noted by Trautmann and O'Rourke [7] for characterizing the force vs. displacement relationship. In the analysis, the following hyperbola was fitted from the force vs. displacement relationship:

$$F = \frac{Y}{\alpha + \beta Y} \quad (2)$$

where $1/\alpha = \lim_{Y \to 0} (dF/dY)$ corresponds to initial force vs. displacement stiffness curve and $1/\beta = \lim_{Y \to \infty} (F)$ corresponds to $(F)_{ult}$ [17]. By extrapolation, the hyperbolic curve was followed to find the location of $(F)_{ult}$ or the asymptote. $(F)_{ult}$ was then multiplied by a reduction factor of 0.9. The value found from this multiplication process is considered F_{max}. Yovichin [40] summarized and reported the corresponding values of each test. Once F_{max} and Y_{max} are defined for each test, the normalization process of dimensionless force vs. dimensionless displacement could occur. The normalization process allowed for each soil type to be presented in two different ways, as discussed in the next section.

5. Methodology

5.1. Single-Normalization

The hyperbolic stress-strain relationship was first developed by Kondner [38] and further discussed and used by Wang and Duncan [23] and Trautmann and O'Rourke [7]. First, the dimensionless force vs. displacement was fitted and plotted on a transformed axis, Y'/F' and Y'' where

$$Y' = \frac{Y}{D} \quad (3)$$

$$F' = \frac{F}{\gamma_d H_c D L} \quad (4)$$

Each parameter used in Equations (3) and (4) are defined previously. Plotting the dimensionless force vs. displacement on a transformed axis allows the hyperbolic curve to be represented by a linear line following the rectangular hyperbola:

$$F' = \frac{Y'}{A + BY'} \quad (5)$$

The parameters describing the hyperbolic equation have physical significance and are shown in Figure 1b, labeled as *A* and *B*. The *A* parameter is the reciprocal of the initial tangential modulus, whereas the *B* parameter is the reciprocal of the asymptotic value of F_{ult}, as discussed in the previous section.

The single-normalization process allows the data to be represented by a linear line on the transformed axis plot. This linear line is then fitted with a linear best-fit line. Furthermore, the best-fit line will represent the best-fit hyperbola [23]. Similar graphs shown in Figure 1b are generated for each test's data, by plotting the dimensionless force and displacement on a transformed axis. The *A* and *B* parameters for each test previously mentioned, are then found and recorded.

5.2. Double-Normalization

After single-normalization occurred, the hyperbolic curve is then converted to a double-normalized force vs. displacement graph, by plotting on a transformed axis Y''/F'' and Y'' [38] (Figure 1b) as

$$F'' = \frac{F'}{F'_{max}} \quad (6)$$

$$Y'' = \frac{Y'}{Y'_{max}} \tag{7}$$

where F'_{max} and Y'_{max} are the maximum values of F' and Y', respectively. The double-normalized force vs. displacement graph can then be approximated by Equation (8).

$$F'' = \frac{Y''}{A + BY''} \tag{8}$$

The average value of Y'_{max} is 0.054 H_c, 0.036 H_c, and 0.025 H_c for medium, dense, and very dense sand, respectively. The double-normalization process followed a similar approach in finding the *A* and *B* parameters, as described in single-normalization (Equation (5)). Equations (3) and (4) are similar to Equations (6) and (7), with the maximum values, F'_{max} and Y'_{max}, playing an influential role in the double-normalization process, whereas single-normalization does not take into account the maximum force or displacement. The other researchers [9,41] also focused their attention on double-normalization when reporting the soil–pipe force vs. displacement relationship.

5.3. Outliers

After the *A* and *B* parameters are generated for each data set, the outliers are identified. The *A* and *B* parameters for single- and double-normalization for each experiment, based on soil type, are input individually and a box plot is generated. A box and whisker plot is a graphical representation of the data as shown in Figure 6. The distance between the upper (Q3) and lower quartiles (Q1) is the inter quartile range (IQR). This range is classified as 50% of the results. The minimum and maximum values are considered the upper and lower whiskers; however, if these values exceed 1.5 IQR they are deemed outliers. These outliers are further classified as mild or extreme, >1.5 IQR and >3.0 IQR, respectively [42,43]. In this study, the mild classification is used and the *A* or *B* parameters greater than 1.5 IQR are neglected for both normalizations. The outlier analysis is useful to find the representative *A* and *B* values for each soil type.

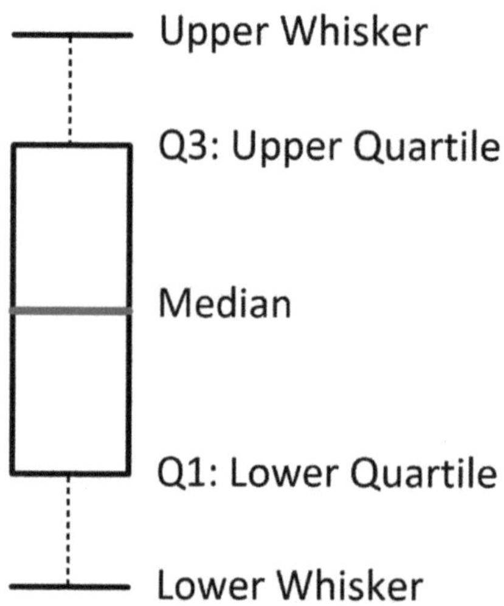

Figure 6. Box and Whisker Plot.

It turned out that there were no outliners for very dense sand. For medium sand, five tests performed by Robert [5] and Li [36] were found to be outliers. For dense sand, out of a total of eighteen tests from four researchers [25,26,30,36], four tests performed by Li [36] were found to be outliers.

6. Results and Discussion

Excluding the outliers, the single- and double-normalized graphs for medium sand are generated and displayed in Figure 7a,b, respectively. The single-normalization appears to show the data beginning at the origin. This is due to the values of the y-intercept (*A* parameter) being relatively small. The double-normalization graph, however, allows each experiment to be more visible with less congestion, causing a larger range for the *A* parameter to be produced. All the experiments shown in the figure exhibit a linear trend. By the nature of each normalization process, the maximum Y'/F' and maximum Y' varies for each individual experiment in the single-normalized graph, whereas the double-normalization process forces the maximum Y''/F'' and maximum Y'' to be 1.

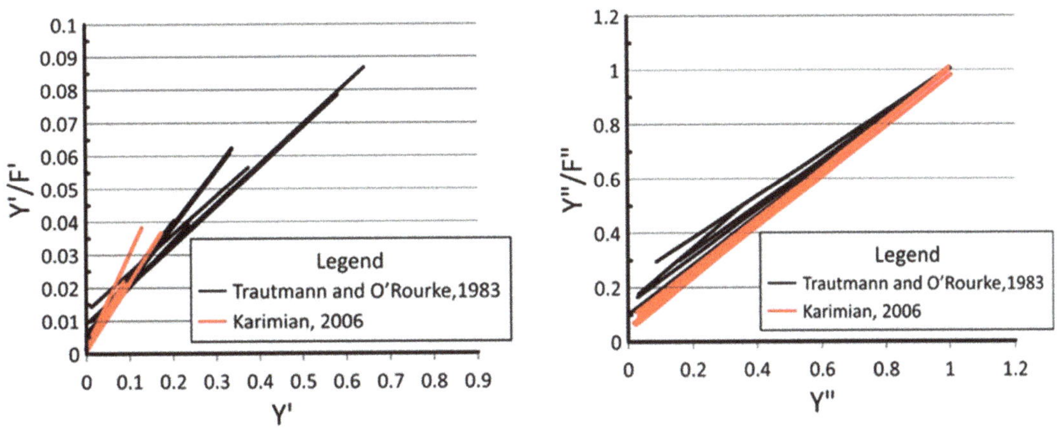

(a) (b)

Figure 7. Diagram of (**a**) Single- and (**b**) Double-normalization for medium sand. Experiments performed by [7,28].

The *A* parameter for single-normalization obtained from Karimian [28] ranges from 0.0005 to 0.0014 with a γ_d = 16 kN/m^3 and H_c/D varying from 1.92 to 2.75. The *B* parameter from Karimian [28] bound between 0.07 and 0.17. Trautmann and O'Rourke [7] had a slightly higher *A* parameter ($0.002 \leq A \leq 0.0074$) and a smaller *B* parameter ($0.07 \leq B \leq 0.11$); however, the γ_d was 16.4 kN/m^3 and the H_c/D ranged from 3.5 to 11. When comparing the two researchers' data separately, the *A* and *B* parameters for single-normalization did not overlap, yet this occurrence did not happen with the double-normalization case.

The double-normalization process allowed for the data from each researcher, regardless of the independent variable (e.g., γ_d, H_c/D), to be well converged. For example, the *A* parameter for Trautmann and O'Rourke [7] is distributed between 0.06 and 0.22, and the *B* between 0.78 and 0.96. Karimian's [28] ranges are similar to Trautmann and O'Rourke [7], from 0.05 to 0.10 and from 0.89 to 0.95 for *A* and *B* parameters, respectively. The double-normalization process allows each respective parameter to correspond to the other test results. The maximum and minimum values of *A* and *B* parameters for medium sand are summarized in Table 4. The single-normalized *A* parameters are less than the double-normalized, and this is also the case for the *B* parameters. These differences in the *A* and *B* parameters are inherently visible in Figure 7. Detailed *A* and *B* parameters for each test are reported in Yovichin [40].

Table 4. Maximum and minimum values of medium sand.

Parameter	Single-Normalized		Double-Normalized	
	A	*B*	*A*	*B*
Maximum	0.0074	0.17	0.22	0.96
Minimum	0.0005	0.07	0.05	0.78

The results for dense sand are visualized in Figure 8 and the maximum/minimum values of the *A* and *B* parameters are summarized in Table 5. The H_c/D for each experiment varied from 2.5 to 10.5. Both the single- and double-normalization graphs, plotted on a transformed axis, display a linear trend for most of the experimental data. For single-normalization, the *A* parameter was approximately between 0.0024 and 0.0073, and the *B* parameter was between 0.03 and 0.06. The double-normalization parameters have a larger range, with *A* being between 0.16 and 0.57 and *B* between 0.38 and 0.81.

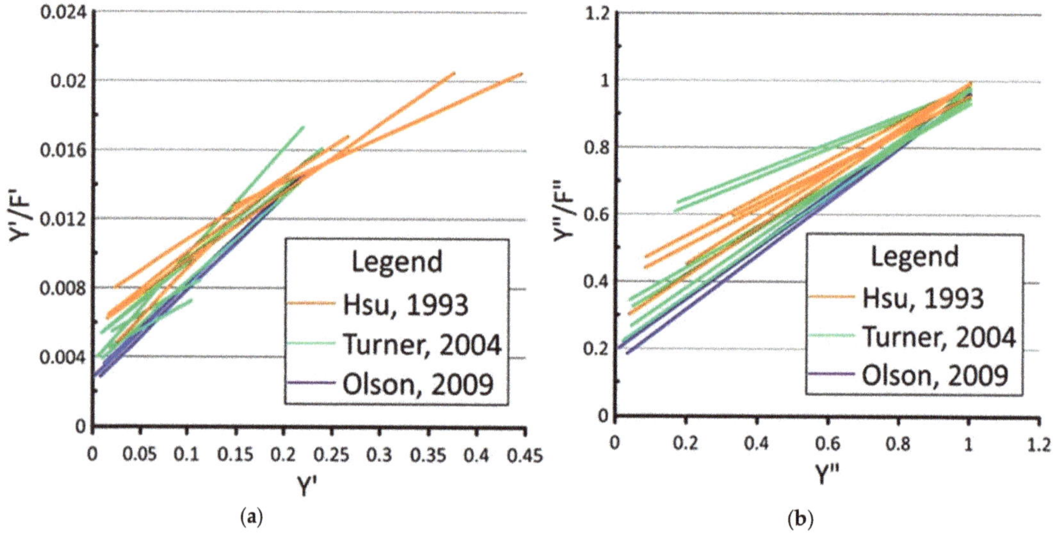

Figure 8. Diagram of (**a**) Single- and (**b**) Double-normalization for dense sand. Experiments performed by [25,27,30].

Table 5. Maximum and minimum values of dense sand.

Parameter	Single-Normalized		Double-Normalized	
	A	*B*	*A*	*B*
Maximum	0.0073	0.06	0.57	0.81
Minimum	0.0024	0.03	0.16	0.38

Tests from Trautmann and O'Rourke [7] and Minh and Zhang [18], classified as very dense sand, are shown in Figure 9. The single-normalization graph is spread out until a dimensionless displacement $\leq$0.075. In the double-normalization, the test data converge into a single linear line. The H_c/D varied from 1.5 to 11. It is found that as H_c/D increases, the *A* parameter increases, whereas the *B* parameter decreases for both single- and double-normalization. Table 6 below illustrates the ranges of the *A* and *B* parameters, regardless of the H_c/D. The single-normalization *A* parameters are between 0.0008 and 0.0038, whereas for double-normalization *A* flocculates from 0.10 to 0.30. The *B* parameters are from 0.03 to 0.11, and 0.65 to 0.91 for single and double, respectively.

Table 6. Maximum and minimum values of very dense sand.

Parameter	Single-Normalized		Double-Normalized	
	A	*B*	*A*	*B*
Maximum	0.0038	0.11	0.30	0.91
Minimum	0.0008	0.03	0.10	0.65

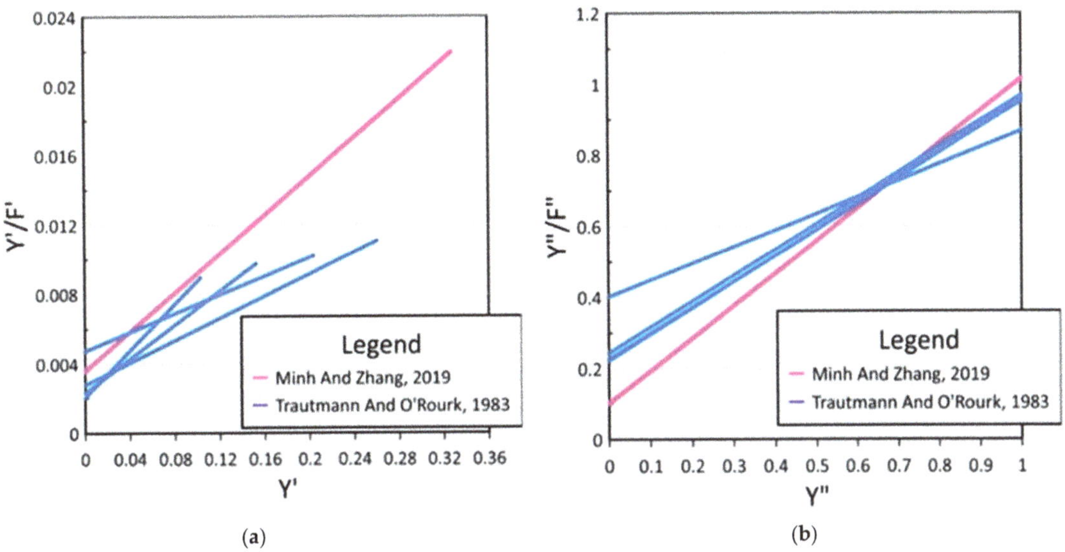

Figure 9. Diagram of (**a**) Single- and (**b**) Double-normalization for very dense sand. Experiments performed by [7,18].

7. Compiled Data and Validation

In general, medium sand has a higher *B* parameter when compared to the other soils. As unit weight increases, the *B* parameter decreases; however, the *A* parameter remains relatively the same. The correlation between *A* and *B* changed significantly with respect to double-normalization. The *A* and *B* parameters for double-normalization, plotted on the x- and y-axis, respectively, are shown in Figure 10. All the data points found exhibit a linear trend, even though the unit weight varied for each experiment. A linear regression line with a high R^2 correlates the *A* and *B* parameters with the negative sloped line as shown in Figure 10. The data points in dense sand have the highest range (distance between maximum and minimum) of *A* and *B* parameters. The medium sand *B*-parameter range was less than very dense sands, whereas the *A*-parameter range was slightly greater. To the best of the authors' knowledge, the correlation shown in Figure 10 between the *A* and *B* parameters has not been reported elsewhere.

The average *A* parameter for single-normalization is converged to 0.003, regardless of soil type. The average *B* parameters for medium, dense, and very dense sand are 0.11, 0.04, and 0.06, respectively. With respect to double-normalization, the *A* parameters for medium, dense, and very dense sand are 0.10, 0.33, and 0.15, respectively, and *B* parameters are 0.91, 0.64, and 0.84, respectively. For medium sand, Trautmann and O'Rourke [7] reported the *A* and *B* parameters for double-normalization as 0.1 and 0.9, respectively, which are in agreement with the values found in this research. Trautmann and O'Rourke [7], however, did not mention parameters for dense soils. The results reported by Trautmann and O'Rourke [7] for very dense sand were 0.25 and 0.75 for *A* and *B* parameters, respectively.

In this study, the *A* and *B* parameters for very dense sand are slightly different with 0.21 and 0.76, respectively.

To validate Figure 10, a double-normalized force vs. displacement curve from an independent test performed by Turner [27] is compared with one using the methodology described above. The soil properties of the test are $H_c/D = 5.5$, $D = 117$ mm, $\gamma_d = 16.7$ kN/m^3, and $L = 1200$ mm. The lateral maximum dimensionless force is obtained from Jung et al. [15] and converted to F'_{max}. Using the average value of *A* parameters for dense sand (Section 7), the *B* parameter is calculated using Figure 10. Then using Equations (6)–(8) and Y'_{max} as reported in Section 5.2, the force vs. displacement curve is approximated and plotted in Figure 11. The force vs. displacement curve from Turner [27] is also plotted in Figure 11. As shown in Figure 11, the curve from Turner [27] and that from Equation (8) agree very well. The favorable comparison implies that the correlation shown in Figure 10 is suitable to approximate a force vs. displacement relationship for later pipe movement in dry soil.

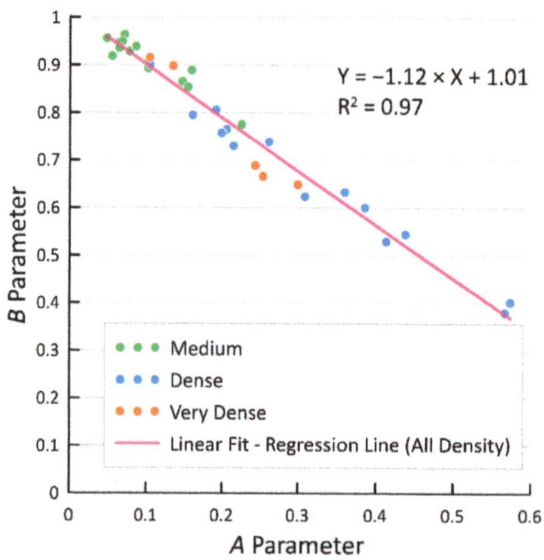

Figure 10. Correlation of A and B parameter for double-normalization.

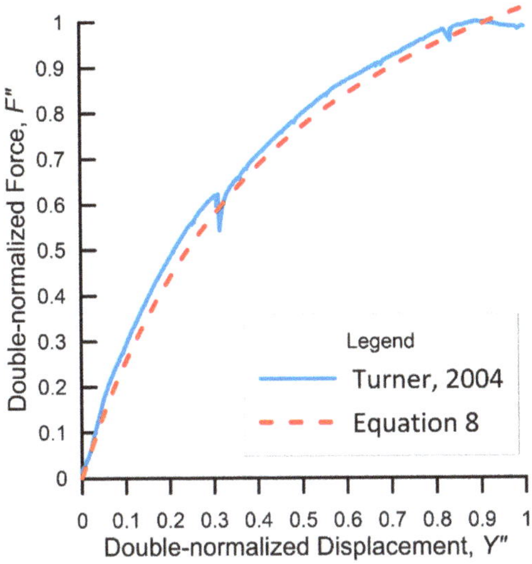

Figure 11. Validation of Figure 10 and Equation (8). Experiments performed by [27].

8. Conclusions

A force vs. displacement behavior for lateral pipe movement was conducted on three types of soil: medium, dense, and very dense sands. This study focused on experiments performed by several researchers involving $H_c/D \geq 1.5$, where the *A* and *B* parameters for single- and double-normalization plotted on a transformed axis were investigated. An extensive literature review was designated to understand how the tests were conducted, what types of tests were suitable for this case study, and how the parameters would be found. The minimum and maximum *A* and *B* parameters are reported for all respective soil types. For validation, the results found by Trautmann and O'Rourke [7] regarding *A* and *B* parameters were compared and agree for medium and very dense sand.

An effort was made to obtain a correlation between the *A* and *B* parameters. The correlation was not visible for single-normalization. However, there was a negative linear relationship with a high statistical significance for double-normalization. The variety of tests from different researchers allows for a wide range of soils to be represented. This allows for the *A* and *B* parameters to be generalized. Due to the generalization, this study can help engineers replicate a soil–pipe interaction without a sophisticated modeling technique. The geotechnical data collection and analysis techniques used in this study may be further used in investigating other types of soil–structure application such as mining and could ultimately limit the potential risk of geotechnical failures.

9. Limitations of the Study

This report is a systematic review study. The data were collected from other researchers within this field of study. While conducting the literature review, the authors were able to find twenty-eight tests for loose sand having $\gamma_d < 16$ kN/m^3 from several researchers [5,25,44–46]. However, the results showed low R-squared values of *A* and *B* parameters for both single- and double-normalization. Therefore, the methodology described in this paper is not applicable for loose sand. In addition, all pipes in this research have a diameter smaller than 500 mm and the majority are 100~150 mm. Therefore, the method proposed in this paper may not be suitable for pipelines greater than 500 mm in diameter. Additional large-scale testing is recommended to help quantify the 500 mm or larger pipe condition.

Author Contributions: R.D.Y.III: data curation, formal analysis, investigation, software, validation, writing—original draft preparation; J.K.J.: Conceptualization, methodology, supervision, validation, writing—review and editing; W.-J.L.: data curation, investigation, software, visualization. All authors have read and agreed to the published version of the manuscript.

Funding: This research received no external funding.

Data Availability Statement: Not applicable.

Acknowledgments: Key Laboratory of Urban Security and Disaster Engineering of Ministry of Education at Beijing University of Technology, Li, and other researchers are acknowledged for their support. The former and current department heads of the Civil and Environmental Engineering at Virginia Military Institute, Riester and Newhouse, are also very much appreciated for their kind support. The first author thanks the Youngstown State University for offering the University Research Council Scholarship, and the other authors thank the Virginia Military Institute for SURI Scholarship support.

Conflicts of Interest: The authors declare no conflict of interest.

References

1. Jung, J.K.; O'Rourke, T.D.; Argyrou, C. Multi-directional force-displacement response of underground pipe in sand. *Can. Geotech. J.* **2016**, *53*, 1763–1781. [CrossRef]
2. Soleimani, N.; Davidson, R.A.; Davis, C.; O'Rourke, T.D.; Nozick, L.K. Multihazard Scenarios for Regional Seismic Risk Assessment of Spatially Distributed Infrastructure. *J. Infrastruct. Syst.* **2021**, *27*, 04021001. [CrossRef]
3. O'Rourke, T.D. Geohazards and Large, Geographically Distributed Systems. *Geotechnique* **2010**, *60*, 505–543. [CrossRef]
4. Jung, J.K.; Koo, D.H.D.; Zhang, K. Verification of the pipe depth dependent model using a finite element analysis. *Tunn. Undergr. Space Technol.* **2014**, *39*, 34–40. [CrossRef]
5. Robert, D.J.; Soga, K.; O'Rourke, T.D. Pipelines subjected to fault movement in dry and unsaturated soils. *Int. J. Geomech.* **2016**, *16*, C4016001. [CrossRef]
6. Nguyen, K.T.; Asimaki, D. Smooth nonlinear hysteresis model for coupled biaxial soil-pipe interaction in sandy soils. *J. Geotech. Geoenvironmental Eng.* **2020**, *146*, 04020035. [CrossRef]
7. Trautmann, C.H.; O'Rourke, T.D. *Behavior of Pipe in Dry Sand under Lateral and Uplift Loading*; Geotechnical Engineering Report 83-7; Cornell University: Ithaca, NY, USA, 1983.
8. American Society of Civil Engineers. *Guidelines for the Seismic Design of Oil and Gas Pipeline Systems*; American Society of Civil Engineers: New York, NY, USA, 1984.
9. Trautmann, C.H.; O'Rourke, T.D. Lateral force-displacement response of buried pipe. *J. Geotech. Eng.* **1985**, *111*, 1077–1092. [CrossRef]
10. Trautmann, C.H.; O'Rourfce, T.D.; Kulhawy, F.H. Uplift force-displacement response of buried pipe. *J. Geotech. Eng.* **1985**, *111*, 1061–1076. [CrossRef]
11. American Lifelines Alliance. *Seismic Design Guidelines for Water Pipelines*; American Lifelines Alliance in Partnership with the Federal Emergency Management Agency: Washington, DC, USA, 2005.
12. Wijewickreme, D.; Karimian, H.; Honegger, D. Response of buried steel pipelines subjected to relative axial soil movement. *Can. Geotech. J.* **2009**, *46*, 735–752. [CrossRef]
13. Argyrou, C.; O'Rourke, T.D.; Stewart, H.E.; Wham, B.P. Large-scale fault rupture tests on pipelines reinforced with cured-in-place linings. *J. Geotech. Geoenviron. Eng.* **2019**, *145*, 04019004. [CrossRef]
14. Honegger, D.G.; Gailing, R.W.; Nyman, D.J. Guidelines for the seismic design and assessment of natural gas and liquid hydrocarbon pipelines. In Proceedings of the International Pipeline Conference, Calgary, AB, Canada, 29 September–3 October 2002; Volume 36207, pp. 563–570.
15. Jung, J.K.; O'Rourke, T.D.; Olson, N.A. Lateral soil-pipe interaction in dry and partially saturated sand. *J. Geotech. Geoenviron. Eng.* **2013**, *139*, 2028–2036. [CrossRef]
16. Ansari, Y.; Kouretzis, G.; Sloan, S.W. Physical modelling of lateral sand–pipe interaction. *Géotechnique* **2021**, *71*, 60–75. [CrossRef]
17. Yimsiri, S.; Soga, K.; Yoshizaki, K.; Dasari, G.R.; O'Rourke, T.D. Lateral and upward soil-pipeline interactions in sand for deep embedment conditions. *J. Geotech. Geoenviron. Eng.* **2004**, *130*, 830–842. [CrossRef]
18. Minh, N.H.; Zhang, D. Soil Responses to Monotonic and Cyclic Lateral Displacement of a Buried Pipe. *J. Pipeline Syst. Eng. Pract.* **2019**, *10*, 04019012. [CrossRef]
19. Morshed, M.A.; Roy, K.; Hawlader, B. Modeling of Buried Pipelines in Dense Sand for Oblique Movement in Vertical–Lateral Plane. *J. Pipeline Syst. Eng. Pract.* **2020**, *11*, 04020050. [CrossRef]
20. Roy, K.; Hawlader, B.; Kenny, S.; Moore, I. A comparative study between lateral and upward anchor-soil and pipe-soil interaction in dense sand. In Proceedings of the International Pipeline Conference, Calgary, AB, Canada, 26–30 September 2016; Volume 50266, p. V002T02A016.

21. Pike, K.; Kenny, S.; Hawlader, B. Numerical and constitutive model development to aid design against pipeline geohazards. *J. Pipeline Eng.* **2014**, *13*, 201–209.
22. Jung, J.K.; O'Rourke, T.D.; Olson, N.A. Uplift soil–pipe interaction in granular soil. *Can. Geotech. J.* **2013**, *50*, 744–753. [CrossRef]
23. Wong, K.S.; Duncan, J.M. *Hyperbolic Stress-Strain Parameters for Nonlinear Finite Element Analyses of Stresses and Movements in Soil Masses;* College of Engineering, Office of Research Services, University of California: Berkeley, CA, USA, 1974.
24. Rowe, R.K.; Davis, E.H. The behaviour of anchor plates in sand. *Geotechnique* **1982**, *32*, 25–41. [CrossRef]
25. Hsu, T.W. Rate effect on lateral soil restraint of pipelines. *Soils Found.* **1993**, *33*, 159–169. [CrossRef]
26. Audibert, J.M.; Nyman, K.J. Soil restraint against horizontal motion of pipes. *J. Geotech. Eng. Div.* **1977**, *103*, 1119–1142. [CrossRef]
27. Turner, J.E. Lateral Force—Displacement Behavior of Pipes in Partially Saturated Sand. Master's Thesis, Cornell University, Ithaca, NY, USA, 2004.
28. Karimian, S.A. Response of Buried Steel Pipelines Subjected to Longitudinal and Transverse Ground Movement. Ph.D. Thesis, University of British Columbia, Vancouver, BC, Canada, 2006.
29. Das, B.M.; Seeley, G.R. Load-displacement relationship for vertical anchor plates. *J. Geotech. Eng. Div.* **1975**, *101*, 711–715. [CrossRef]
30. Olson, N.A. Soil Performance for Large Scale Soil-Pipeline Tests. Ph.D. Thesis, Cornell University, Ithaca, NY, USA, 2009.
31. Robert, D.J.; Soga, K.; O'Rourke, T.D.; Sakanoue, T. Lateral load-displacement behavior of pipelines in unsaturated sands. *J. Geotech. Geoenviron. Eng.* **2016**, *142*, 04016060. [CrossRef]
32. Robert, D.J. Soil-Pipeline Interaction in Unsaturated Soils. Ph.D. Thesis, University of Cambridge, Cambridge, UK, 2010.
33. Fern, J.; Soga, K.; Robert, D.J. Shear strength and dilatancy of partially saturated sand in direct shear tests. In Proceedings of the TC105 ISSMGE International Symposium on Geomechanics from Micro to Macro, Cambridge, UK, 1 September–4 September 2014; pp. 1391–1396.
34. Fern, J.; Soga, K.; Robert, D.J.; Sakanoue, T. Shear strength and dilatancy of unsaturated silica sand in triaxial compression tests. In Proceedings of the 14th International Conference of International Association for Computer Methods and Recent Advances in Geomechanics, Kyoto, Japan, 22–25 September 2014; pp. 535–540.
35. British Standards Institution (BSI). *Methods of Test for Soils for Civil Engineering Purposes;* BS 1377-2; BSI: London, UK, 1990.
36. Li, J.L. Test Study on Ultimate Soil Bearing Capacity of Buried Pipeline—Sand Lateral Interaction. Master's Thesis, Beijing University of Technology, Beijing, China, 2016.
37. Duncan, J.M. Hyperbolic stress-strain relationships. In *Limit Equilibrium, Plasticity and General Stress-Strain in Geotechnical Engineering;* American Society of Civil Engineers: New York, NY, USA, 1980; pp. 443–460.
38. Kondner, R.L. Hyperbolic stress-strain response: Cohesive soils. *J. Soil Mech. Found. Div.* **1963**, *89*, 115–143. [CrossRef]
39. Fellenius, B.H. The analysis of results from routine pile load tests. *Ground Eng.* **1980**, *13*, 19–31.
40. Yovichin, R. Hyperbolic Representation of Force Versus Displacement Relationship for Lateral Pipe Movement in Dry Soil. Master's Thesis, Youngstown State University, Youngstown, OH, USA, 2018.
41. Duncan, J.M.; Chang, C.Y. Nonlinear analysis of stress and strain in soils. *J. Soil Mech. Found. Div.* **1970**, *96*, 1629–1653. [CrossRef]
42. Larsen, R.D. Box-and-whisker plots. *J. Chem. Educ.* **1985**, *62*, 302. [CrossRef]
43. Massart, D.L.; Smeyers-verbeke, A.J.; Capron, X.; Schlesier, K. Visual presentation of data by means of box plots. *LC GC Eur.* **2005**, *18*, 215–218.
44. Daiyan, N.; Kenny, S.; Phillips, R.; Popescu, R. Investigating pipeline–soil interaction under axial–lateral relative movements in sand. *Can. Geotech. J.* **2011**, *48*, 1683–1695. [CrossRef]
45. Almahakeri, M.; Fam, A.; Moore, I.D. Longitudinal bending and failure of GFRP pipes buried in dense sand under relative ground movement. *J. Compos. Constr.* **2013**, *17*, 702–710. [CrossRef]
46. Burnett, A. Investigation of Full Scale Horizontal Pipe-Soil Interaction and Large Strain Behaviour of Sand. Ph.D. Thesis, Queen's University, Kingston, ON, Canada, 2015.

mining

MDPI

Article

Geotechnical Issues in Decommissioning Surface Lignite Mines—The Case of Amyntaion Mine in Greece

Michael Kavvadas [1], Christos Roumpos [2,*], Aikaterini Servou [2] and Nikolaos Paraskevis [2]

[1] Professor Emeritus, School of Civil Engineering, National Technical University of Athens (NTUA), 157 80 Zografou, Greece; kavvadas@central.ntua.gr

[2] Mining Engineering and Closure Planning Department Public Power Corporation of Greece, 104 32 Athens, Greece; servoy_aikaterini@upnet.gr (A.S.); nikolasparaskevis80@gmail.com (N.P.)

* Correspondence: c.roumpos@dei.gr; Tel.: +30-697-979-9291

Abstract: Recent changes in the fossil-fuel energy sector require coal mining industries to plan for the future, including developing procedures for decommissioning and closure associated with mines. In surface coal mining, the geotechnical issues of decommissioning include the long-term stability of the pit slopes, particularly as the pit is gradually filled with water. This paper investigates such slope stability issues, with emphasis on the conditions prevailing in the Amyntaion surface lignite mine, in Western Macedonia, Greece. Analytical and numerical methods have been developed and used to estimate the temporal evolution of the overall safety factor, as the water level in the pit rises, creating a lake. It is shown that until the water level in the lake reaches a critical depth of approximately 15–35% of the final equilibrium condition, the safety factor against the overall slope instability decreases slightly (by about 3% in the case study, and up to 5–10% in other conditions) compared to its value at the end of exploitation. At higher lake levels, the safety factor increases significantly, as the beneficial effect of the lake water pressure acting on the slope overcomes the adverse effect of pore water pressure rise inside the slope. In typical mines, the critical water depth is achieved within a few years, since the surface area of the pit is smaller at deeper levels; thus, more favorable slope stability conditions are usually reinstated a few years after mine closure, while the small reduction in safety during the initial stages after closure is inconsequential. The paper investigates the parameters influencing the magnitude of the small reduction in the short-term safety factor and produces normalized graphs of the evolution of the safety factor as the lake water level rises. The results of the analyses can be used in preliminary closure studies of surface coal mines.

Keywords: mine closure; slope stability; pit lake; water levels; post-mining

1. Introduction

With international policies for the transition to a decarbonized economy, increasing greenhouse gas (GHG) emission costs, the use of lignite (brown coal), and the local fossil fuel for electricity production in Greece becomes gradually uneconomical. As a result, large-scale exploitation of surface lignite mines in Greece is gradually reduced and plans are being developed for decommissioning several lignite mines in an environmentally friendly way. As earth materials for backfilling the deep mine pits (several kilometers in size and up to 200 m deep) are not available by opening new mines (and using their spoils for filling older mines), the present mine pits will be gradually filled with water from precipitation and groundwater seepage, creating artificial lakes and new aquatic ecosystems. However, either in the case of filling the under-closure mining area with materials or with water, the slope stability investigations are considered of primary importance [1]. It is thus important to investigate the long-term environmental effects of these lakes, including the stability of the mine slopes, as the water level in the pit lakes rises gradually.

Mine closure activities include a set of procedures from concept planning to a detailed closure plan, post-closure monitoring activities, and post-closure land re-use [2,3]. These activities also include the determination of end uses, which depend on the water quality, slope stability, and safety issues. Possible end-uses could be for wildlife (flora and fauna), recreation uses (swimming, hunting, diving, fisheries), or primary production (agricultural areas, irrigated crops). Therefore, a successful mine closure should be designed with a transdisciplinary approach combining geological, hydrogeological, hydrological, geochemical, and ecological aspects [4]. The creation of pit lakes constitutes one of the most environmentally friendly and sustainable post-closure land uses, especially in areas favored by ground morphology and a high groundwater table. However, if not properly designed, pit lakes can constitute an environmental risk, because the average inclination of their slopes is usually designed for exploitation conditions with lowered groundwater table, and not for post-closure pit lake conditions, where the natural groundwater table is re-instated [5,6]. The average slope inclination of lignite mines in Greece ranges between 1:4 and 1:7, steeper than the slope inclination of typical natural lakes. Thus, evaluation of the long-term slope stability is required, combined with careful monitoring of ground movements as the water level rises in the pit.

The international literature mentions several successful and unsuccessful cases of pit lakes, showing that there is a lack of established design criteria for this purpose. According to [7], successful lakes are those that meet their purpose (e.g., fish and wildlife habitat, aquaculture, recreational parks, etc.) and those certified by regulators for use. On the other hand, unsuccessful lakes are those that do not follow their planned objectives, usually due to poor water quality (not meeting regulatory requirements). Cases reported in the literature show that mine decommissioning plans depend strongly on local conditions. In the Golden Cross mine in New Zealand, backfilling of the pit was required to face the geotechnical instabilities, submerge the geochemically active formations, and avoid the formation of a lake with poor-quality water [8]. Reference [8] also estimates the geotechnical hazard of Victor and Voorspoed mines (in Northern Ontario, Canada) considering elevation changes of the pit lakes. Reference [9] examined the geotechnical stability and pit lake filling for the decommissioning of the Victor mine in two key phases: active and post-closure. The filling rate of the lakes was calculated, considering that better slope stability conditions are achieved when the lake is filled more rapidly. Geotechnical monitoring was performed in stages, based on a hydrogeological plan, to continuously ensure the safety of the pit slopes, as they are loaded with the rising water level in the lake.

The Sleeper mine (25 miles northwest of Winnemucca in Humboldt County, NV, USA) commenced its closure in 1996, including backfilling the unstable sector of weak rock, dewatering, regrading alluvial materials, and improving water quality. The pit lake of Oaks mine was designed to collect contaminated groundwater, thus preventing contamination of the regional aquifer [10]. In the Zloczew surface lignite mine (situated about 52 km west of Belchatow in central Poland), numerical analyses using the Shear Strength Reduction Method were used to assess slope stability during the operation of the mine. This method was considered more precise than limit equilibrium models in calculating the factor of safety of the mine slopes during operation [11]. However, the design for decommissioning used a limit equilibrium model in combination with probabilistic methods as more suitable in assessing the long-term failure risk of the mine slopes [12]. The closure of Pumarabule and Mosquitera mines in Spain, included a groundwater risk assessment, and risk mitigation strategies evaluation in a performance and economic aspect [13]. In an iron ore mine in Bosnia and Herzegovina, a resilience approach was applied, regarding some crucial parameters for the closure planning strategies. The same study showed that this approach contributes to the safe closure and ensures the economic viability of the mining companies [14].

Many studies of decommissioning surface mines focus on the water quality of the created pit lakes. In references [15,16], a geochemical model was applied, while in [17] the authors compared the physical characteristics and mining conditions for three pit lakes and concluded that the formation of pit lakes is an environmentally sustainable tool of mine closure, strongly influenced by water balance and quality. Water balance is considered a critical parameter because it controls the stability of the lake when the inflows are equal to or less than the outflows [18]. By controlling this equilibrium in such infrastructures, is of great significance for the overall sustainability of the operation. Particularly, in the study of Gaagai et al. [19] a flooded dam simulation was elaborated for a risk assessment, in the mitigation context. Water quality is also an important parameter because it affects the potential life of several species in the created aquatic ecosystem.

In Greece, most surface lignite mines are operated by the Public Power Corporation (PPC). Large-scale exploitation techniques using large bucket-wheel excavators and conveyor belts result in 150–250 m deep pits developed as a sequence of benches (60–150 m wide) and drops (15–25 m high) with an overall inclination (v:h) between 1:4 and 1:7, depending on slope stability considerations of each case. Figure 1 (taken in 2016) shows the benches and drops of the SW exploitation front (excavation face) of the ca. 200 m deep Amyntaion mine of the Ptolemais Field in Western Macedonia, with an average slope inclination of about 1:6. This mine was decommissioned in 2020, with pit lake level rising since then, and forms the case study of the present paper.

Figure 1. Typical exploitation of surface lignite mines in Greece using large bucket-wheel excavators and conveyor belts. The photo (taken in 2016) shows the SW exploitation front (average slope inclination about 1:6) of the recently decommissioned ca. 200 m deep Amyntaion mine.

Based on the above literature review, it appears that there is a lack of established methods and criteria for assessing the long-term stability of surface mining slopes after closure, as a function of the rising water level in the pit lake. The present study aims to contribute to filling this research gap, using a simplified analytical method to study the parameters controlling this problem and a more detailed numerical analysis for the slopes of the Amyntaion mine as the pit lake rises. Specifically, the paper investigates the long-term slope stability of surface lignite mines by calculating the evolution of the safety factor as the level of the lake rises, compared to its present value (when the mine is still in operation, and the bottom of the pit is kept dry to suit exploitation requirements).

2. The Amyntaion Mining Area

2.1. Historical Data—Reclamation Planning

Public Power Corporation (PPC) of Greece operates several mines in the large lignite fields of Western Macedonia using mainly continuous (bucket wheel excavators—spreaders—belt conveyors) and occasionally non-continuous (trucks—loaders) mining techniques. In the Amyntaion area, three surface lignite mines have been exploited in the last 35 years: (a) the Anargyroi mine between 1984 and 2010, with 49 million tons of lignite production and 173 million m^3 total excavations, (b) the Amyntaion mine between 1989 and 2020, with 179 million tons of lignite production and 1595 million m^3 total excavations, and (c) the smaller, but higher quality, Lakkia mine, between 2013 and 2021, with 4 million tons of lignite production and 49 million m^3 total excavations. A total of 232 million tons of lignite has been mined in the Amyntaion area, with total excavations of 1817 million m^3 and an overall stripping ratio (steriles to lignite) of 7 m^3 per ton of lignite.

Figure 2 shows a plan of the Amyntaion area mines in the Amyntaion flood plain, extending from SW to the NE along the four natural lakes shown in the figure. The piezometric regime of the flood plain practically reaches ground level and has a very mild inclination from SW (Cheimaditida and Zazari lakes at elevation ca. +595) towards the NE (Petron Lake at elevation ca. +570) over a distance of about 15 km.

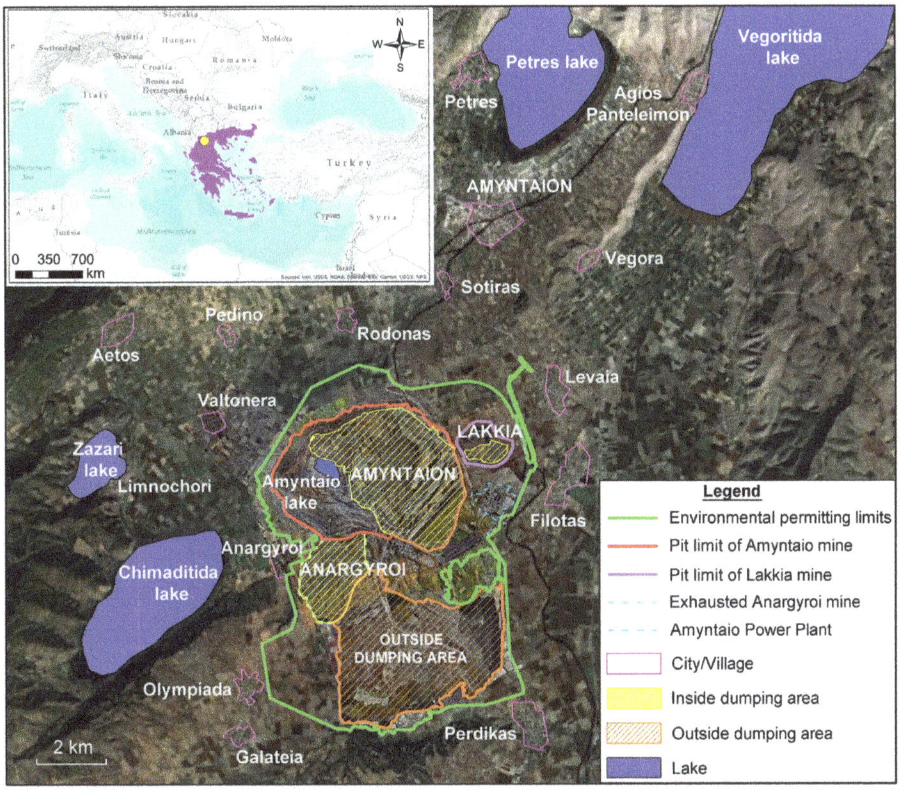

Figure 2. Plan of the Amyntaion area mines (Anargyroi, Amyntaion, Lakkia). The case study (Amyntaion mine) is marked in red. The green boundary shows the limits of the exploitation license.

Figure 3 shows the elevation distribution in the mining area and the internal and external spoil dumping areas through the hillshade effect. The brown (higher elevation) area in the south is the external spoil dumping site. The bluish color shows the deeper part of the present Amyntaion mine. Yellow sections A and B show the sections analyzed for slope stability.

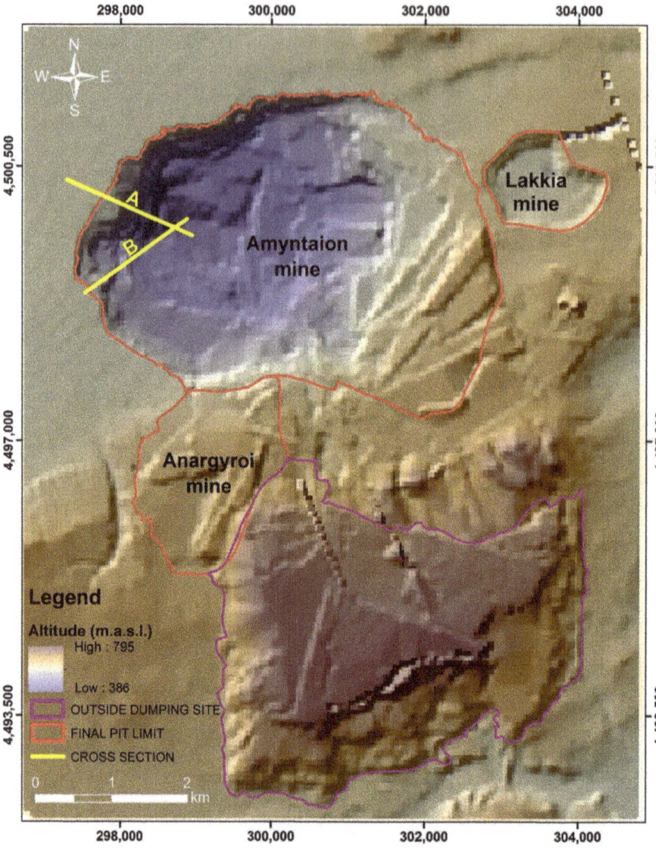

Figure 3. Digital Elevation Model (December 2021) of the Amyntaion mining area, showing the three main mines (Anargyroi, Amyntaion and Lakkia). Elevations are shown in hillshade effect.

Figure 4 shows the total annual excavation volumes (full lines) and lignite production (dashed lines) in the three mines of the Amyntaion area. The green line corresponds to the case study (Amyntaion mine).

Figure 5 shows the progress of the excavation work (location of the upper exploitation front) and the corresponding maximum depths of the Amyntaion mine over time. The red points show cases of slope failures that occurred through time in the mining area, whereas the orange dashed polygon shows the landslide area of the 10 June 2017. Figure 6 shows two typical cross-sections (I-I and II-II) of the Amyntaion mine (section locations are shown in Figure 5).

In the initial exploitation stages, sterile materials from the Anargyroi and Amyntaion mines were deposited in the external dumping area to the south of the mines; with the progress of mining excavations, spoils were deposited inside the pits, backfilling the mines. Complete backfilling of the Anargyroi pit with materials from the Anargyroi and the Amyntaion mines was achieved in 2016. In 2013, extraction started in the smaller (but better quality) Lakkia mine, with spoils transported for environmental reclamation of the Anargyroi and Amyntaion mines. Up to the present time, an area of about 8 km^2 has been reclaimed for forestry and 1 km^2 for agriculture.

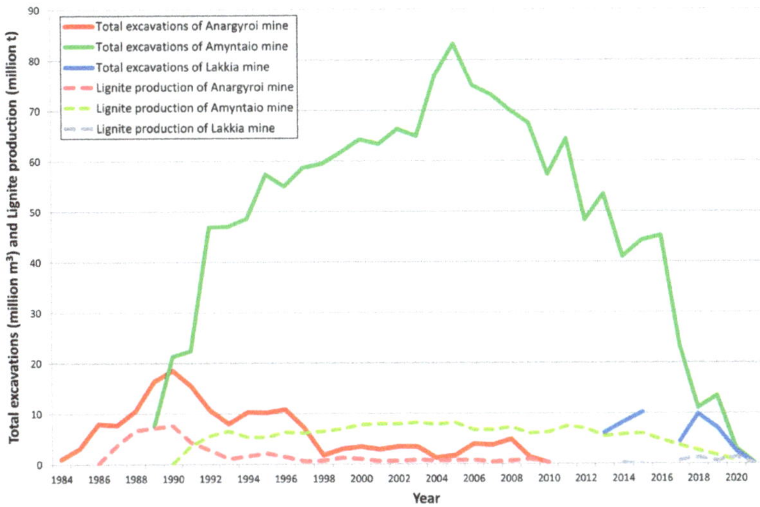

Figure 4. Total excavations (full lines) and lignite production (dashed lines) of the Amyntaion area mines per year. The green line corresponds to the case study (Amyntaion mine).

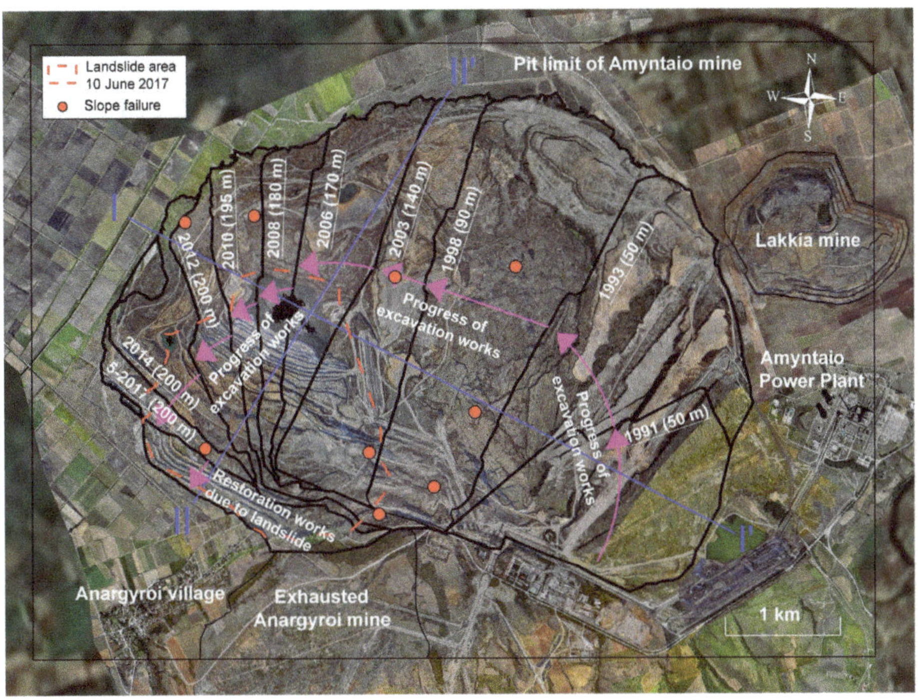

Figure 5. Progress of excavation works and corresponding maximum depths of the Amyntaion mine. Blue lines show cross sections I-I and II-II.

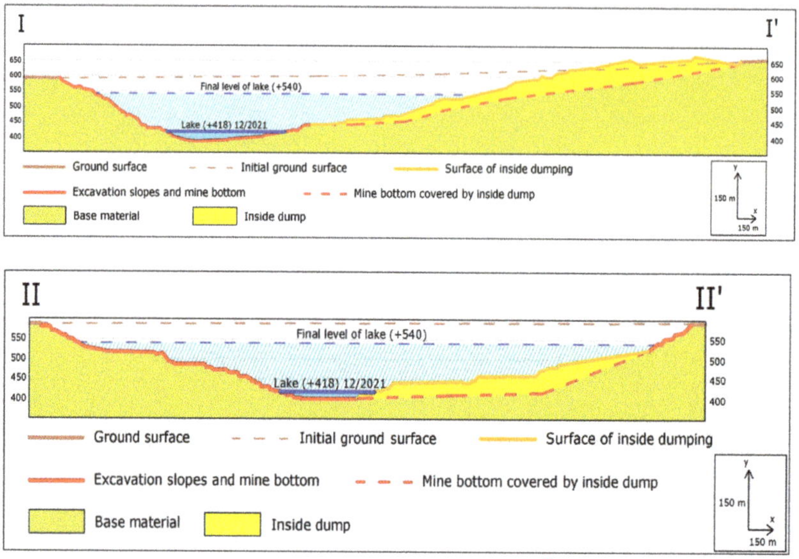

Figure 6. Sections (**I-I′**) and (**II-II′**) of the Amyntaion mine (locations are shown in Figure 5). The recent lake level is presented with continuous blue line, while the predicted lake level is presented with dashed blue line.

In June 2017, a significant landslide (about 80 million m^3) occurred on the southwestern exploitation front of the Amyntaion mine, despite its very mild slope inclination (milder than 1:6). According to the geotechnical study for the rehabilitation of the affected area [20], which practically coincided with the decision to decommission the Amyntaion mine as part of the national decarbonization policy, the failed slope was regraded to eliminate abrupt drops created by the failure and the mine was decommissioned in July 2020. Planning for land re-use after decommissioning, combined geotechnical and geochemical studies to examine the feasibility of creating an aquatic ecosystem, by allowing the natural filling of the mine pit from precipitation and groundwater seepage.

2.2. Geological, Geotectonic, and Hydrogeological Setting

Figure 7 shows a simplified geological map of the Amyntaion basin with the mining area in the red envelopes (see also Figure 3). The Amyntaion basin is a deep tectonic trough (up to 500 m deep) in the SW-NE direction in the Pelagonic geotectonic zone, consisting of an upper Palaeozoic bedrock (metamorphic rocks including schists, crystalline schists, and gneiss) and a Middle Triassic—Lower Jurassic carbonate cover (marbles, crystalline limestones, and tectonic breccias). In the Neogene, the trough was filled with lacustrine-marshy sediments (including the lignite-bearing horizons) and was subsequently covered with fluvial-lacustrine sediments in the Quaternary.

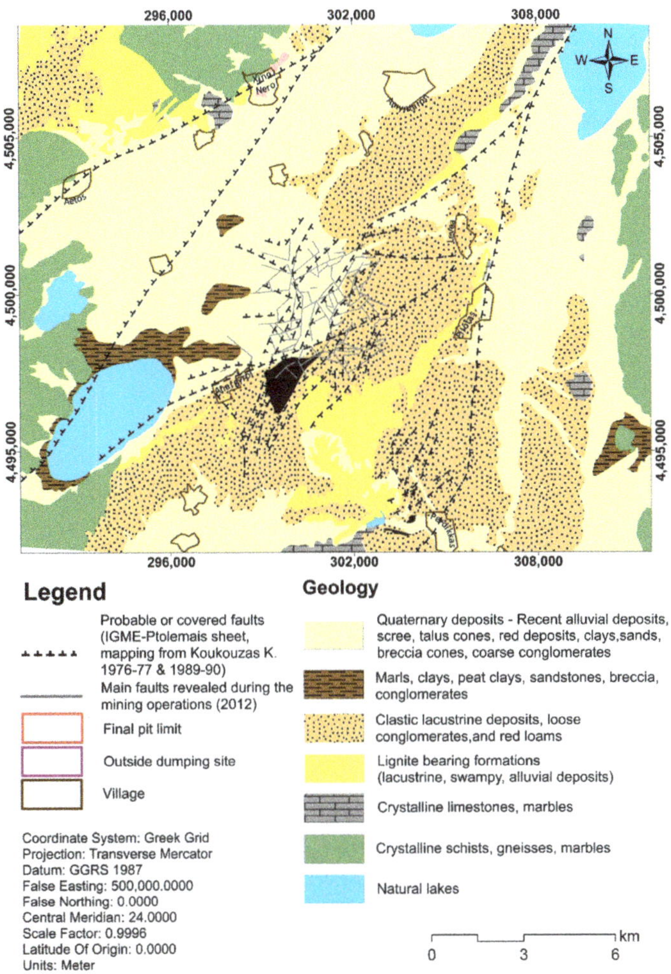

Figure 7. Simplified geological map of the Amyntaion basin which includes the mining area.

According to the geologic map of Greece [21], the Neogene sediments include an upper lignite-bearing (Ptolemaida) formation of Pliocene age (about 50–100 m thick) which also includes lacustrine-marshy deposits of grey to grey-green silty clays and clayey sands, and a lower lignite-bearing (Komnina) formation of upper Miocene-lower Pliocene age (over 200 m thick), which also includes lacustrine-marshy deposits of silty clays and sandy layers with intercalations of sandstones, above a thick base of steriles including horizons of clay-marls, sandy clays, and sands.

The Quaternary deposits, mainly of Low-Middle Pleistocene age, have a thickness of about 80–120 m and include an upper (Perdika) formation (including fluvial-lacustrine sediments comprised of intercalations of fine sand with alternating layers of sandy clays, clays, and marls, as well as lenticular intercalations of weakly cemented conglomerates created by small-sized pebbles) and a lower (Proastion) formation which overlies the Neogene deposits with unconformity and includes alternating horizons of loose silty to clayey sands and conglomerates cemented with red clay [22].

Since the Middle-Upper Miocene, extensional neotectonic activity prevails in the area, shaping the large Florina–Ptolemaida basin by major normal faults in the NW–SE direction;

later, the second group of NE-SW faults created secondary troughs, one of which is the Amyntaion basin [23–26]. Six major fault zones control the Amyntaion lignite mining area, and they constitute decisive factor for the overall mine stability, as their complex character cause fragmentation to the geological formation and decrease of their cohesion [27].

The Amyntaion flood plain is intensely cultivated. Irrigation during the dry (summer) season is achieved by about 600 deep pumped wells (shown in Figure 8), which penetrate the upper (quaternary), higher permeability, aquifer to depths up to about 120 m.

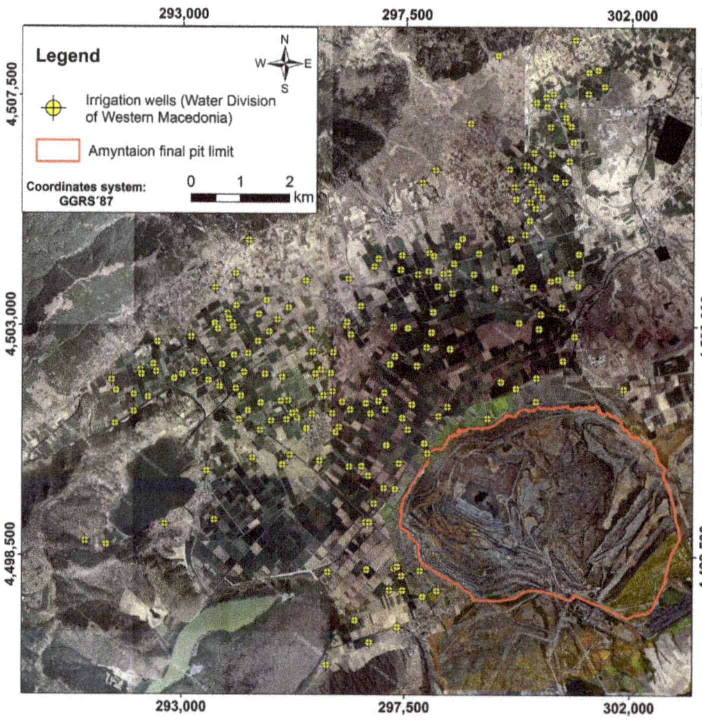

Figure 8. Distribution of the ca. 600 nos. irrigation wells in the Amyntaion flood plain (yellow shapes). The Amyntaion mine is shown in the red envelope.

Changes between limnic and telmatic environments are responsible for the lignite formations of the Amyntaion-Ptolemaida deposit. This deposit occurs in an alternating nature between lignite and interburden [28].

The ca. 200 m deep Amyntaion mine uses peripheral pumped wells to reduce groundwater seepage on the exploitation slopes. These wells penetrate the upper (more permeable) aquifer to a depth up to 120 m. The underlying lignite and intermediate sterile zone are relatively impermeable and do not need dewatering. Calculations and measurements show that the ca. 100 m deep lowering of the upper aquifer along the perimeter of the mine influences a zone up to about 500 m beyond the crest of the mine. In the rest of the flood plain, groundwater table lowering is due to the many deep wells for irrigation purposes. Since 2019, when mining operations stopped, pumping from the peripheral dewatering wells of the mine was discontinued, and pumping from the bottom of the pit stopped when the mine was decommissioned in July 2020.

According to the planning studies for the closure of the Amyntaion mine, new land uses will be created in the area surrounding the pit lake, such as recreational parks, wildlife, industrial areas, and photovoltaic parks. According to the references [29,30] the expected rise of the water level in the Amyntaion artificial lake from the initial (2020) level of +390, corresponding to the bottom of the pit, up to the final equilibrium level (about +540) after more than 100 years. It is observed that the lake development is distinguished into three stages; the early one, where the lake begins to be filled at a big rate ($\approx$10 m per year for the first 10 years), since the surface area of the pit is smaller at deep levels, the second phase of a more normal rate of filling and the third where the raise rate has begun to be stabilized $\approx$0.1 m per year). The expected trend evolution was calculated through a spatiotemporal forecasting model, which considers the crucial parameters of rainfall and temperature and by applying linear (autoregressive integrated moving average) and non-linear methods (artificial neural networks) [29,30].

3. Stability of Surface Lignite Mining Slopes after Closure

This section develops a simplified analytical tool for the calculation of the evolution of the safety factor of surface lignite mining slopes after closure, as the water level in the pit lake rises.

References [31,32] describe the geotechnics of surface lignite mining slopes in Western Macedonia. Typical ground profiles include a thick (50–120 m) zone of sterile overburden consisting of Quaternary and Neogene sediments overlying the exploitable deposits (30–130 m thick), giving a total depth of the mine up to about 250 m. The exploitable deposits include lignite seams, each several meters thick, separated by sterile interlayers consisting of hard clays/marls, usually with medium (15–35%) to high (35–50%) carbonate content. The sterile interlayers have a thickness varying between a few millimeters and several tens of centimeters (occasionally even several meters). Steriles with lower carbonate content (less than 15%) usually have high plasticity (PI > 30–40%), low peak friction angle (18–22 degrees), and an even lower residual friction angle (15–20 degrees). Steriles with medium to high carbonate content usually have lower plasticity (PI = 10–30%), higher peak friction angle (22–26 degrees), and residual friction angle in the range of 16–24 degrees.

Experience of the study area shows that instabilities of the exploitation fronts and the permanent (final) slopes are usually initiated by, or associated with, the development of a planar sliding surface along with a weak (i.e., having lower shear strength) sub-horizontal lignite-to-clay/marl interface, located either slightly above or (more often) slightly below the base of the slope. Sliding along such a surface is usually initiated by the significant horizontal stress release as the excavation front is advanced, in conjunction with the high stiffness contrast between stiffer lignite zones and much softer intermediate steriles. The elastic shear deformation along such interfaces increases with depth (due to the larger stress release) and, in deep mines, it may exceed the peak shear strength, leading to even larger deformations as the shear strength gradually drops towards the (lower) residual value. If the residual shear strength is not sufficient to provide the required shearing resistance for the stability of the sliding mass, slope instability occurs. Due to the above sliding mechanism, movement in such slopes starts at a slow velocity and gradually accelerates, eventually leading to failure. Local experience shows that the velocity of the movement is not usually a reliable criterion for incipient instability (via a critical velocity), as long as the velocity is practically stable or varies in a consistent manner (e.g., influenced by rainfall). It appears that the acceleration rate of slope movement is a more reliable criterion in accessing upcoming instability: sustained velocity changes per day approaching the velocity of the previous day are usually alarming.

The inclination of the above sub-horizontal interfaces is critical for slope stability; in deep mines, even a mild dip towards the toe of the slope can reduce the safety factor significantly, compared to a horizontal interface and, even more, compared to interfaces with reverse inclination (towards the slope interior). In some cases, when the lignite seams are highly tectonized (as usually occurs close to the edges of the tectonic trough), the above weak interfaces are disturbed by successive faults and become discontinuous; in such cases, slope instabilities occur along a more or less circular or multi-linear failure surface which may locally align with weak interfaces, where kinematically possible.

Finally, the rate of filling of the pit lake is also critical in slope stability. Rapid filling of the pit lake is always beneficial for stability (compared to slower filling rates) because, in the rapid filling, the stabilizing effect of the lake water pressure acting on the slope occurs faster than the destabilizing effect of the rising pore water pressure inside the slope). Unfortunately, this condition cannot be controlled in most cases, because the expected filling rate of the lake may be delayed by unusual dry seasons and, most importantly, the rate of increase of the pore water pressure in the slope may be accelerated by the presence of undetected high permeability zones. Thus, the present analysis assumes (conservatively) that the piezometric regime in the slope corresponds to steady-state seepage conditions for each pit lake level, i.e., filling of the pit lake is sufficiently slow that pore pressure equilibrium is achieved in the slope.

Based on the above mechanisms, the present investigation studies the effects of the following parameters on slope stability conditions after mine closure with pit lake development:

1. The height (H) and average inclination of the slope (angle β, from crest to toe). Although the overall slope consists of several benches and drops, overall slope stability is controlled by the average slope inclination rather than the size of the intermediate benches and drops.
2. The inclination (angle β') with respect to the horizontal of the lignite-bearing layers and intermediate steriles in the slope. In slopes with few tectonic faults, potential failure surfaces usually follow such interfaces because of their lower shear strength. In contrast, in heavily tectonized materials, failure surfaces are more or less circular, crossing interfaces, since aligning with them is not kinematically feasible.
3. The shear strength (especially the residual strength) parameters (c', φ') of the ground along the failure surface, with emphasis on the residual strength along with weak interfaces between lignite-bearing layers and underlying steriles (in slopes with few tectonic faults).
4. The temporal evolution of the piezometric levels in the slope, as the water level in the pit lake (depth Hw) rises, in conjunction with the stabilizing effect of the pressure of lake water acting on the lower part of the slope.

Various methods have been used in the literature for the analysis of mining slope stability during exploitation. Analytical methods (e.g., [32]), produce a relatively simple formula for the safety factor of the slope and thus can illustrate better the effect of various input parameters. Numerical methods using suitable software (e.g., [33]) can model complicated geometric and material characteristics more accurately but lack insight on the sensitivity of the results on the various input parameters. Most of these methods, however, do not study the effect of the rising water level in the pit lake after mine closure, because they are developed for the stability of mining slopes during operation. The following analytical method is developed to study the overall stability of mining slopes as the water level in the pit rises. The method produces an analytical formula for the safety factor as a function of the water level in the pit lake and can be used for preliminary assessments of mine closures.

Figure 9 shows the geometry and the forces acting on a typical sliding mass (OAB) of a surface lignite mining slope (height H, average inclination angle $\beta \rightarrow \tan(\beta) = 1/h$). Assuming the experience, such slope failures usually emerge close the toe of the slope (O) and develop along with a sub-horizontal lignite-marl interface (OA) with angle $\beta'(\ll \beta,\ \tan\ \beta' = h')$ with respect to the horizontal. Experience also shows that stability conditions are more adverse, as the inclination of OA increases (h' increases). In some cases, the material is intensely tectonized (i.e., line OA is crossed by several tectonic faults), and the development of a planar failure surface (OA) is not possible kinematically. In such cases, failure occurs along a quasi-circular surface and the below analytical method is not applicable.

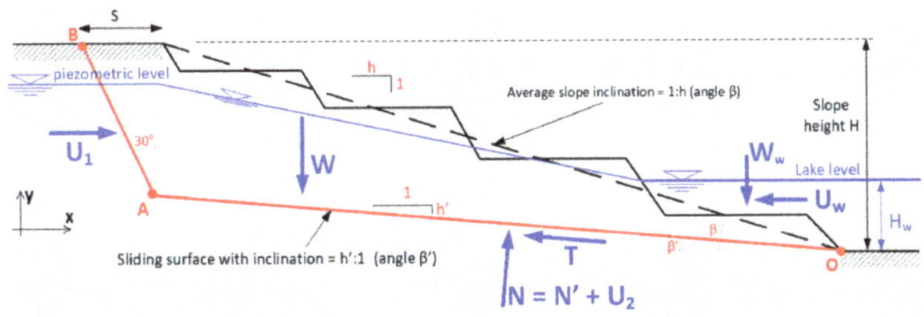

Figure 9. Geometry and forces acting on a typical sliding mass of a lignite mining slope.

The sliding surface is assumed to reach the top of the slope, at distance (S) from the crest, with a planar transition (AB). It is assumed that a tension crack develops along (AB), only a water thrust (U_1) acts on (AB) and that (AB) has an inclination of 30 degrees to the vertical plane (the inclination of (AB) has a small effect on the stability calculations). The blue line in the figure shows the assumed piezometric level along the failure surface (OAB). A linear transition of the piezometric surface is assumed between the crest of the slope (ambient piezometric level) and the pit lake level. It is noted that other aquifers which may develop at higher elevations in the slope are irrelevant in the slope stability analysis.

Geometrical considerations give the horizontal (X) and vertical (Y) projections of lines (AB) and (OA):

$$X(AB)= X_1 = \frac{(1-hh')H-h'S}{\sqrt{3}-h'} \qquad Y(AB)= Y_1 = \sqrt{3}X_1 \tag{1}$$

$$X(OA)= X_2 = \frac{\left(\sqrt{3}h-1\right)H+\sqrt{3}S}{\sqrt{3}-h'} \qquad Y(OA)= Y_2= h'X_2 \tag{2}$$

$$(OA) = L = \sqrt{(X_2)^2+(Y_2)^2} \tag{3}$$

The tension crack (AB) is assumed to be filled with water (unit weight γ_w) at a fraction λ_1 (between 0 and 1) of its total height Y_1. Thus, the water force on (AB) is:

$$U_1 = \frac{1}{2}\gamma_w(\lambda_1 Y_1)^2 \tag{4}$$

Considering that the pore water pressure at A is: $u_A = \lambda_1\gamma_w Y_1$ and the pore water pressure at O is: $u_o = \gamma_w H_w$, with linear variation in between, the pore water force (U_2) acting on the base (OA) of the sliding mass is:

$$U_2 = \frac{1}{2}\gamma_w(H_w+\lambda_1 Y_1)L \tag{5}$$

where L is the length of the sliding surface OA

The lake water forces W_w and U_w acting at the toe of the slope are:

$$U_w = \frac{1}{2}\gamma_w(H_w)^2 \tag{6}$$

$$W_w= hU_w$$

where H_w is the height of the lake's water depth

The weight (W) of the sliding mass is:

$$W = \frac{1}{2}\gamma_w [S(Y_1 + Y_2) + X_2 Y_1 - X_1 Y_2] \tag{7}$$

Force equilibrium of the sliding mass along axes normal and parallel to its base (OA) gives:

$$N = (W + W_w)\cos\beta' + (U_w - U_1)\sin\beta' \tag{8}$$

$$T = (W + W_w)\sin\beta' + (U_1 - U_w)\cos\beta' \tag{9}$$

Finally, the shearing resistance (T_u) along the base (OA) of the sliding mass is the sum of the cohesive and frictional components:

$$T_u = c'L + (N - U_2)\tan\varphi' \tag{10}$$

where (φ', c') are the effective friction angle and the effective cohesion along the base of the sliding mass (along with an interface between a lignite zone and the underlying stiff plastic clay-marl layer).

Equations (8) and (9) to (10) can be used to calculate the overall safety factor (FS) of the slope:

$$FS \equiv \frac{T_u}{T} = \frac{c'L + [(W + W_w)\cos\beta' + (U_w - U_1)\sin\beta']\tan\varphi' - U_2\tan\varphi'}{(W + W_w)\sin\beta' + (U_1 - U_w)\cos\beta'} \tag{11}$$

and the value of the safety factor (FS_o) before the lake level rises ($H_w = 0$):

$$FS_o \equiv \frac{T_u}{T} = \frac{c'L + [W\cos\beta' - U_1\sin\beta']\tan\varphi' - U_2\tan\varphi'}{W\sin\beta' + U_1\cos\beta'} \tag{12}$$

Note: In Formulae (11) and (12), the pore water force (U_2) is different, as it depends on the depth (H_w) of the pit lake (see Equation (5)).

Application of Formulae (11) and (12) for various combinations of geometrical (H, H_w), h, h', λ_1, S) and material (γ_w, c, φ) parameters shows that while the values of (FS) and (FS_o) depend on all input parameters, the ratio (FS/FS_o) is weakly dependent on input parameters other than the ratio (H_w/H) and the inclination angle (β'). The reason for the weak dependency of the ratio (FS/FS_o) on the various geometric and material parameters is that the effect of these parameters is expressed in both components (FS and FS_o) in the same way (i.e., tends to increase or decrease both values) and thus their influence of the ratio FS/FS_o is significantly reduced.

Using the above analytical method, Figure 10 plots the calculated safety factor (FS) of a slope with height H and lake water depth (H_w), normalized with the safety factor (FS_o) of the same slope for $H_w = 0$, versus the ratio (H_w/H) for all examined cases in the following ranges: H = 150–250 m, h = 4–6, $\lambda_1 = 0.90$, $S/H = 0.10$, $\gamma_w = 10$ kN/m^3, c = 3–5 kPa, φ = 17–25 deg). For each case, the combinations of the various parameters were such, that the safety factor during operation of the mine was in the range $FS_o = 1.20 - 1.35$, the typical range of safety factors in exploitation slopes for surface lignite mines in Greece.

The curves in the figure correspond to various inclinations of the base (OA) of the failure surface (β' = 0–4 deg). The figure shows that as the water level in the lake rises, the safety factor reduces, reaching a minimum value when H_w/H = 0.2 to 0.35. The reduction of the safety factor (FS) with respect to FS_o is smallest (about 10%) when the base of the failure surface is horizontal (β' = 0) and increases up to about 25% when this inclination reaches 4 degrees. Subsequently, the safety factor increases monotonically, exceeds the value FS_o when $H_w/H > 0.40$ to 0.65, and reaches large values when the slope becomes fully submerged.

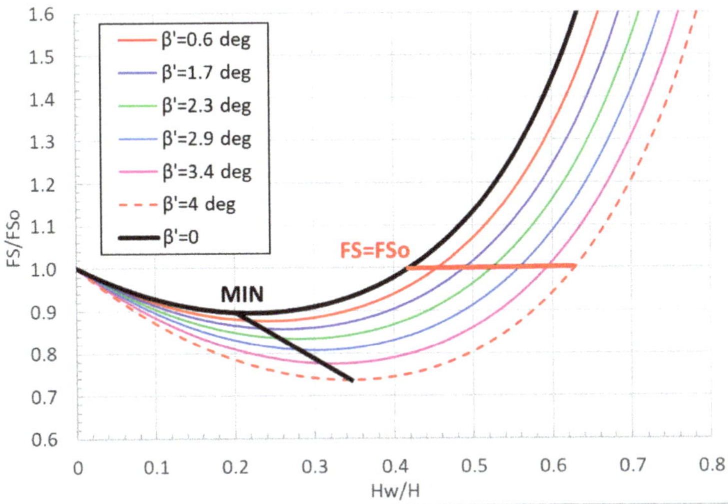

Figure 10. Variation of the safety factor (FS) against overall slope instability with water depth (H_w) in the lake. The safety factor (FS) is normalized with the safety factor (FS_o) of the same slope when $H_w = 0$. H is the height of the slope and β' is the slope of the base of the failure surface.

In real lignite slopes, the reduction of the safety factor (FS) at low levels of the pit lake is expected to be smaller than the above simplified analysis, because the base of the sliding mass (several hundred meters long) is rarely perfectly planar and uniformly inclined, as assumed in the analysis. The next section of the paper presents the results of more realistic numerical analyses which show that the maximum reduction of the safety factor does not exceed 5% of the initial value (FS_o) in more realistic cases of the geometry of the failure surface, where the failure surface crosses some material boundaries rather than following a single interface.

4. Stability of the Amyntaion Mine Slopes after Closure

The stability of the Amyntaion mine slopes was analyzed at two representative Sections A and B (locations shown in Figure 3) as the water level in the pit lake rises. Both Sections are close to the SW corner of the mine because the depth of the mine is maximum there, the steepest slope is at Section A and the significant landslide of 2017 occurred close to and along the direction of Section B. Figure 11 presents the geological profiles of Sections A and B. Both Sections are crossed by several tectonic faults (shown in red) because the slopes are close to the edge of the tectonic trough, which is defined by several tectonic faults. Since the deposits are highly tectonized, potential failure surfaces are not expected to develop along with a single weak lignite-sterile interface, as examined analytically in the previous section of the paper. On the contrary, potential failure surfaces are expected to be more or less circular, cutting through the various deposits shown in the geological sections.

Table 1 shows the strength parameters of the relevant geological formations, based on past experience from back analyses on similar slopes (either failed or approaching failure) and the literature [34]. The piezometric surface along the examined failure surfaces was assumed to be very close to the ground surface (at a depth of about 10 m).

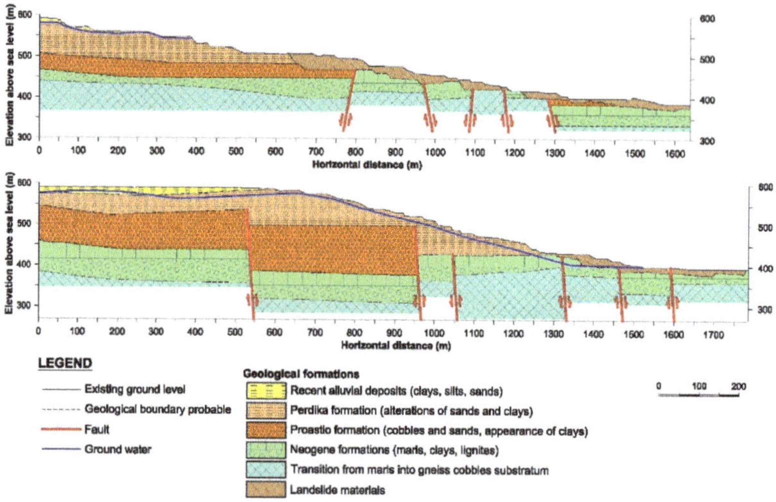

Figure 11. Geological Sections A (**top**) and B (**bottom**). Section A: Height 200 m (from +390 to +590) and average inclination 1:4. Section B: Height 200 m (from +400 to +600) and average inclination 1:7 (due to the 2017 landslide that accumulated most of the Fill material shown on the surface of the slope).

Table 1. Geotechnical parameters used in the stability analyses.

Geological Formation	Cohesion c (kPa)	Friction Angle φ (°)
Quaternary deposits (upper and lower)	7.5	26.5
Neogene deposits (including lignites)	10	30
Base marl	10	24

The numerical stability analyses of the two slopes (A and B) with variable pit lake levels and corresponding equilibrium piezometric levels in the slope were performed with a 2D limit equilibrium method (method of slices) using the computer software SLIDE2 (2D Limit Equilibrium Analysis for Slopes), [35].

Figures 12 and 13 show the critical failure surfaces and the corresponding calculated safety factors at the end of mine operation (when the piezometric level was at the bottom of the pit) and at an elevated water level in the pit lake when the safety factor is at a minimum (before starting to increase).

Figure 14 plots the variation of the calculated safety factor (FS) with depth (H_w) of the pit lake for Sections A and B and a "predicted" (average) curve. For the calculations, the same input parameters were used for the two sections. The safety factor (FS) is normalized with the safety factor (FS_o) of the same slope when Hw = 0. Both sections show similar responses, despite the very significant difference in the initial safety factor (FS_o= 1.294 in Section A and FS_o= 1.879 in Section B) which is mainly due to the difference in the two slope inclinations (1:4 and 1:7). The safety factor decreases slightly (by about 2%) up to lake depth about H_w = (10–15%) H and then increases monotonically, confirming that the overall safety of the slope increases as the water depth in the lake increases. The small initial decrease and the subsequent significant and monotonic increase of the safety factor of the slopes with increasing lake water level are due to the fact that the piezometric level along the critical failure surface at the end of the operation of the mine is high (almost up to ground level) due to the low permeability of the Neogene deposits and the underlying base marl. Thus, when the water level in the pit lake rises, the destabilizing increase of the pore water pressure along the failure surface is relatively small, compared to the stabilizing effect of the lake water thrust acting at the toe of the slope.

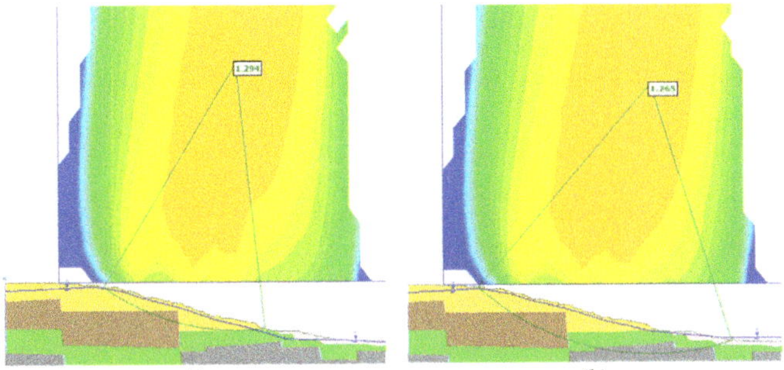

(a) (b)

Figure 12. Stability analysis of Section A. (**a**) End of exploitation, with lake level at the base of the pit (+390). Safety factor FS_o = 1.294. (**b**) Minimum safety factor FS = 1.265 when the water depth inside the pit is Hw = 20 m (level +410). The initial safety factor (FS_o) is reduced by 2.2%.

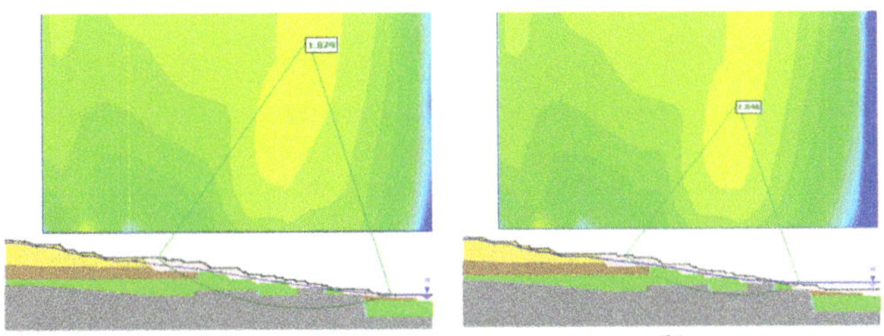

(a) (b)

Figure 13. Stability analysis of Section B. (**a**) End of exploitation, with lake level at the base of the pit (+400). Safety factor FS_o = 1.879. (**b**) Minimum safety factor FS = 1.846 when the water depth inside the pit is H_w = 20 m (level +420). The initial safety factor (FS_o) is reduced by 1.8%.

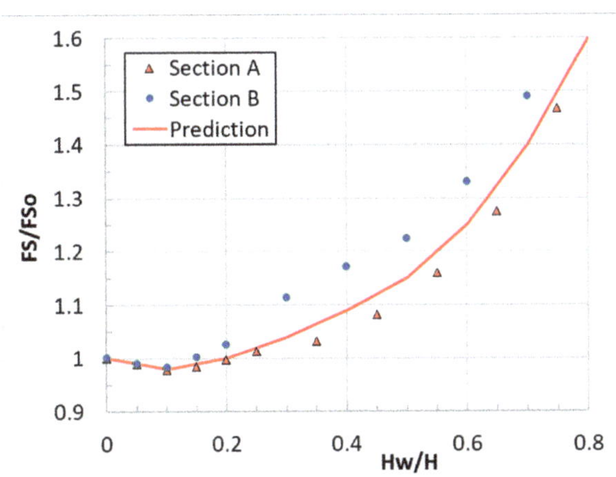

Figure 14. Variation of the safety factor (FS) of Sections A and B, against slope instability with pit lake water depth (H_W). The safety factor (FS) is normalized with the initial safety factor (FS_o) of the same slope when H_W = 0. H is the total slope height. The curve "Prediction" plots an average curve of the two Sections that can be used for preliminary predictions.

5. Conclusions

This paper investigates the evolution of the safety factor of surface lignite mining slopes after mine closure, as the water level in the pit lake rises. A simplified analytical model calculates the safety factor of such slopes, assuming that the base of the failure surface develops along a weak lignite-to-sterile interface, located close to the base of the slope. The analysis shows that the safety factor (FS) of the slope can be expressed in terms of the initial safety factor (FS_o) before the water level starts to rise in the lake, the ratio (H_w/H) of the depth of the lake to the height of the slope and the inclination (β') of the weak interface at the base of the slope. Parametric analyses have shown that the effect of the remaining geometrical and material parameters can be lumped into the initial safety factor (FS_o) with reasonable accuracy. In the initial stages of water filling, the safety factor of the slope decreases mildly and then increases significantly. The simplified analytical model shows that the minimum safety factor is controlled by the inclination (β') of the weak interface at the base of the slope and ranges between 10–25% of the initial safety factor (FS_o). In real slopes, the reduction of FS is expected to be appreciably smaller, because the base of the sliding mass is rarely perfectly planar and uniformly inclined, as assumed in the simplified analysis.

The same trends are confirmed by numerical stability analyses of two characteristic slopes of the Amyntaion mine in Western Macedonia, Greece, which was decommissioned in 2020 and the water level in the pit lake is rising ever since. The numerical analyses show that the maximum reduction of the safety factor does not exceed 5% of the initial value (FS_o) and occurs at lake depths about 10–15% of the slope height. Subsequently, the safety factor quickly recovers to the initial value (FS_o) and then increases rapidly, almost doubling when the slope is practically submerged. The small initial decrease and the subsequent significant and monotonic increase of the safety factor of the slopes with increasing lake water level are due to the fact that the piezometric level along the critical failure surface at the end of the operation of the mine is high (almost up to ground level) due to low ground permeability. Thus, the destabilizing pore water pressure rise along the failure surface is relatively small, compared to the stabilizing effect of the lake water acting on the slope.

Both methods of analysis (analytical and numerical) have used the conservative assumption that the piezometric regime in the slope corresponds to steady-state seepage conditions for each pit lake level, i.e., filling of the pit lake is sufficiently slow that pore pressure equilibrium is achieved in the slope. This assumption is conservative, since faster filling of the pit lake is beneficial for stability, because the stabilizing effect of the lake water pressure acting on the slope occurs faster than the destabilizing effect of the rising pore water pressure inside the slope. Unfortunately, this condition cannot be controlled in most cases, because the expected filling rate of the lake may be delayed by unusual dry seasons and, most importantly, the rate of increase of the pore water pressure in the slope may be accelerated by the presence of undetected high permeability zones.

Author Contributions: Conceptualization, C.R. and M.K.; methodology, M.K. and C.R.; software, M.K., C.R., A.S. and N.P.; validation, M.K., C.R., A.S. and N.P.; formal analysis, M.K., C.R., A.S. and N.P.; investigation, M.K., C.R., A.S. and N.P.; resources, M.K., C.R., A.S. and N.P.; data curation, M.K., C.R., A.S. and N.P.; writing—original draft preparation, M.K. and C.R.; writing—review and editing M.K., C.R., A.S. and N.P.; visualization, M.K., C.R., A.S. and N.P.; supervision M.K. and C.R.; project administration, M.K. and C.R. All authors have read and agreed to the published version of the manuscript.

Funding: This research received no external funding.

Conflicts of Interest: The authors declare no conflict of interest.

References

1. Tao, Z.; Li, M.; Zhu, C.; He, M.; Zheng, X.; Yu, S. Analysis of the Critical Safety Thickness for Pretreatment of Mined-Out Areas Underlying the Final Slopes of Open-Pit Mines and the Effects of Treatment. *Shock Vib.* **2018**, *2018*, 1306535. [CrossRef]

2. De Graaf, P.; Beale, G.; Carter, T.; Dixon, J. Geotechnical Guidelines for Open Pit Closure—A New Publication by the Large Open Pit (LOP) Project. In *Mine Closure 2021: Proceedings of the 14th International Conference on Mine Closure*; Fourie, A.B., Tibbett, M., Sharkuu, A., Eds.; QMC Group: Ulaanbaatar, Mongolia, 2021; pp. 217–230. [CrossRef]
3. Agboola, O.; Babatunde, D.E.; Isaac Fayomi, O.S.; Sadiku, E.R.; Popoola, P.; Moropeng, L.; Yahaya, A.; Mamudu, O.A. A Review on the Impact of Mining Operation: Monitoring, Assessment and Management. *Results Eng.* **2020**, *8*, 100181. [CrossRef]
4. McCullough, C.; Schultze, M.; Vandenberg, J. Realizing Beneficial End Uses from Abandoned Pit Lakes. *Minerals* **2020**, *10*, 133. [CrossRef]
5. Doupé, R.G.; Lymbery, A.J. Environmental Risks Associated with Beneficial End Uses of Mine Lakes in Southwestern Australia. *Mine Water Environ.* **2005**, *24*, 134–138. [CrossRef]
6. McCullough, C. Key Mine Closure Lessons Still to Be Learned. In *Mine Closure 2016: Proceedings of the 11th International Conference on Mine Closure*; Fourie, A.B., Tibbett, M., Eds.; Australian Centre for Geomechanics: Perth, Australia, 2016; pp. 325–338. ISBN 978-0-9924810-4-9. [CrossRef]
7. Vandenberg, J.; Mccullough, C. *Global Review of Pit Lake Case Studies*; Presentation of Golder Associates Inc. British Columbia MEND ML/ARD Annual Workshop: Vancouver, BC, Canada, 2018. Available online: https://bc-mlard.ca/files/presentations/2018-8-VANDENBERG-MCCULLOUGH-review-pit-lake-case-studies.pdf (accessed on 22 April 2022).
8. de Graaf, P.; Desjardins, M.; Tsheko, P. Geotechnical Risk Management for Open Pit Mine Closure: A Sub-Arctic and Semi-Arid Case Study. In *Mine Closure 2019: Proceedings of the 13th International Conference on Mine Closure*; Fourie, A.B., Tibbett, M., Eds.; Australian Centre for Geomechanics: Perth, Australia, 2019; pp. 211–234. [CrossRef]
9. Desjardins, M.; de Graaf, P.; Beale, G.; Rougier, M. Geotechnical Risk Management for Victor Mine Closure. In *Slope Stability 2020: Proceedings of the 2020 International Symposium on Slope Stability in Open Pit Mining and Civil Engineering*; Dight, P.M., Ed.; Australian Centre for Geomechanics: Perth, Australia, 2020; pp. 399–414. [CrossRef]
10. Botham, N.; Kelso, C.; Annegarn, H. *Best Practice in Acquiring a Mine Closure Certificate—A Critical Analysis of the De Beers Oaks Diamond Mine, South Africa*; Australian Centre for Geomechanics: Perth, Australia, 2011; pp. 401–410. [CrossRef]
11. Bednarczyk, Z. Slope Stability Analysis for the Design of a New Lignite Open-Pit Mine. *Procedia Eng.* **2017**, *191*, 51–58. [CrossRef]
12. Apostu, I.-M.; Lazar, M.; Faur, F. A Suggested Methodology for Assessing the Failure Risk of the Final Slopes of Former Open-Pits in Case of Flooding. *Sustainability* **2021**, *13*, 6919. [CrossRef]
13. Fidalgo Valverde, G.; Duda, A.; Iglesias Rodríguez, F.J.; Frejowski, A.; Todorov, I. Groundwater Risk Assessment in the Context of an Underground Coal Mine Closure and an Economic Evaluation of Proposed Treatments: A Case Study. *Energies* **2021**, *14*, 1671. [CrossRef]
14. Komljenovic, D.; Stojanovic, L.; Malbasic, V.; Lukic, A. A Resilience-Based Approach in Managing the Closure and Abandonment of Large Mine Tailing Ponds. *Int. J. Min. Sci. Technol.* **2020**, *30*, 737–746. [CrossRef]
15. Anim, F.; Nyankson, E.; Nyame, F.K. Analysis of Mine Water from Four Decommissioned Pits in South-Western Ghana–Implications for Remediation Programmes for Mine Closure. *Water Policy* **2017**, *19*, 957–977. [CrossRef]
16. Triantafyllidis, S.; Psarraki, D. Implementation of Geochemical Modeling in Post-Mining Land Uses, the Case of the Abandoned Open Pit Lake of the Kirki High Sulfidation Epithermal System, Thrace, NE Greece. *Environ. Earth Sci.* **2020**, *79*, 518. [CrossRef]
17. Johnstone, A.C. Are Pit Lakes an Environmentally Sustainable Closure Option for Opencast Coal Mines? *J. S. Afr. Inst. Min. Metall.* **2021**, *121*, 1–5. [CrossRef]
18. Sakellari, C.; Roumpos, C.; Louloudis, G.; Vasileiou, E. A Review about the Sustainability of Pit Lakes as a Rehabilitation Factor after Mine Closure. *Mater. Proc.* **2021**, *5*, 5052. [CrossRef]
19. Gaagai, A.; Aouissi, H.A.; Krauklis, A.E.; Burlakovs, J.; Athamena, A.; Zekker, I.; Boudoukha, A.; Benaabidate, L.; Chenchouni, H. Modeling and Risk Analysis of Dam-Break Flooding in a Semi-Arid Montane Watershed: A Case Study of the Yabous Dam, Northeastern Algeria. *Water* **2022**, *14*, 767. [CrossRef]
20. Kavvadas, M. *Report for the Slope Instability of Southwestern Side of PPC Amyntaion Field Mine, Near the Side of Anargyroi Village*, Public Power Corporation: Athens, Greece, 2017; *Unpublished Work.*
21. Institute of Geology and Mineral Exploration (IGME). *Geological Map of Greece, Ptolemais Sheet*; Scale 1:50.000. Geological mapping by Koukouzas, K. Lignite Investigation Program of NW Macedonia, implemented on behalf of Public Power Corporation (PPC) in cooperation with the Primary Energy Resources Division of IGME and the Lignite Exploration Division of PPC Program for compiling the Geological Map of Greece in scale 1:50.000; Department of Geology and Geological Mapping of IGME, 1997.
22. Koukouzas, C.; Kotis, T.; Ploumidis, M.; Metaxas, A. *Coal Exploration of Anargiri-Amynteon Area, Mineral Deposit Research*; I.G.M.E: Athens, Greece, 1979; Volume 9, pp. 1–69.
23. Pavlides, P. Neotectonic Evolution of the Florina-Vegoritida-Ptolemaida Basin (SW Macedonia). Ph.D. Thesis, Aristotle University of Thessaloniki, Department of Geology, Thessaloniki, Greece, 1985.
24. Pavlides, S.B.; Simeakis, K. Neotectonics and active tectonics ἱη low seismicity areas of Greece: Vegoritis (NW Macedonia) and Melos isl complex (Cyclades)-comparison. In *Annales Géologiques Des Pays Helléniques*; Aristotle University of Thessaloniki: Thessaloniki, Greece, 1987; Volume XXXIII, pp. 161–176.
25. Mountrakis, D.; Pavlides, S.; Zouros, N.; Astaras, T.; Chatzipetros, A. Seismic Fault Geometry and Kinematics of the 13 May 1995 Western Macedonia (Greece) Earthquake. *J. Geodyn.* **1998**, *26*, 175–196. [CrossRef]

26. Mountrakis, D.; Tranos, M.; Papazachos, C.; Thomaidou, E.; Karagianni, E.; Vamvakaris, D. Neotectonic and Seismological Data Concerning Major Active Faults, and the Stress Regimes of Northern Greece. *Geol. Soc. Lond. Spec. Publ.* **2006**, *260*, 649–670. [CrossRef]
27. Pavlides, S.; Chatzipetros, A.; Lazos, I. The Role of Geological Faults in Mine Stability: Amynteon Mine, Western Macedonia (Greece) as a Case Study. In *Proceedings of the 14th International Symposium of Continuous Surface Mining, ISCSM*; Department of Geology, Aristotle University of Thessaloniki: Thessaloniki, Greece, 2018.
28. Mavridou, E.; Antoniadis, P.; Khanaqa, P.; Riegel, W.; Gentzis, T. Paleoenvironmental Interpretation of the Amynteon–Ptolemaida Lignite Deposit in Northern Greece Based on Its Petrographic Composition. *Int. J. Coal Geol.* **2003**, *56*, 253–268. [CrossRef]
29. Louloudis, G.; Kasfikis, G.; Mertiri, E. *Investigation of Spatiotemporal Development of Amyntaion Pit Lake*, Public Power Corporation: Athens, Greece, 2020; *Unpublished Work.*
30. Louloudis, G.; Louloudis, E.; Roumpos, C.; Mertiri, E.; Kasfikis, G.; Chatzopoulos, K. Forecasting Development of Mine Pit Lake Water Surface Levels Based on Time Series Analysis and Neural Networks. *Mine Water Environ.* **2021**. [CrossRef]
31. Kavvadas, M. Stability and Movements of Open-Pit Lignite Mines in Northern Greece. In Proceedings of the 18th International Conference on Soil Mechanics and Geotechnical Engineering (ICSMGE), Paris, France, 2–6 September 2013.
32. Kavvadas, M.; Roumpos, C.; Schilizzi, P. Stability of Deep Excavation Slopes in Continuous Surface Lignite Mining Systems. *Geotech. Geol. Eng.* **2020**, *38*, 791–812. [CrossRef]
33. Mikroutsikos, A.; Theocharis, A.I.; Koukouzas, N.C.; Zevgolis, I.E. Analytical versus Numerical Analysis on Slope Stability of Surface Lignite Mines. In *The Evolution of Geotech-25 Years of Innovation*; CRC Press: London, UK, 2021; pp. 161–167. ISBN 978-1-00-318833-9.
34. Anagnostopoulou, S.; Boumpoulis, V.; Lampropoulou, P.; Servou, A.; Depountis, N.; Sabatakakis, N. The Behavior of the Highly Weathered and Partially Decomposed Flysch in the Reactivation of Landslide Phenomena in Greece. In *IAEG/AEG Annual Meeting Proceedings, San Francisco, California, 2018-Volume 1*; Shakoor, A., Cato, K., Eds.; Springer International Publishing: Cham, Switzerland, 2019; pp. 119–124. ISBN 978-3-319-93123-4. [CrossRef]
35. *SLIDE2; 2D Limit Equilibrium Analysis for Slopes*, version 5.010; Rockscience Inc.: Toronto, ON, Canada, 2004.

mining

Review

An Overview of Slope Failure in Mining Operations

Peter Kolapo [1,*], Gafar Omotayo Oniyide [2], Khadija Omar Said [3], Abiodun Ismail Lawal [2], Moshood Onifade [4] and Prosper Munemo [5]

1 Department of Mining Engineering, University of Kentucky, Lexington, KY 40506, USA
2 Department of Mining Engineering, Federal University of Technology, Akure 340100, Nigeria; gooniyide@futa.edu.ng (G.O.O.); abiodunismail18@yahoo.com (A.I.L.)
3 Department of Mining and Mineral Processing Engineering, Taita Taveta University, Voi 80300, Kenya; skhadijaomar@gmail.com
4 Department of Civil and Mining Engineering, University of Namibia, Ongwediva P.O. Box 3624, Namibia; lonifade4@gmail.com
5 Department of Mining and Mineral Processing Engineering, Manicaland State University of Applied Science, Mutare P.O. Box 7001, Zimbabwe; munemoprosper@gmail.com
* Correspondence: peter.kolapo@uky.edu

Abstract: The primary aim of every twenty-first century mining operation is to extract as much ore as possible in a safe and economical manner. Failure in mine excavation occurs when the shear stress acting on the rock is greater than the shear strength of the rock mass. The stability of rock slopes in open-pit mine and quarry operations is extremely important from both economic and safety points of view because unstable slopes can result in the loss of human life and damage to properties. This paper presents an overview of several case studies of slope failure in mining operations and explains various modes of failure in rock slopes, as well as factors that influence the stability of slope walls. With the aim of enforcing the importance of monitoring and evaluating slope stability in mining, both linear equilibrium and numerical modeling techniques were reviewed to elaborate their importance in designing stable slopes. In addition, the process of slope failure was discussed, and key signs of failure were indicated. In an effort to prevent mines from experiencing the hazards of slope failure, this study reports previous work performed in determining slope failure and the current state-of-the-art models, which entail the integration of analytical methods with artificial intelligence techniques. This innovation would help overcome the drawbacks of conventional prediction techniques that are cumbersome and ambiguous.

Keywords: slope failure; in situ; shear stress; rockmass; mining; factor of safety

1. Introduction

The stability of slopes in open-pit mining operations and quarries is extremely important from both economic and safety points of view. The stability of rock slopes entails the design of safe, economical, and functional excavated slopes to attain equilibrium conditions of natural slopes [1]. It is generally accepted that during the design of a stable slope, a proper understanding of the geological processes, such as stratigraphy, weathering, geomorphology, petrography, and earthquakes, is necessary. The most significant structures that influence the stability of slopes are joints, bedding planes, and the intersection of joints, faults, and shear zones [2].

Instability in rock slopes can be harmful and could result in the loss of human life and damage to basic properties. Failure in slope occurs as a result of the downward movement of materials due to the effects of gravity. However, it is assumed that the sliding of a rock slope will take place where there is an intersection of joint sets. Failure of rock mass is inevitable when the shear stress is greater than the shear strength of the rock [3]. However, failure of slope walls depends on some activities such as cracking of rock mass, weathering, increase in pore pressure, presence of decomposed clay rock filling materials, leaching, increase in water permeability, strain softening, and change in groundwater dynamics, which causes an increment in shear stress [4]. Therefore, the nature and behavior of the rock mass must be well-understood to ensure that the design of a pit wall remains stable for the life of the mine while extracting as much ore as safely and economically as possible [5].

The most common mode of failure in rock slopes is a plane failure. This type of failure happens when the angle of a structural discontinuity plane such as a bedding plane is smaller than the slope angle and greater than the angle of friction of the discontinuity surface [6]. Moreover, the water forces acting along the potential failure plane can also destabilize the slope. According to Wang and Niu [7], the dynamic loading and surcharge forces are other factors that contribute to the driving force that causes failure in rock slopes. The plane failure is influenced by factors such as the geometry, groundwater conditions, dynamic loading, potential failure plane characteristics and surcharge conditions [8]. However, the presence of major structural features such as faults, major joint planes, and unfavorably oriented bedding planes may also have a significant influence on the stability of the slope.

Over the last few decades, there have been significant advances in slope stability research works that investigated the causes of slope failure and factors that can trigger failure in slope. These factors are categorized by Sha [4] as internal and external factors. The internal factors that can affect the stability of a sloping wall include the mineral composition of the rock, rock types, and geotechnical and structural strengths. In addition, environmental factors such as earthquakes, rainfall, and weathering that can reduce the strength of the rock mass are also categorized as internal factors, while the external factors are mainly caused by human activity [9].

The increase in mineral demand in the 21st century has without a doubt compelled the expansion of mining operations globally, which has resulted in the extraction of minerals in larger capacities and deeper levels. However, it is important that these operations are conducted in a safe and economical manner. Therefore, this means that competent designs and techniques should be adopted to support the ever-growing mining capacities. Despite the analysis of slope stability using conventional techniques, mines are still experiencing slope failures, which have proven to be catastrophic and expensive. Therefore, this requires a better understanding of slope failure and the development of accurate prediction models to forecast this hazard before it occurs. Therefore, this study reports the application of slope stability evaluation techniques using linear equilibrium and numerical modeling, as well as the mechanism of slope failure. Additionally, the common factors known to affect slope stability are discussed, together with the various slope failure cases recorded globally. Last but not least, the application of artificial intelligence in predicting mine slope failure is reviewed, together with recommendations to improve prediction modelling.

Instability and Rock Mass Failure in Slopes

Unstable strata in rock slopes can lead to rock mass movement and cause harmful incidents that can affect mining operations and loss of ore reserves. Consequently, it can result in the premature closure of mines. Nicholas and Sims [9] reported the risk associated with slope failures in mining operations. These ranged from the loss of equipment to loss of reserves or mine closure and, in the worst cases, loss of life. Hustrulid et al. [10] argued that instability in rock mass slopes is largely caused by mining activities such as rock drilling, blasting, and the use of heavy machines. Similarly, Read and Stacey [11] indicated that the presence of groundwater, slope design, complex geology, discontinuities on rock mass, and mining operations are factors that affect the stability of rock slopes in mines. According to Eberhardt [12], the most common factors that influence the stability of rock slopes are redistribution of in situ stresses, complexity in geology, anisotropy and inhomogeneity of the rock materials, pressure pores and seismic loading. Similarly, Stacey and Swart [13] reported that the effects of blasting and groundwater are the only two major significant factors that control the stability of slopes. In addition, they indicated that blasting can cause ground vibration that in turn may have a significant influence on the stability of the highwall.

Practical experience in the design of rock slope projects has demonstrated the importance of having a robust design. The basic parameters to be considered in the design are height, overall slope angle, and area of failure surface, as shown in Figure 1 [14]. There is a correlation between the height and stability of the slope wall; the slope stability decreases with the increment in the height of the wall. Similarly, the slope needs to be steeped to minimize the amount of waste rock mined, thereby reducing mining costs. However, the economic effect of steeping a slope is that some portion of the waste material will be mined but the slope will be stabled [5].

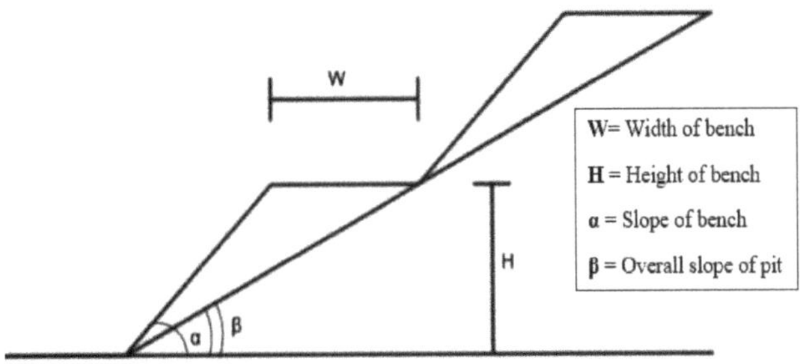

Figure 1. Design parameters in rock slope (Adapted from [14]).

The design of rock slopes is a major challenge in open-pit mines as it requires knowledge of the geological structure, lithology, and geotechnical properties of the rock mass [11]. It is necessary to determine the optimum design parameters to be able to evaluate all geotechnical materials before creating a slope. Read and Stacey [11] diagrammatically illustrated the design process of a rock slope in Figure 2, which is grouped into five stages, namely: models, domains, design, analysis, and implementation stages. Prior to the implementation of the design, the site investigation and data play important roles in evaluating the economic viability and stability of a rock slope. Furthermore, if predictions have been made that slope failure will occur, it is very important that the structural characteristics of the rock mass be considered. However, at the early stage of the design, data from the site can be obtained during geological exploration, which provides information on the strength of the rock mass, deformability properties, and geological structure and highlights the presence of major planes of weakness. These parameters could also be used to predict slope stability. Typical samples, as reported by Simataa [5] from a site investigation as shown in Figure 3, are selected for laboratory testing to determine the mechanical properties of a rock mass.

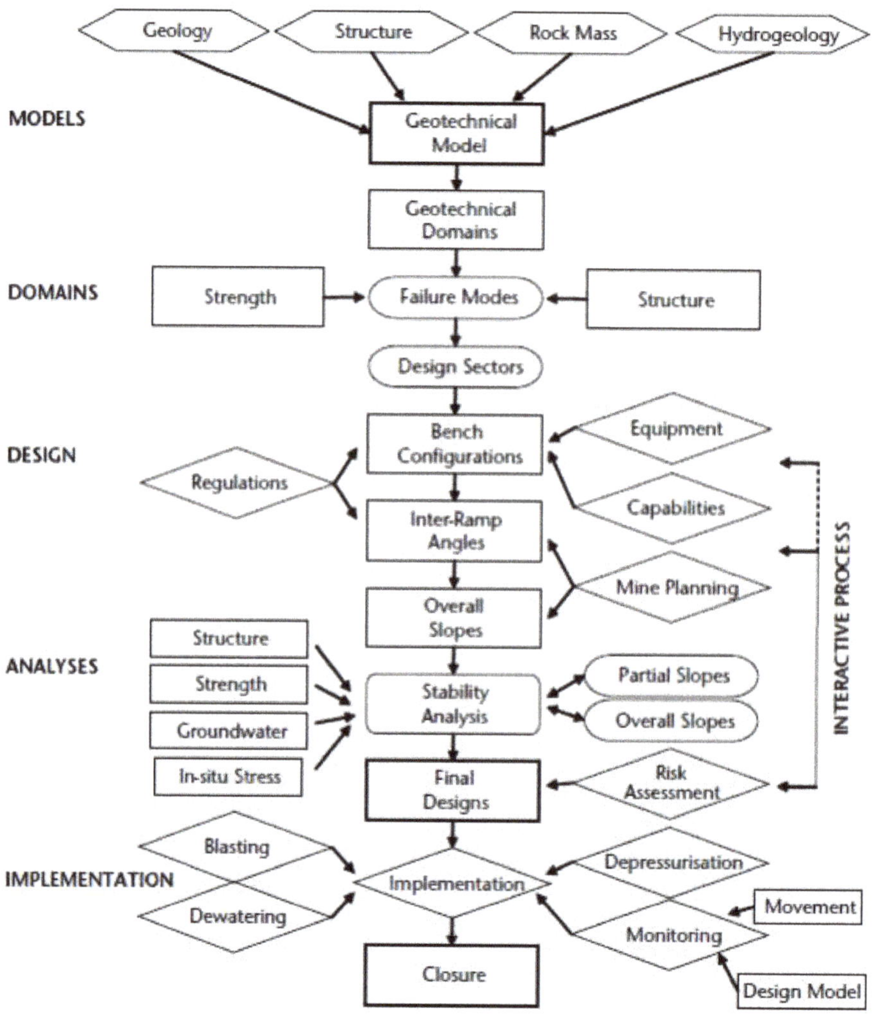

Figure 2. Rock slope design procedure (Adapted from [11]).

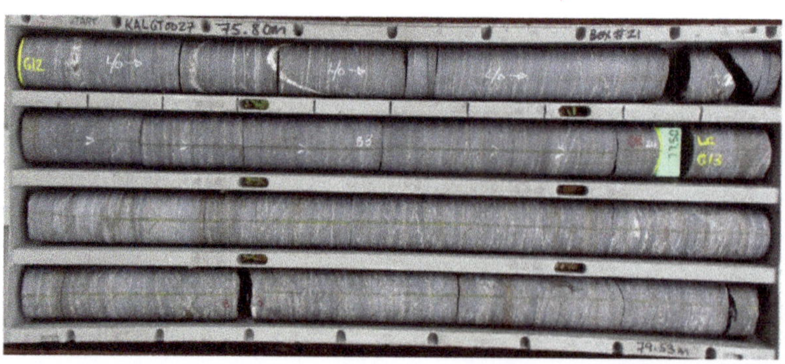

Figure 3. Core logs from borehole drilling during site investigation (Adapted from [5]).

2. Evaluation of Slope Stability in Mining

Failure of rock slope occurs when excess loading shear stress in a rock mass is redistributed and the load exceeds the strength of the rock. The shear strength of a rock mass plays an important role in the stability of the rock mass. Therefore, factors that tend to change the shear strength must be taken into consideration during the design as these factors may have an overriding influence on the stability of the slope. According to Abramson [15], failure starts in the rock slope from a single point and propagates to the entire rock mass. According to the study by Eberhardt et al. [16], during slope failure simulations, their model suggested that in the absence of triggering events, the strength reduction and progressive failure in a rock mass are attributed to causes of failure in the rock mass.

Over the years, a number of researchers have conducted research on the evaluation of slope stability. However, numerous methods were proposed, but only two major techniques, namely the limit equilibrium method (LEM) and numerical analysis (deformation analysis), are commonly used to evaluate the performance of rock slopes' stability [17,18].

2.1. Limit Equilibrium Method (LEM)

Hoek et al. [19] stated that the LEM has been available for more than 25 years and can be considered as a reliable slope design tool. The LEM approach is based on evaluating the applied forces and the strength of the ground. This method has been the most popular technique used in estimating the stability of a slope in geotechnical engineering [20]. In addition, Duncan et al. [21] described the LEM as the procedure used to calculate the shear strength of the ground against some factors causing the shear stresses. In general terms, the factor of safety (FOS) of a slope is described as the ratio of strength to the stress load. In a potential failure surface, the resisting force (strength of the rock mass) could be compared against the driving force. The balance between the shear stress acting along the potential failure surface and the strength conditions of the rock mass also describes the FOS.

According to Fleurissen and Cojean [22], if the calculated FOS is greater than 1, this shows that the strength of the rock mass exceeds the stress; thus, the slope is stable. In the case where the FOS is united, then there is an equal chance of failure or stability of the slope. However, when the FOS is less than or equal to 1, this shows that the stress exceeds the strength; hence, the slope is unstable. This shows that the calculated value of FOS determines an average value for the slip surface. The stability conditions for wedge stability analysis are presented in Table 1.

Table 1. Stability condition analysis for wedge failure (Adapted from [23]).

Calculated FOS	Stability Condition	Recommended Action
FOS > 2.0	Stable	None
1.0 < FOS < 2.0	Marginal	Analyse stability rigorously
FOS < 1.0	Unstable	Revise design or stabilise

According to the study reported by Simataa [5], the strength of LEM lies in its ability to evaluate the sensitivity of stability using various input parameters, the simplicity of the component and, most importantly, the experience of the geotechnical engineers acquired over the years in estimating FOS. Similarly, the engineer must have an in-depth knowledge of the limitations of the method. For instance, LEM can become inadequate when there is internal deformation in rocks, such as the presence of microcracks, brittle fracture, extension strains and liquefaction of the weaker soil layer [24]. In a complex scenario, computerized codes and software can be utilized to further analyze the stability conditions of the slope.

The LEM can be combined with other geotechnical analysis methods, such as the Mohr-Coulomb, failure criterion to assess the initiation of failure in slopes to the worst credible scenario. In this approach, it is assumed that the shear strength of the rock along the potential failure zones is controlled by linear (Mohr-Coulomb) or non-linear relationships in between the shear strength and the normal stress on the failure surface, which is defined by FOS [24]. However, the LEM is so fast that it can make thousands of FOS calculations, while the numerical modelling method takes a longer period to estimate just one FOS [8,25].

Despite the advantages of LEM in analyzing the stability of slopes, there are still some limitations reported by Ceryan et al. [26], as stated below.

(a) In estimating the stability of slopes, if the movement of rock mass is detected, the LEM approach cannot estimate the impact of such movement on the overall stability [27];
(b) LEM is restricted to the evaluation of slope stability with simple problems, such as providing little insight into the slope failure mechanism [12];
(c) LEM can only identify the onset of slope failure. Complex rock slope stability problems associated with in situ stresses, such as the geometry of the slope, pre-pressure and seismic loading, require a continuum-mechanics-based numerical modeling approach [28].

2.2. Numerical and Mathematical Modeling Method of Slope Stability Analysis

The numerical modeling method of analyzing instability in rock slopes has offered solutions for complex scenarios. The design of open-pit mines often involves complexities as a result of inherent geological conditions that can be too cumbersome for the conventional LEM to solve. The numerical modeling approach can be applied in such cases for the simulation of potential rock slope failure mechanisms and to carry out a comprehensive rock slope investigation [29]. However, many researchers have applied computational tools, such as PLAXIS 2D, FLAC 3D by Itasca Consulting Group Inc and Phase2 from Rocscience Inc., for such cases.

Over the years, the advancements in technology and high computing software tools have brought about the introduction of computer codes and computational tools that can provide more comprehensive and reliable slope stability analyses for geotechnical engineering. The two most commonly used numerical methods are the finite element method (FEM) and finite difference method (FDM) [30]. The shear strength reduction is the most widely used approach for performing FEM in slope analysis [31]. Hammah et al. [32] conducted a study on the principle of systematical reduction in the shear strength of materials by FOS and computed the FEM models of the slope until deformations were unacceptably high. The FOS in the numerical simulation methods was calculated using the shear strength reduction techniques that relate the existing strength to the limit equilibrium strength [26].

In the numerical modeling approach, slopes are divided into elements. The split elements are modeled with the stress–strain relationship and deformation properties of the slope to predict the behavior of slopes. The boundary conditions are defined, and the numerical modeling software is able to determine FOS and predict the displacement of rock mass that will take place during failure [26]. Similarly, a study by Chiwaye [8] reported that numerical software is capable of calculating FOS using the shear strength reduction techniques. The calculated FOS determined by the numerical modeling approach is usually equal to or slightly less than that of the conventional LEM.

According to Eberhardt [12], the numerical method of analyzing slope stability can be categorized into two approaches, viz., continuous and discontinuous modeling. The continuous computer codes assume that the material is continuous throughout the body. This technique is used to analyze slopes that are massive, have weak strata, are heavily fractured and have soil-like rock masses [32]. However, with this technique, the discontinuous surface does not form with the continuous modeling and it is impossible to have after failure analysis. In addition, discontinuities inside the rock mass cannot be modeled clearly [33]. On the other hand, the discontinuous modeling approach is applicable if the stability of the slope is governed by a joint bounded blocks or intact deformation [34]. During modeling, the model treats the rock slope as an arrangement of deformable blocks [12]. However, the discontinuous modeling requires the geometry of the discontinuities, shear strength of the rock, groundwater characteristic, in situ stress condition, and the intact constitutive criteria as the input parameters [8,29]. Nonetheless, of the capabilities of the discontinuous modeling approach, some shortcomings are associated with this method. Chiwaye [8] acknowledged that modeling of rock mass with discontinuous methods requires a representative of the discontinuity geometry. A study by Stead et al. [29] reported that both continuous and discontinuous computer codes can be used to analyze various failure modes in rocks but are most suitable for rocks with complex translation or rotational instabilities where failure brittle fracturing, internal yielding and shearing are present.

The main advantage of the numerical approach in analyzing slope stability is to predict the deformation analysis (stress–strain distribution), which may be used in the interpretation of slope behavior [10]. In addition, Hammah et al. [32] stated that numerical modeling has two major advantages; firstly, the ability of models to compute deformation and displacement of a rock mass and the proficiency of the method to process more sophisticated and complex problems than LEM. Secondly, numerical models can analyze complex geometries, simulation of stages in excavation and the influence of stress field conditions and groundwater seepage on the stability of the slope.

Recent studies have shown that the application of geomechanical parameter modeling of rock slopes is another approach to conducting stability analysis of geotechnical structure in slopes. Ground properties such as the cohesion and friction angle are the most used parameters that are required for the modeling [34]. The modeling of geomechanical parameters has been widely used by a great number of researchers to predict the stability factor and effects of the geomechanical structure of rock slopes. Lei et al. [35] applied the discrete fracture network (DFN) to model a couple of geomechanical properties of natural discontinuity rocks. Additionally, four copulas, that is, the Gaussian, Plackett, Frank, and Number 16 copulas, were used by Tang et al. [34] to model the dependence structure between cohesion and friction angle. The outcomes of the model were employed to construct the joint probability density function of cohesion and friction angle.

Similarly, Ahmad et al. [36] used Tree Augmented Naïve Bayes (TAN) to develop a Bayesian belief network to create seven nodes (unit weight, slope stability, slope angle, pore pressure ratio, slope height, cohesion, and internal friction angle) that represent parameters such as the "slope geometry", "geomaterial shear strength" and "water condition" to predict the slope stability. In another study by Haghshenas et al. [37], a meta-heuristic algorithm, that is, Harmony Search (HS) algorithm and K-means algorithm, was used to determine a clustering analysis that will highlight whether a slope will be stable or fail. Additionally, to develop a slope engineering geological model framework, an algorithm model was introduced by Huang et al. [38] to evaluate decision-making in slope management. This model was developed from soft set theory and fuzzy soft set theory. A probabilistic approach was applied in complex anisotropic rock masses to model a factor of safety in Figueredo et al. [39]. The study made use of a series of stochastic simulations that provided scenarios for the failure moment, and a factor of safety close to 1 was obtained.

The study by Tugelbayeva [40] developed prediction models for describing dynamic behaviors of non-uniform media and structures to solve non-stationary issues in rigid bodies. The study developed mathematical models to determine mechanical processes utilizing explicit finite-difference techniques for explaining finite-difference techniques for solving partial differential equations and a numerical solution method using integral transformation. The study analyzed wave propagation with a cavity lying on an elastic foundation under dynamics loads from the surface overlying the cavity. The study utilized discontinuity decay for analyzing stress-strain variables. Analysis of the results proved that the models can be used to design and assess geomechanical features in mineral deposits [40]. Another study by Nemirovisky and Tyrymov [41] utilized mathematical modeling of a deformable body to calculate stress–strain conditions of layered rock mass with an overlying mountainous.

3. Process and Mechanisms of Slope Failure

The mechanisms of slope failure from a stationary hillside to an active slide imply that there is either an external force that triggers the failure process, or the shear stress is greater than the strength of the ground. In most cases, the downward movement of rock mass occurs very suddenly and most of the time, it is not just one cause that results in failure [42]. A combination of factors may influence the failure mechanisms in slopes. The redistribution of excess loading of the shear stress triggers the downward movement on sections of rock slope, especially where the shear stress exceeds the strength of the rock mass [5]. The study by Eberhardt et al. [16] suggested that in the absence of any triggering event, degradation in the strength of the rock mass may result in progressive failure, but subject to a time-dependent mechanism. Wyllie and Mah [43] categorize the mode of failure in rock slope into four classes as shown in Figure 4. These include plane failure, wedge failure, rotational and toppling failure.

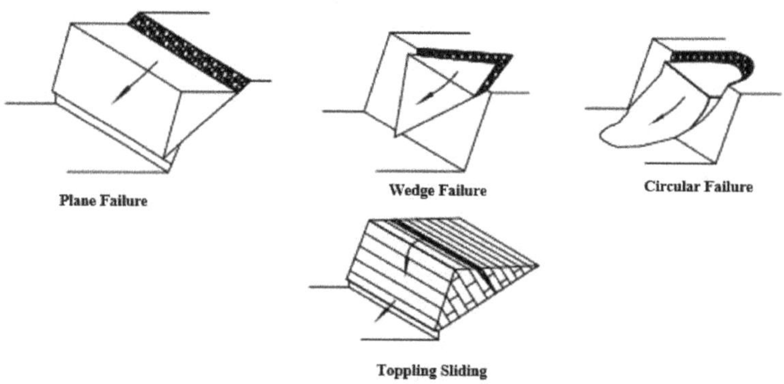

Figure 4. Classes of failure in rock slope (Adapted from [44]).

3.1. Plane Failure

Plane failure occurs when the discontinuity striking approximately parallel to the slope face and dipping at a lower angle intersects the slope face to enable the slope to slide [24]. This failure involves the rock slope downward and outward movement along a gently undulating surface or sliding [45]. When there is a potential plane failure, it will be evident as the bedding plane strikes parallel to the slope and dips out of the slope. In such conditions, there will be a lack of confinement, which will trigger planar failure in the slope. Kliche [45] and Simataa [5] in their studies agreed that planar failure is likely to occur when pre-existing joints are striking parallel to the slope and also when the dip is less than the slope angle as shown in Figure 5. In practice, the lack of lateral confinement in rock slopes will lead to planar failure. This type of failure is common in slopes that have convex designs where the direction is parallel to the strike of the weak planes. Planar failure can be limited to benches or areas of the pit with adverse geometry or the structures that strike perpendicular to the slope face, and thus allow a planar slide to occur. One of the major contributing factors in planar failure is the presence of groundwater, which causes the stability of the slope to deteriorate when there is temporary groundwater pressure, especially during heavy rainfall.

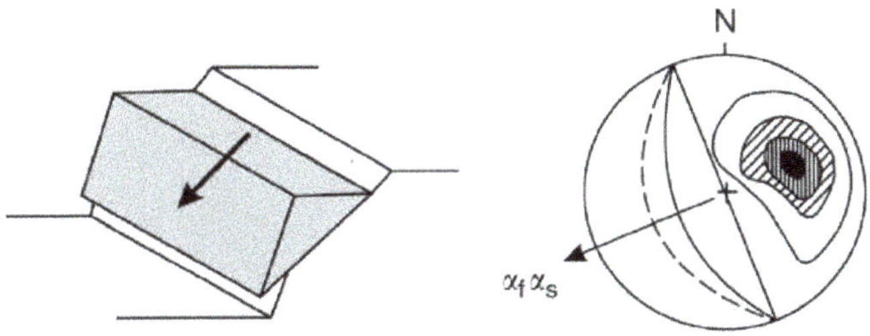

Figure 5. Pre-existing joints dipping out of slope face and striking parallel to the face (Adapted from [44]).

Wyllie and Mah [43] improved on the conditions reported by Hoek and Bray [46] and suggested the conditions that must be satisfied for planar failure to occur as follows:

(a) The strike of the plane of weakness must be within +/−20° of the crest of the slope;
(b) The toe of the failure plane must daylight between the toe and the crest of the slope;
(c) The dip of the failure plane must be less than the dip of the slope face and greater than the angle of internal friction of the failure plane;
(d) The upper end of the sliding surface either intersects the upper slope or terminate in tension cracks;
(e) Release surfaces that provide negligible resistance to sliding must be present in the rock mass to define the lateral boundaries of the slide.

3.2. Wedge Failure

Wedge failure occurs as a result of the intersection of two or more discontinuities that lead to the formation of a tetrahedral failure block, which could slide out when the angle of inclination of the line is greater than the internal angle of friction along the discontinuities [5]. When two discontinuities hit obliquely across the slope face, the rock resting on these discontinuities will slide down the line of intersection, as shown in Figure 6.

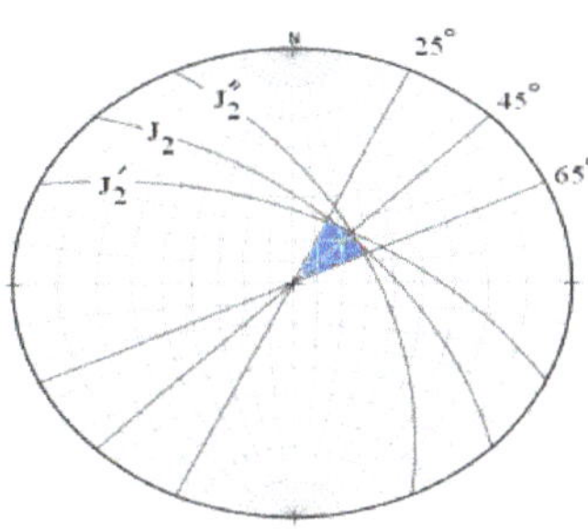

Figure 6. Plot of intersection of three joint sets on Stereonet dipping out of slope face (Adapted from [47]).

Wedge failure has a high tendency of being the most commonly experienced failure mechanism in rock slope [48]. The stability analysis of a wedge failure is one of the main components in the design of a competent bench face angle–height configuration. The stability evaluation will include large wedges, which could influence the stability of the overall slope, and joint planes, which could link up and affect the stability of the bench or even the ramp.

Hudson and Harrison [49] reported that wedge failure is inevitable in rock slope under the following conditions:

(a) When the line of intersection of two discontinuity planes associated with the potentially unstable wedge is daylighting on the slope plane;
(b) When the dip of the slope exceeds the dip of the line of intersection of the two discontinuity planes associated with the potentially unstable;
(c) When the line of intersection of the two discontinuity planes associated with the potentially unstable wedge must be such that the strengths of the two planes are reached.

Low [50] argued that there are four classes of failure modes for a wedge, namely:

(a) Sliding along the line of intersection of both planes forming the block;
(b) Sliding along plane A only;
(c) Sliding along plane B only;
(d) A floating type of failure.

In evaluating the stability of wedge failure in rock slope, one of the most rapid and convenient methods is the use of stereonet. The stereonet has widely been used to examine the kinematic feasibility of wedge failure, i.e., the type of sliding that is likely to occur when there is intersection of two or more major planes. However, the method cannot be used to determine the actual FOS but can be estimated from the geometry of the wedge, the strength of each plane and the water pressure [43]. Moreover, the application of a friction stability chart is another rapid method to check the stability of two-plane wedges. In this approach, the chart only considers the frictional strength on the plane of weakness while ignoring the cohesion and water pressure. When using the friction method, the calculated FOS value must be greater than 2; a slope with FOS less than 2 is regarded as potentially unstable [43]. In using the stereonet approach to determine the stability of wedges, great circles for the planes of weakness are plotted on the stereonet by considering the shear strength on the weakness planes. Similarly, the slope designer must also consider the frictional strength using a friction circle.

3.3. Circular Failure

Circular failure is most common in soil and deeply weathered or closely fractured rock. Coates [51] stated that circular failure is experienced in continuum slopes with highly jointed or weak rock mass, as shown in Figure 7. However, circular failure can also be experienced in hard rock [52].

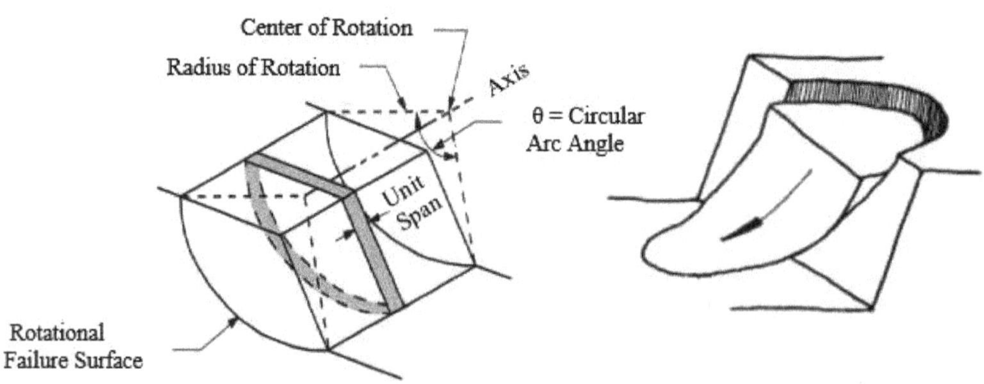

Figure 7. Rotational failure mode and circular failure mode in slopes (Adapted from [5]).

In weak strata such as soil or deeply weathered rock, the circular failure is defined by a single discontinuity surface but will tend to follow a circular path [53]. However, if the failure surface is curvy, it helps to prevent the extension of tension cracks at the upper ground surface [5]. The result of Sjöberg [54] investigations indicated that it might be difficult to conduct numerical simulations of rock slopes in large scales. Sjöberg [55] also managed to carry out a model study of circular failure which showed that circular failure occurs in six (6) stages, as shown in Figure 8. The six stages are explained below:

(a) Elastic displacement is caused by the removal of rock material during mining activities;
(b) Yielding commences at the toe and spreads upwards as more material is removed or as a result of mining to a new and critical slope height;
(c) Accumulation of shear strain at the toe of the slope will progress upward;
(d) When failure surface is developed, the slope will start showing some displacements, which can be tracked if there is a good monitoring system in place;
(e) Slope fails with time with larger displacement starting from the toe;
(f) When failure occurs, the failing mass can slide away from the slope.

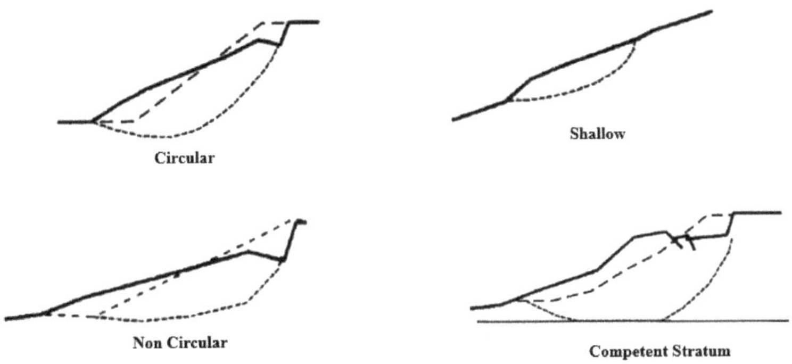

Figure 8. Mechanics of rotational and compound failure in rock slope (Adapted from [56]).

Kliche [45] reported that most circular failures in homogenous materials such as fills, highly jointed rock slopes, and constructed embankments are aggravated by water intrusion. Gundewar [24] classified circular failure into three (3) categories based on the area that is affected by the failure, namely: slope failure, toe failure, and base failure. Slope failure happens when the arc of the rupture surface meets with the shape above the toe of the slope. Toe failure occurs when the arc of the rupture meets with slope at the toe, while base failure occurs when the arc of the failure passes below the toe and into the base of the slope. In most cases, base failure materializes when the strata below the base are softer than the soil above the base.

3.4. Toppling Failure

Toppling failure happens in slopes with joint sets that are vertical or nearly vertical. Kliche [45] reported that this type of failure occurs when there is a mass movement where the weight of the vector of a rock block resting on an inclined plane falls outside the base of the block. The failure is triggered once joint sets are disturbed; the slope will collapse from a small to large size, as shown in Figure 9 [57]. The movement of the failed slope in toppling is characterized by the downslope overturning, through rotation and flexure of blocks with steep discontinuities [5]. Moreover, boulders from the upper bench face may bounce off and over benches to pose a hazard at lower levels. In the area where toppling is potentially identified, the bench should be sufficiently wide to ensure that boulders cannot be bounced over the crest.

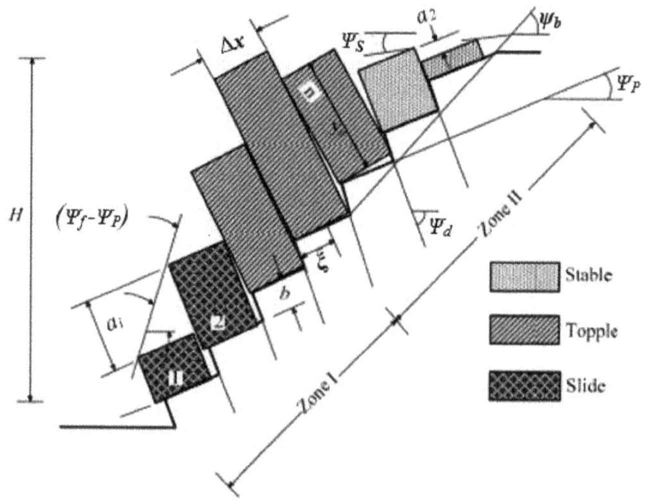

Figure 9. Geometry model of toppling slope failure (Adapted from [57]).

In major toppling instability, there will be a long warning period with noticeable events such as progressive development of tension cracks adjacent to the crest of the slope. However, there will be sufficient time to implement stabilizing measures or to evacuate the area before failures occur. The potential for small toppling failure could be sudden from bench faces, which represents a hazard during mining activities such as drilling, charging and mucking operations. According to the study reported by Sjöberg [55], the following conditions have to be satisfied in order for toppling to occur:

(1) The joint sets must dip relatively steeply into the slope and must be able to slip relative to each other;
(2) The rock mass must be able to deform substantially for toppling to have room to develop;
(3) The tensile strength of the rock mass must be low to allow a tensile bending failure at the base of the toppling columns.

4. Factors Affecting Rock Slope Stability

Many factors have a significant influence on the stability of rock mass slopes. Some of these factors are taken into account when analyzing the stability of slope so that their effects can be included on a quantitative judgment basis, while some factors are not considered during stability analysis. The most important factors that affect the stability of slopes in both hard and soft rocks are slope geometry, geological structures, groundwater, lithology, cohesion, and angle of internal friction, effects of blasting and mining method, and equipment used [24]. Similarly, Chiwaye [8] pointed out some factors as sources of uncertainty that lead to the failure of rock slopes, as presented in Table 2.

4.1. Slope Geometry

The geometry of slopes in open-pit mines and quarries plays a significant role in controlling the stability of the slope. Slope geometry is described as the basic slope design parameters that comprise the height, overall slope angle and area of failure surface [53]. The study reported by Chaulya and Prasad [58] indicated that the basic geometrical slope design comprises parameters such as the overall slope angle, bench height and surface area, as shown in Figure 10. The stability of the slopes decreases with an increase in the height of the pit wall and slope angle.

Table 2. Sources of uncertainty in rock slope failure.

Slope Aspect	Sources of Uncertainty
Geometry	Topography Geology/Structures Groundwater surface
Rock mass Properties	Strength Deformation Hydraulic conductivity
Loading	In situ stresses Blasting Earthquakes
Failure Prediction	Model reliability

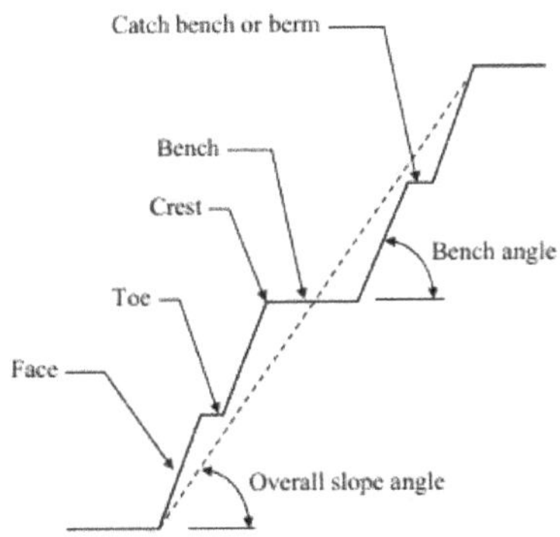

Figure 10. Schematic diagram showing open-pit bench slope parameters (Adapted from [58]).

The overall slope angle plays an important role in the stability of the slope. An increase in the slope angle can increase the chance of failure occurring in the rock slope. Ideally, the overall rock slope at 45° should be considered safe and stable in open-pit mines [14]. Similarly, the curvature of the slope also influences the stability of the wall. The design of convex pit wall structures should be discouraged as it has effects on the stability of a slope. Planar failure is more likely to take place when the slope geometry is convex in a plan, i.e., the plan shape of slopes affects the potential for failure and the effects of failure. Several studies [46,59–62] have shown that geometry has a significant effect on the stability of slopes. The result of the study reported by Zhang et al. [63] showed the effects of curving surface (different geometry) slopes on the FOS and the stability of a 3D slope. In general, a curving slope surface on a steep slope shows more significant influence than that of a gentle slope.

In addition, the relationship between the curvature and slope height of the wall influences the stability of the slope. In most cases, if the radius of curvature in a concave slope is less than the height of the slope, the slope angle can be 5° to 10° steeper than conventional slope stability analysis. On the other hand, for convex slopes where the radius of curvature is lesser than the slope height, the slope angle should be 5° to10° flatter than the conventional ones. The majority of natural and excavated slopes are not infinite in the plane; most of them are either concave or convex in configuration [46] most especially in open-pit mines where the curvature of the pit walls can affect the stability angle of the slope [64]. According to Wines [60], the stability of concave slopes is more than that of a straight slope. Due to the lack of confinement and effects of the side resistance in convex slopes, the potential failures are structurally controlled, and the stability of convex slopes is often less than that of straight slopes.

The relationship between the radius of curvature and the height of the slope is another factor that is used in the stability analysis of slopes. When the radius of curvature of a concave slope is less than the height of the slope, the slope angle can be 10° steeper than the angle suggested from the conventional stability analysis [46]. Similarly, the ratio of the curvature radius to the height due to an increase in the lateral pressure causes an increase in the stability of a slope [64]. However, practical experiences have shown that concave geometry is the most widely used type of curvature design of slope geometry, most especially in open-pit mines, narrow long pits, circular pits and in the wall at the end of the pits [60].

4.2. Geological Structure

This is an important factor to be considered in evaluating various factors that influence the stability of slopes. It is essential, not only from a safety but also from an economic point of view. A wrong approach to geological conditions during design and operations can have severe consequences in mining operations. Geological structures include fault, joints, presence of bedding plane and intra-formational shear zones that influence the stability of a rock slope. In most cases, geological structures, also known as discontinuities, control the type of failure that will take place in the rock slopes. The properties of the discontinuities include roughness, orientation, and persistence and mineral infillings such as clay material present in between rocks play important role in the stability of the rock slope. For instance, the joint orientation (dip and the dip direction) can be used to predict the mechanism of failure that will occur. The intersection of discontinuities such as joints and fractures in rocks results in instability and accidental falling of rock blocks in rock mass as a result of the actual strength of the in situ rock mass being less than the strength of the intact rock [8,61]. These discontinuities are usually several sets occurring at different directions, which causes separation in rock mass into discrete and interlocking pieces. However, intersection of discontinuities causes a reduction in the shear and tensile strength of the rock, resulting in sliding along the structural plane [65].

The geological structure and strength of rocks have a significant influence on the mechanical properties of the rock. According to Liu et al. [66], the mechanical properties of the ground depend on different factors such as stone content, rock shape, rock distribution and bonding strength. In most cases, the mechanical properties serve as input parameters for evaluating the stability of slopes and to understand the characteristics of a rock mass [67].

The initiation of tensile and shear failure in rock mass often leads to instability of rock slopes, which usually occur as a result of response from several factors, such as temperature and insolation [68], seismic loading [69], weathering [70] and precipitation [71]. In underground excavations, the rock in the roof behaves oppositely, where there are overhanging and threatening blocks or slabs remaining in their position because of the strength initiation of discontinuities, which are mainly arising from the rock bridges [72]. However, results from investigations revealed that rock bridges significantly increase with the shear strength of individual incipient rock discontinuities, especially when they are under constant normal stiffness boundary conditions [73]. The intersection of joints in rock masses results in the formation of discrete blocks with variable geometries, especially when the discontinuities are not fully continuous [74].

The main geological structures known to influence slope stability include direction of dip, intra-formational shear zones, joints, discontinuities and faults. Slope failure my occur due to failure along the structural discontinuities of intact zones or along surfaces formed close to discontinuities [75,76]. In addition to that, dip direction affects slope stability especially when the dip direction of the strata of discontinuity is similar to slope dip direction with an angle of the strike of less than 20°, which consequently triggers failure from the toe zone. Furthermore, slope failure is promoted when seepage occurs from beneath the ground into the strata, which lubricates the rock mass and consequently triggers material failure in the upper zones, leaving a planar surface [77]. Additionally, in zones where the rocks have weathered, the infiltration of water into the sediments increases pore pressure, alters the degree of saturation, and compromises the shear strength of rock mass leading to failure [78]. The density of slope material has considerable implications on its stability, as a very low density promotes water seepage into the rock mass, while a high-density limits infiltration [79].

4.3. Groundwater

Groundwater opens up the little cracks and activates the force on blocks and wedges that cause instability of slopes in rocks. The water pressure on potential failure planes reduces the normal stresses across planes, which in turn reduces the frictional strength on these planes. Consequently, it increases the thrust and driving forces, which in turn controls the stability of the slope.

Unusual ingression and distribution of groundwater can enlarge the spacing between joints, which can result in the failure of rocks when subjected to gravitational forces. According to Simataa [5], when the tension crack in the rock mass is filled with water, there will be a linear increase with depth and total force, as shown in Figure 11. From Figure 11, Z is the depth of the tension crack, W represents the weight of the wedge rock, U is the uplift force due to water pressure on failure surface, V is the water pressure in the tension crack, and (φ) represents the angle of internal friction of sliding. It is assumed that the water moves in between the tension crack and the base of the block, which causes distribution along the base of the block. The distribution of water becomes an uplift force U, which reduces the normal force acting across the surface.

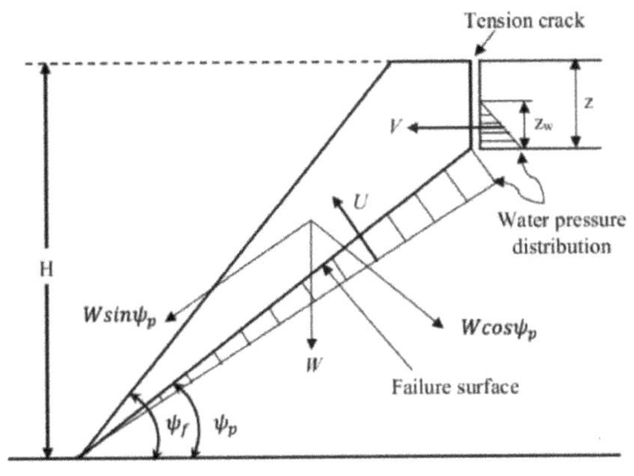

Figure 11. Effects of water pressure in a crack (Adapted from [80]).

According to Liu and Li [81], the ingression of water is one of the factors that triggers landslides. Based on statistics, over 90% of landslide and slope failures are caused by water-related issues. The outcome of their investigations revealed that water seepage, rainfall, and water level fluctuations are the major mechanisms of landslide failure.

Prakash [53] stated that the presence of groundwater alters the property of the rock and reduces the normal effective stress and alters rock strength parameters such as cohesion and friction. Moreover, the effects of the groundwater pressure and transient flow of water within the rock and soil affect the pore pressure conditions, strength, and deformation behavior of the rock [17]. Additionally, the presence of water pressure within the rock enlarges the rock pores, which decreases the compressive strength, particularly where confining stress has been reduced.

There are various computer codes used to analyze and assist in assessing groundwater pressure. Numerical modeling techniques such as FLAC/FLAC 3D, UDEC/3DEC, and PHASE2 are capable of performing coupled hydro-mechanical analyses [8]. This numerical modeling software can predict pore pressure drops due to volumetric expansion associated with the excavation of a pit. However, modeling of groundwater pressure and expansions with numerical modeling software is cumbersome.

Hydrogeological conditions of the rock mass influence slope stability, which normally manifests through ground seepage from an underground water aquifer, surface water, or precipitation from stormwater or snow thaw. The influence of groundwater on slope failure can generally be experienced in three major avenues. To begin with, the presence of water in the rock mass creates hydrostatic pressure, which tends to lift present rock blocks, thereby posing lateral pressure on present discontinuity planes. Consequently, the imposed lateral pressure reduces plane resistance against sliding, as well as the shear strength through a reduction in normal stress on the discontinuity plane, which loosens rock blocks, resulting in slope failure. Furthermore, depending on the nature of the groundwater, such a condition may alter the physical and chemical compositions of discontinuity fillings, thereby reducing the shear friction and ultimately leading to slope failure [82].

4.4. *Lithology*

The lithology of rock mass is an essential factor that influences the failure process of rock slopes in open-pit and quarry operations. In most cases, the physical characteristics of the rock mass show the potential areas that are prone to failure. The mineral constituents forming the rock control the mechanical properties and the behavior of the rock mass and also determine the strength of the rock mass [14,83]. McNeilly et al. [84] stated that the porosity, mineralogy, density, and degree of cementation are the rock properties that can influence the strength of rock. The mineralogy dictates the rock type, strength of the rock, color and other properties of the rock. According to Yasir et al. [85], the strength properties of the rock increase with the degree of cementation. The mineral composition and the physical properties of the rock mass affect the strength of the rock, thus influencing the stability of the slope. In the design of rock slope, pit high wall that contains weathered rocks or alluvium rock materials will have low shear strengths and a high propensity to further weathering through erosion. Despite the roles cohesion and frictional angle play in the stability of rock slope by opposing the effects of the magnitude of the shear stress, there are some limitations of these strength parameters. In most cases, the estimated value of these parameters contains errors during calculations that influence the uncertainty in the result [86]. Likewise, Sullivan et al. [87] stated that the friction angle is influenced by the grain sorting, grain size and grain angularity, while the chemical bonding and cementation also affect the cohesion.

4.5. *Cohesion and Angle of Internal Friction*

Rock strength properties such as cohesion (c) and angle of internal friction (φ) are used in evaluating the FOS of rock slopes during the design stage. The shear strength is a key mechanical characteristic of rock–soil related to slope stability assessment. The cohesion and friction angle values are used to describe the shear strength of the rocks. These parameters are determined in the laboratory through a triaxial compression test. The angle of internal friction in a rock defines the ability of the rock to withstand shear stress [53]. However, the particle size of the rock affects the angle of internal friction; i.e., the larger the particle size, the larger the angle of internal friction. In addition, high water content, undercutting slopes, ground vibration, and alternating expansion by wetting and contraction by dryness of water reduce the strength of cohesion in a rock mass.

The cohesion and friction angle of rock–soil aggregates are influenced by the particle size distribution characteristics and other conditions, especially the water content [88] However, the lower the cohesion in a rock mass, the lower the stability of the slope, and the higher the internal frictional angle of rock, the higher the slope wall will be. Therefore, it is necessary to evaluate the cohesion and the angle of internal friction from the core specimen during site investigation to determine whether the slope will be stable or not. In analyzing the stability of the slope both in rock and soil, the cohesion and the angle of internal friction must be taken into account as these properties control the shear strength of the rock [89].

4.6. Blasting

Poor blasting in open-pit mining operations is detrimental to rock slope stability, not only due to blasting-induced vibrations but more largely because the rock behind the slope face can be fragmented and loosened [5]. The ground vibration that occurs as a result of poor blasting causes redistribution of stresses in the rock slope, which eventually leads to a dynamic acceleration of materials, which in turn causes instability in the slope plane [14]. Similarly, the effects of poor blasting largely result in the creation of discontinuities such as fractured zones, faults, joints, and fissures in rock slopes [90]. Poor blasting reduces the cohesion and creates the potential for ingress of water, which causes further loosening of the rock mass.

Another implication of poor blasting is that it reduces the bench-face angle and eventually leads to ground vibration that could potentially cause failure in a rock mass [53]. During blasting operations, natural cracks and fractures in a rock mass structure are extended by additional stresses induced by the blasting, and therefore the shear strength of the rock mass is significantly reduced, thereby causing instability of the rock mass.

Over the course of detonation, seismic waves are produced, which spread in the form of a stress wave within a rock mass. The effects of this distributed seismic wave on the geometry of a slope will produce vertical and horizontal constant acceleration that will result in an unstable slope. When the stress wave is higher than the tensile strength of the rock, there may be potential damage of the rock mass as a result of seismic waves, elastic vibration and elastic stress waves caused by poor blasting and the presence of fracture zones, thus leading to slope instability and other damages. However, a combination of the effects of poor blasting with the external loading can also create a surcharge on the crest of the benches.

4.7. Mining Method and Equipment Usage

In mining operations, the surrounding rock mass around excavation develops deformation due to changes in the in situ stress field conditions. During excavation, natural slopes may face deformation as a result of the reduction in shear strength, which can lead to slope failure [91]. The rock mass movement may continue if no solutions are implemented on the cut slope. The overall slope stability must be considered when selecting mining methods and equipment usage. Similarly, the state of in situ stress field conditions described by the magnitude and orientation of the principal stress must be taken into account during the excavation process [92]. In the open-pit mining method, the highwall slopes are designed to be steeped due to the increase in slope height that is prone to the buckling mode of failure [53]. Additionally, the movement of heavy equipment during mining for mine haulage and other operational equipment such as rigs give rise to an increase in surcharge, which in turn increases the forces that trigger the downward movement of the slope.

4.8. Stresses on Slope

The presence of field stress conditions in rock masses has a significant influence on the stability of a slope [93]. The in situ field-stress condition of a rock mass can be used to investigate the behavior of slopes regarding the height of the wall and provide a summary of several failures. Although the impact of in situ stress fields can be less significant in small slopes, it has to be considered for the design of a very large slope. The advantages of using the stress analysis approach to evaluate the stability of slope include modeling of the excavation sequence, modeling of deformation, modeling of water pressure and flow in the rock mass and modeling of the progressive development of yield and failure zones. However, the stress can be advantageous as it provides confinement to the rock mass [94]. The confinement assists in stabilizing the slope. Hoek et al. [19] stated that the lateral stresses during rock excavation have an important influence on the stability of the slope. At high confinement conditions, most of the joints reach their residual aperture, and as a result, the stiffness of the rock is no longer affected by the increment in the confining stress [95].

The presence of discontinuities in rocks influences the shear strength, which in turn determines the stability properties of the rock mass. The effects of discontinuities (geological structures) significantly influence the load condition of the rock mass. Notably, natural discontinuities such as joints and faults are evident in rocks that influence the mechanical behavior of the rock and also create stress in the anisotropy, heterogeneity and scale effects. The mechanical properties of a jointed rock mass are controlled by the geometry and properties of the discontinuities. Additionally, the physical properties of discontinuities such as friction, compressive strength, weathering and the presence of infilling materials (clay) affect the mechanical behavior of the rock mass. This is more reason why the joint network and other geological structures reduce the strength of the rock mass and increase the deformability of the rock. Therefore, it is important to investigate the properties of joints to be able to accurately determine the stability factor of slopes. Moreover, knowing the joints network will enable a safe and economic design of the slope and stability analysis [96]. The nature of the jointing, such as their orientations, length, the distance between joints and joint spacing, also influence the stability of slopes.

During site investigation, discontinuities are classified according to how they are formed. In most cases, discontinuities with the same category usually have the same properties in terms of their shear strength property, which can be used to predict the stability conditions of the excavation. Evaluation of joints network and the strength properties can serve as the guide for designing a stable slope and other rock engineering applications. It is assumed strength properties of jointed rock masses depend on the properties of the intact rock pieces and the freedom of the pieces of the slide and rotate under different stress conditions. Ryan [97] stated that the strength of jointed rock masses depends on the strength of the joints, the strength of the intact rock locks between the joints and the degree of interlocking between the rock blocks. For instance, in jointed rock masses, it is expected that the water pressure in the joints will continue to increase and dissipate faster than the pore pressure of the intact rocks, most especially in rocks where the porosity and permeability are low.

In slope stability analysis, the joints are too many to be taken into account individually; thus, an approximate property is usually considered. Other influences of discontinuities are represented by the shape, the block size and the strength of the rock mass. Likewise, the importance of taking into account joint conditions in the rock mass is that the joint information is used to highlight the failure mechanisms as well as the failure intact between joints [98]. According to Cundall et al. [99], the compressive strength of the rock is the best described property of the rock against shearing under confined conditions. Recently, the advancements in technologies have brought about accuracy in the method of predicting slope stability in jointed rock masses. The availability of various computer programs can simulate different stability scenarios in jointed rock masses. However, the application of computer programs such as the finite element method (FEM) and the discrete element method (DEM) have proven to be effective in analyzing slope stability. These approaches make use of the shear strength reduction (SRR) methods to determine the stability factor in a jointed rock mass. However, to use the computer programs for jointed rock mass, it is necessary to consider the joint patterns to obtain an adequate representation of real rock configuration. That is, when all the joint components are considered in the model, it is possible to predict the slope failure mechanism and joint movement that can influence the sliding of the rock mass.

5. Slope Failure in Mining Operations

Slope stability is a crucial consideration in the management of mining operations as slope failure compromises the economical and safety aspect of production. Due to the increase in demand for mineral resources and the invention of more sophisticated mining methods as well as machinery, most mines are designed to reap more resources from deeper or steeper mines. Such mines have a higher angle of inclination, which make them more susceptible to slope failure. Slope failure can lead to injury of personnel, damage of mine machinery, and disruption of operations, which all negatively impact the mining performance. In order to prevent such a hazard from occurring, it is imperative that extensive geotechnical studies are conducted to ensure that FOS is within a tolerable range (>1) far from failure. In cases where the FOS is very low, more attention should be given to monitoring such slopes to prevent slope failure.

Slope failures are common in open-pit mines and become unavoidable as excavation is becoming deeper and, as such, more difficult to manage. A review of existing failed slopes around the globe provides a better understanding of the failure modes and factors to be considered in controlling the mechanisms that trigger mass movement of materials in slope engineering [55]. Mitigation of slope failure risk requires an in-depth understanding of structural geology, rock mass properties, influence of groundwater pressure and other external forces in the area [100]. Failure in rock slopes occurs due to many reasons, which have been discussed earlier in this paper. During excavation, stripping excavation walls are required to be as steep as possible for operational efficiency. Failure may occur if the slope design is over steeped [101], although most failures in open-pit mines are triggered by fracture and shear on existing defects. Structures inside the rock mass interact in different ways as the rock deforms, and thus affect the general behavior of the rock. In addition, understanding the in situ stress field pattern is critical in understanding the deformation processes in open-pit mines. For instance, mining of ore using open-pit mining methods expose the surface and the rock adjacent to the pit wall; therefore, it becomes unconfined in the direction normal to the slope face over large areas [102].

When a rock mass comprises blocks separated by joints, shear zones and bedding planes, sliding on multiple discontinuity sets as well as tensile and shear failure are bound to occur [103]. For instance, the report of the slope failure that occurred at Yanqianshan iron mine in China confirmed that the eastern part of the slope collapsed as a result of deformed strata in the area [104]. The collapsed area comprises of eight build-up stages that are represented by actions from point (a) to point (g). These stages are classified as a cave rock zone, cracking zone, toppling zone, and sliding, as shown in Figure 12. Based on the report of the investigations, the process of initiation and development of failure zones were categorized into three stages: (i) overlaying area above the goaf, (ii) initiation of the collapsed strata sliding into the pit to form a small landslide and (iii) a large landslide occurring as a result of mining activities and creeping of rock that triggered the mass movement in the northeastern phyllite slope. The disturbance produced by underground mining activities initiated the potential sliding body on the northern phyllite slope and the retaining wall structure gradually tended towards instability.

Similarly, another slope failure was reported in Western Macedonia in Northern Greece as a result of sliding along the sub horizontal direction, unfavorably sloping as shown in Figure 13. From the analyses, the stability was governed by the interface between lignite and an underlying stiff presence of plastic clay or a marl layer that are very close to the bottom of the slope where there is a significant amount of stress. Based on the study reported by Zevgolis et al. [105], some factors such as groundwater conditions, pit geometry and shear strength of the critical interfaces between clays and lignite influenced the stability of the slope.

In addition, a massive landslide at the Grasberg gold and copper mine in Indonesia, which forced the pit to suspend operations, was reported by DTE [106]. In October 2003, a fatal accident happened at the southern wall of the open-pit mine, which collapsed and claimed the life of eight people and injured another five people. This incident moved 2.3 million tons of rock and mud down the slope, thereby engulfing mineworkers and heavy equipment. Similarly, a rock mass slippage happened in the year 2000 at the same mine when the overburden was washed by heavy rainfall into Lake Wanagon [107]. The main cause of material movement was attributed to ingress of groundwater from a nearby water table, coupled with an abnormally high rainfall season.

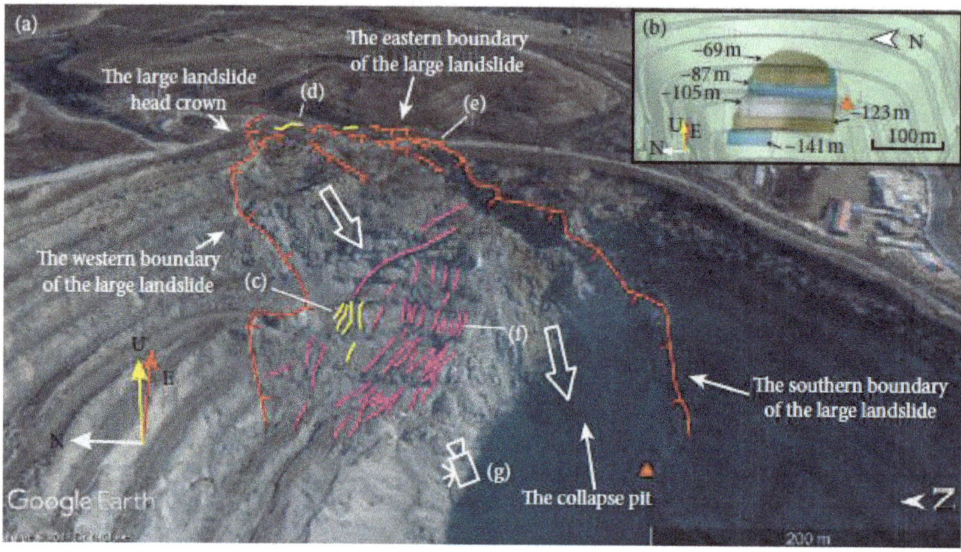

Figure 12. Image of collapsed phyllite materials in the northeastern slope (Adapted from [104]). (**a**) cracks initiation point (**b**) location of the mining area (**c**) Tension cracks from the western boundary of the landslide (**d**) Initial collapse area (**e**) Tension crack at the eastern boundary of the landslide (**f**) Collapsed pits (**g**) Google Earth monitoring technique.

Figure 13. Sliding portion of lava lignite mine in Greece (Adapted from [105]).

Recently, another landslide happened at the Gamsberg mine located in the northern Cape Province in South Africa. The open-pit mine is operated by Vedanta zinc international (VZI), which currently produces 40,000 tons of ore per month [108]. According to Petley [109], the landslide incident that occurred at the south pit of Gamsberg was described as a "geotechnical failure". According to the report, eight mine workers were rescued after the incident, while two people were still missing at the time when the report was published. The failure happened across the road that provides an access ramp. The incident appears to have occurred above the access ramp, but the debris reached the pit floor, as shown in Figure 14. After the failed slope, Vedanta Zinc International (VZI) suspended all mining-related activities at Gamsberg to carry out a proper investigation into the slope failure.

Figure 14. Image of Gamsberg mine landslide (Adapted from [109]).

Another notable incident of slope failure in mining was the movement of the Chiquicamata slope in Chile, which was triggered by an earthquake [110]. In this study, the authors stated that the mass movement progressed at a more or less steady rate until it fell in 1968. However, the kinematics of the slope movements were not clear but seemed to involve load transfer from the north block to the south block dominated by discontinuities. However, due to the presence of discontinuities, the cohesion that was supposed to prevent the movement of the slope was quite small, with little intact rock. The failure started from the north, i.e., the upper part of the slope, to the south block, which is the toe of the slope, as shown in Figure 15.

Possibly the largest landslide of all time was the event that took place at Kennecott Utah copper's Bingham Canyon Mine in the United States of America. This open-pit mine is regarded as the largest man-made excavation in the world, which measures 1 km deep by 4 km wide [111]. The Bingham Canyon landslide occurred on 10 April 2013 in the canyon open-pit copper mines, causing a massive movement of the upper half of the northern pit wall. The slide filled the mine floor with thick debris, as shown in Figure 16. Before the incident, the geotechnical surveillance teams and mine operators were fully aware of the instability and evacuated the mine workers and equipment from the unstable zones. Thus, no fatalities or injuries were recorded. According to Pankow et al. [112], the seismograph showed that the landslide was triggered by several small earthquakes. Six days after the landslide event, 16 additional seismic events were detected in the mine area. The study by Hibert et al. [113] indicated that the slide caused the onset of the mobilization of the second slide at a higher elevation.

Figure 15. A large scale image of some portion of Chiquicamata slope (Adapted from [1]).

Figure 16. Image of landslide at Kennecott Utah Copper's Bingham Canyon Mine (Adapted from [112]).

The failure mechanisms driving the instability and causes of failure are presented in Table 3.

Table 3. List of failed mine slopes.

Country	Location	Names of Mining Company	Mode of Failure	Causes of Failure	Sources
Botswana	Central District	Letlhakane mine	Toppling	Presence of tension crack formation, crack widening and extension	i
Canada	British Columbia	Afton Mine	Wedge, Toppling and circular failure	Multiple failures occurred as a result of intersection of discontinuities	j, r, a
	British Columbia	Brenda Mine	Toppling	Intersection of joint sets	q, a, j
	British Columbia	Cassiar Mine	Toppling	Presence of shear zones, faults and sets of discontinuities	j, k
	British Columbia	Highland Valley Copper	Toppling	Steeply dipping joints, increase in groundwater pressure and melting of snow	t
	British Columbia	Lonex Pit at Highland valley	Toppling	Groundwater condition, Steeply dipping faults	o
	British Columbia	Highmont	Planar	Structural discontinuities, precipitation, run off, poor quality and low strength rock mass	o
	Vancouver	Island Copper	Wedge and Toppling	Large fault zone passing through a weaker rock mass	l
	Quebec	Jeffrey Mine, Asbestos	Wedge and Planar	Intersection of several thick shear zones and smaller scale discontinuities	t, c
	British Columbia	Nickel Plate Mine	Wedge	Steeply dipping joint sets and faults	t
China	Mongolia	Changshanhao open-pit	Wedge and Toppling	Presence of faults and joints	s
	Shazhenxi	Qianjiangping	Planar	Increase in water level, poor geological structure and continuous rainfall	h
Norway	Hange i Dalane	Tellness Dagbrudd	Wedge	Heavy rainfall	n
Mexico	Calama, Antofagasta	Chiquicamata	Toppling	Presence of fault zones	f, p
Spain	Seville	Aznacollar Mine	Complex	Presence of tension cracks, Heavy rainfall, groundwater pressure	g
Sweden	Kiruna	Kirunavaara	Rotational	Presence of tension cracks	u

Table 3. *Cont.*

Country	Location	Names of Mining Company	Mode of Failure	Causes of Failure	Sources
United States of America	Utah	Bingham Canyon Mine	Rotational, Planar	Rise in water table, fractured rock mass with minor joints and larger fault structure	t, v
	Nevada	Carlin Trend	NA	Presence of wider fault zones and clay infillings	v
	Arizona	Cyprus Bagdad and Sierrita	Toppling	Presence of steeply joint sets	j
	Nevada	Liberty Pit	Wedge	Intersection of joint sets	c
	Nevada	Veteran—Tripp Pit	Wedge	Intersection of faults, presence of clay gouge in fault zones	m
	Nevada	Kimbley pit	Wedge	Presence of flat sipping fault, High water pressure	b
	Arizona	Twin Butes	Toppling	Numerous faults and several joints	c
South Africa	Limpopo	Palabora Mine	Wedge	Presence of faults and set of joints	t, d
	Mokopane	Sandsloot open pit	Planar and Wedge	Presence of set of joints	e
Zambia	Chingola	Nchanga Open Pit	Wedge	Intersection of joint sets, abnormally rainfall, weathering	u

Source codes: a: Blackwell and Calder [114]; b: Blake [115]; c: Brawner [116]); d: Brummer et al. [117]; e: Bye [118]; f: Call et al. [119]; g: Golder Associates UK Ltd [120]; h: Jian et al. [121]; i: Kayesa [122]; j: Martin [123]; k: Martin and Mehr [124]; l: Mathis et al. [125]; m: Miller [126]; n: Nilsen and Hagen [127], o: Newcomen and Martin [128]; p: Rapiman [129]; q: Pritchard and Savigny [130]; r: Reid and Stewart [131]; s: Ren et al. [132]; t: Stacey [133]; u: Sjoberg [54]; v: Zavodni and Mccarter [134].

However, events of landslide and rock slope failure in underground mining environments have not yet been fully researched [18]. In most cases, slope failure in underground mine spaces happens during the transition from open-pit to underground mining methods. The transition from open pit to underground usually happened to exploit minerals from deep ore deposits. When mine deposits start from the shallow subsurface and extend to a great depth, sequential use of open-pit and underground mining is an efficient and economical way to maintain mining productivity.

An example of a failed slope during the transition to underground mining was reported by Brummer et al. [117] for Palabora mine in South Africa. Parabola mine was changing their method of extraction from open-pit to the underground block caving method as the ore extraction was getting deeper, which was no longer economical for the mine. During the transition, there was a failure at the north wall which, is evident by the daylight caving of the zone as a result of intersections of four main faults crossing the pit and three dominant joint sets present at the mine [117].

6. Factors Required in the Design of a Stable Slope

The essence of strengthening a rock slope is to prevent rock mass movement and premature closure of mining operations. When making an attempt to stabilize a rock slope that is prone to failure, many solutions are opened to the geotechnical engineering team to decide what to do. According to Niroumand et al. [135], the method used in preventing slope failure includes a change in geometry, rock–soil drainage, rock material nailing, turfing, shotcreting, geotextiles and the application of retaining the wall. Similarly, Roux et al. [136] applied techniques such as a regular collection of geotechnical mapping and logging data, the collection of laboratory strength test data, adherence to good housekeeping and the use of a hazard plan and evacuation procedure to achieve a stable slope. The study also implemented precautionary measures, such as surveying the actual excavation profile, measuring the change in groundwater level with a piezometer instrument, monitoring the seepage flow from the toe drains and face, measuring the damage caused by blasting to the rock mass behind the mine design line, inspecting the blast face and crests to allow removal of loose materials, and periodically installing bolt supports, as recommended by the geotechnical engineering team.

The study reported by Singh et al. [137] indicated that anchored tensioned rocks were installed, along with scaling and trimming of loose rock blocks, to prevent further slope failure. The study mentioned that shotcrete can be sprayed on the slope face with proper drainage galleries to prevent the development of pore water pressure. These techniques protect the slope from degrading agents such as rainwater and serves as part of an effort to strengthen and improve the ground conditions, which in turn increase the FOS and improve the stability of the slope. Another method to stabilize a rock slope involves the use of a leguminous tree planted on the rock slope to control hydrological factors such as erosion and run-off water that have the potential to trigger rock mass movement in rock slope engineering [138]. The outcomes of the study showed that the Leucaena Leucocephala leguminous plant has the capability to prevent slope failure in open-pit mining operations.

The use of an early warning system (EWS) is another approach to prevent the risk associated with slope stability. Roux et al. [136] confirmed that the installation of smart technologies such as extensometers aid in monitoring and checking the source of rock slope failures. In deeply weathered rock, the presence of cracks and faults are inevitable and need close monitoring. Extensometers and other smart systems can be installed in open-pit mines to monitor cracks and generate early warning systems. An extensometer integrated with other EWS was installed at Navachab Gold Mine in Namibia to give an early warning prior to failure. A study reported by Roux et al. [136] indicated that extensometers set off alarms an hour before failure. A similar study reported by Karam et al. [139] indicated that warning systems in mines should be installed in affected areas and serve as a guide to passive or active countermeasures. The EWS is designed to reduce threat, as shown in Figure 17.

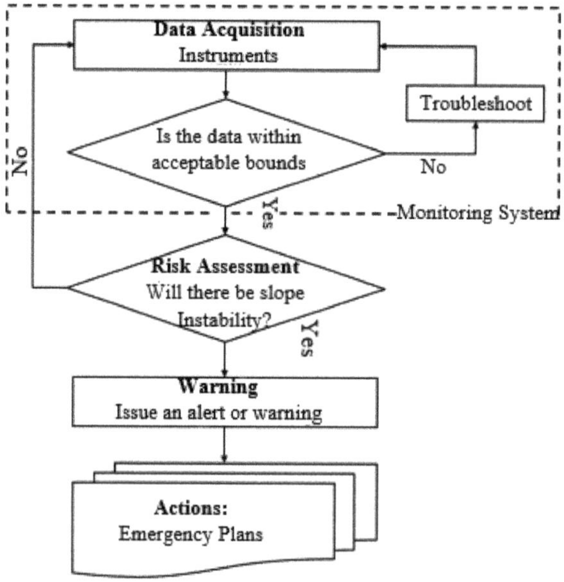

Figure 17. The flow chart for early warning system (Adapted from [139]).

6.1. Methods to Improve Slope Stability

There are several ways of achieving stability or strengthening rock slopes in mining operations. These include the construction of drainage and pump installation, slope monitoring, ground control with improvement in geological structures and the application of reinforcement.

6.1.1. Drainage Construction

Since water is one of the factors that promote failure in rock slopes, mine water management systems should be a priority. The presence of groundwater in open-pit mines often creates problems by reducing the stability of pit slopes. Most of the problems are caused by pre-pressure and hydrodynamic shock from blasting operations [140]. The presence of pore pressure on rock mass reduces the shear strength and seepage pressure, and water in the tension cracks causes an increment in the unit weight, which in turn increases the shear strength of the ground.

A number of researchers confirmed that drainage construction around the mines can reduce the impact and occurrence of groundwater in a mining environment. The three main objectives of drainage systems in mines are to keep working areas dry, stable, and safe, to ensure pit floor workability, and to lower the hydrostatic pressure and increase the effective stress of soil to improve stability [141]. Some of the drainage systems may include the construction of channels, water collection sumps and pump stations with pipelines to divert water from the surface. In addition, drains may be constructed within the surface to remove excess seepage. Hence, it is necessary for the geotechnical engineering team to know the hydrogeological conditions of the mine. This will assist in the selection of pumping systems required in the mine.

6.1.2. Slope Monitoring

Monitoring of rock slopes is necessary to determine how the rock structure behaves during excavation and if they are a threat to safety. This has been a key technique used to assess instability in rock slopes. The slope monitoring approach can be of value to provide information that is useful in data collection, recording and qualitative and quantitative analysis. The effectiveness of any monitoring system depends on the ability to give warnings before rock mass movement or failure takes place [142] Traditionally, there are geotechnical instrumentations (piezometers, crack meters, tiltmeters, borehole extensometers, and stress meters) available for the monitoring of slopes in mining and civil engineering. Despite the contribution of these instruments in the monitoring of rock instabilities, they also have some limitations.

Over the years, there have been improvements in geotechnical monitoring instrumentations, which include the use of remote-sensed technological tools. These instruments are commercially available to monitor rock mass movement in a mining environment and are cost-effective. Examples of these are the integration of synthetic aperture radar (SAR) with optical images and interferometric SAR (InSAR), which are currently being used to characterize instabilities in slopes. Advancement in remote sensing technology systems has brought about differential synthetic aperture radar (DInSAR) with high-resolution image processing, which is also being used to monitor instabilities. Similarly, there is a ground-based radar monitoring instrument known as linear SAR (LISA), which is capable of assessing the deformation of an unstable area in slopes that are characterized by high radar reflectivity [139]. Other advanced geotechnical monitoring instruments are light detection and ranging (LiDAR) and optical satellite images.

6.1.3. Ground Improvement with Enhancement of Geological Structures

The interpretation of the data acquired during site investigation provides information about the structure of the rock mass, strength of the rocks, planes of weakness, and the range of their strengths. This will provide adequate information on the relationship between slope parameters, such as the spatial relationship, face inclination of the rock, and geometrical dimensions of each of the individual elements of a particular sector of the mine shell, such as ramp and geotechnical berm width, bench spill berm width, individual slope height, and slope angle, slope bench, etc.

Apart from the design parameters, the geotechnical engineering team must be familiar with the geology of the mine, i.e., the discontinuities and engineering properties of the rock mass that can influence the stability of the slope. Karama et al. [139] argued that gravitational load is the most important load affecting the stability of the mine and, notably, the shear strength of the ground, which confers major resistance to failure. To prevent further damage of the rock mass, the shear strength of the ground must be enhanced to strengthen and protect the ground. This strengthening can be completed with the application of backfill, rapid yielding properties, the construction of a retaining wall, the spraying of shotcrete, and application of reticulated micropiles.

6.1.4. Installation of Reinforcement Units

The application of reinforcement in rock slope is a viable approach to increase the stability of unstable slopes. The installation of various support systems increases the shear strength of rocks and reduces sliding effects along the slip surface [143]. The essence of reinforcement in weathered rock or soil is to stiffen the base and to reduce shear stress magnitudes. The two most commonly used support systems in rock slope and slope engineering measures are rock bolt fame and anchor cable fame [144]. These reinforcement tools have proven to be effective in resisting the shear stress in rock mass deformation by transferring the load to the bounded rock so that resistance is provided by the rock to balance the deformation. During the installation of bolts, the effects of the inclination angle of bolts must be defined by the designer of the slope for effective reinforcement. According to Sazzad et al. [145], the FOS of slopes increases with an increase in the angle of inclination. However, the FOS value depends on the overall angle of the slope. Therefore, the relationship between the slope angle and inclination angle for a maximum FOS should be enhanced, i.e., the positioning of the bolts is a key factor in reinforcement design. In addition, when installing a bolt to resist dip bedding slope, the failure mechanism and process occurring in such slope need to be considered. A great number of researchers have acknowledged the effectiveness of bolts as reinforcement tools in controlling the movement of discontinuous layers that are subjected to shearing. The installation of rock bolts in unstable zones can be regarded as one of the major techniques used to improve and control the unstable ground in open-pit mines.

6.2. Role of Artificial Intelligence in the Management of Slope Failure As a Reflection on the Current State of the Art

The monitoring and evaluation of slope stability in mining operations is a crucial necessity to avoid potential damage to both the personnel safety and financial base of mining companies. Despite the effort to reduce slope failure, the geotechnical phenomenon is known to be complex, comprising numerous factors such as ground conditions, geological activities and human actions, which make the prediction of slope failure challenging [146,147]. More so, these factors are dynamic and ever-changing, making it cumbersome to continuously measure them. However, the invention of artificial intelligence (AI) has greatly aided engineers in forecasting possible slope failure using detailed analysis rather than solely relying on phenomenological models [148].

The study by Kothari and Momayez [148] compared the prediction performance of inverse velocity (IV) and the artificial neural network (ANN) model to forecast slope failure of an open-cast mine. Twenty-two datasets were collected using radar equipment, and we developed a double-layered feed-forward ANN in the MATLAB environment. The output from the study indicated that ANN prediction was 86% accurate compared to IV. A total of 82% of the slope failure predictions indicated that the slopes were safe, while 18% indicated that the slopes were unsafe. The unsafe predictions were approximated within 5 min before actual failure, proving that the ANN model gave a safer prediction compared to IV. Another study by Chebrolu et al. [149] utilized 46 datasets (23 obtained from wet slopes and 23 obtained from dry slopes) to predict the FOS of a slope using multi-gene genetic programming (MGGP) and multi-adaptive regression spline (MARS). Thirty-two datasets were used in training the models, and fourteen datasets were used in testing the models. Despite both models proving to have an accurate prediction, MGGP has a better prediction performance, with R-values of 84.38% in training and 85.71% in testing, compared to MARS with R-values of 81.25% in training and 85.71% in testing.

To establish an accurate prediction model for determining the slope failure of a mine in Vietnam, a study by Bui et al. [146] integrated the M5-rule and genetic algorithms to develop a novel prediction model that was then compared with conventional AI models, including ANN, support vector regression (SVR), the firefly algorithm (FFA), the imperialist competitive algorithm (ICA), artificial bee colony (ABC), and the genetic algorithm (GA). This study utilized bench height, soil unit weight, cohesion, angle of internal friction, and slope angles as input parameters to forecast the factor of safety against failure. The results of this study indicated that prediction performance from the M5-rules GA model gave the best accuracy due to its enhanced optimization prowess. Another study by Du et al. [150] utilized a ground-based interferometric radar (GB-SAR) to record 150 datasets in the Anjialing open-pit coalmine of China. Twelve input parameters (slope shape coefficient, deformation rate, reverse deformation rate, deformation amplitude, rainfall, temperature, atmospheric pressure, relative humidity, wind speed and direction, and groundwater temperature and level) and one output variable (deformation) were used to develop five prediction models. These models included backpropagation, NN, support vector machine (SVM), recurrent neural network (RNN), adaptive neural fuzzy inference system (ANFIS), and relevant vector machine (RVM). From this result, it was evident that not all models can be reliable in the prediction of slope failure due to high error values. The error values for each model were 4.122 mm for BPNN, 3.612 mm for SVM, 1.660 for ANFIS, 0.578 mm for RNN, and 0.442 mm for RVM. RVM had the lowest root-mean-square error value (RMSE) of 2.64, while BPNN had the highest RSME value of 4.58.

In addition, a study by Ferentinou and Fakir [151] developed a BP-NN model using 141 databases obtained from worldwide cases of surface mines and eighteen (18) input variables, including environmental factors, rock quality, rock mass characteristics, rock stresses, hydraulic profile, presence of geological features, dimensions of slopes, blast design and previous instability occurrence. This study established slope stability indices, where all slopes were examined to predict failure. The results obtained indicated a mean square error value of 0.0001 converging at 98%, proposing the utilization of BP-ANN as a reliable tool to predict slope failure in feasibility studies. To evaluate and monitor the slope stability of open-cast mines, Luo et al. [147] developed a particle swarm optimization-cubist algorithm (PSO-CA) to forecast the factor of safety of a Vietnamese mine. This model used five input variables, including bench height, angle of slope and internal friction, cohesion coefficient, and specific weight of the material. The output from this model was then compared with the output obtained from the prediction of the same variables using SVM, classification and regression tree (CART), and k-nearest neighbor. Performance comparison of the models inferred that PSO-CA had the lowest error values (mean absolute error (MAE) of 0.009 and root mean square error (RMSE) of 0.025 and a high correlation coefficient R-value of 0.981. The SVM, CART, and k-NN produced poor prediction performance with MAE values of 0.014 to 0.038, RMSE 0.030–0.056, and R 0.917–0.974.

The reported studies that have utilized AI in determining factors of safety in mines have proven to be efficient and effective in accurately forecasting slope failure before the hazard occurs. Moreover, the models are endowed with the ability to handle large amounts of data at a given time and execute predictions at a high rate compared to conventional stability analysis techniques. The models are designed using various factors known to affect slope stability in order to map out FOS. Various AI models have been used to predict slope failure in mines, but limited studies have been conducted to compare and validate the most accurate model for such a task. Therefore, in order to establish the most robust, versatile, and reliable model, more comparative studies need to be conducted to improve prediction performance.

7. Concluding Remarks

Stability analysis of slopes in open-pit mines and quarries is extremely important from both economic and safety points of view. The effect of unstable ground cannot be over-emphasized as it may lead to the temporary or permanent closure of the mine depending on the level of damage. Instabilities usually occur where there is a presence of geological discontinuities such as fractures, cracks, faults, unfavorably oriented bedding planes, etc. Rock slope failures are triggered when the shear stress exceeds the shear strength of the rock mass. Several factors, such as geometry of the slope, groundwater condition of the rock, lithology, geological structures, cohesion and angle of internal friction, effect of blasting, mining method and equipment selection, have been declared to influence the failure of rocks. The effects of all these factors must be considered during the planning and designing stage of the mine, not only that these factors have effects on the design, but that they also influence the engineering judgement on the stability of the slope. Several techniques, such as construction of drainages, slope monitoring, application of reinforcement and improvement of geological structures, have been used over the years to improve and maintain the stability of the rock slope.

Advancements in technology have brought about improvement in slope monitoring techniques and provide an understanding of slope dynamics through the assessment of stability over time. The accurate prediction of slope failure is an important task in active open-pit mines in order to avoid slope failure so as to prevent injury to personnel and damage to machinery. When determining the stability of a slope, it is imperative to assess the factors affecting slope stability. Different studies have used different AI models with various input variables to predict slope failure in mining. These studies have reported the prowess of AI models as being fast, reliable in mapping out FOS and sturdy to handle large amounts of data at a given time. Despite the usage of such a model, there is a need for comparing the output of various AI models in order to determine the most accurate model for forecasting slope stability. In addition, various elements known to affect slope stability should be captured and used as input variables to forecast slope stability and compare performance of different AI models.

Author Contributions: Original manuscript writing, review and editing, P.K.; Manuscript Administration, G.O.O.; Review and Manuscript Editing, K.O.S.; Manuscript Administration, A.I.L.; Conceptualization and Manuscript Draft, M.O.; Manuscript Editing, P.M. All authors have read and agreed to the published version of the manuscript.

Funding: This research received no external funding.

Conflicts of Interest: The authors declare no conflict of interest.

References

1. Naghadehi, M.; Jimenez, R.; Khalokakaie, R.; Jalali, S.-M. A probabilistic systems methodology to analyze the importance of factors affecting the stability of rock slopes. *Eng. Geol.* **2011**, *118*, 82–92. [CrossRef]
2. Ghobadi, M. Engineering Geological Factors Influencing the Stability of Slope in the Illawarra Region. Ph.D. Thesis, University of Wollongong, Wollongong, Australia, 1994; p. 1.
3. Gao, W.; Dai, S.; Xiao, T.; He, T. Failure process of rock slopes with cracks based on the fracture mechanics. *Eng. Geol.* **2017**, *231*, 190–199. [CrossRef]
4. Sha, L. Analysis of slope instability factors and protection. *Int. J. Multidiscip. Res. Dev.* **2016**, *3*, 181–182.
5. Simataa, E. *Investigating Slope Stability in Open Pit Mine—A Case Study of the Phyllites Wall at Sentinel Pit*; The University of Witwatersrand: Johannesburg, South Africa, 2019.
6. Raghuvanshi, T. Plane failure in rock slopes—A review on stability analysis techniques. *J. King Saud Univ. Sci.* **2019**, *31*, 101–109. [CrossRef]
7. Wang, X.; Niu, R. Spatial forecast of landslides in three gorges based on spatial. *Sensors* **2009**, *9*, 2035–2061. [CrossRef] [PubMed]
8. Chiwaye, H. A Comparison of the Limit Equilibrium and Numerical Modelling Approaches to Risk Analysis for Open Pit Mine Slopes. Ph.D. Thesis, Faculty of Engineering and the Built Environment, University of the Witwatersrand, Johannesburg, South Africa, 2010.
9. Nicholas, D.E.; Sims, D.B. *Collecting and Using Geological Structure Data for Slope Design*; Hustrulid, W.A., McCarter, M.K., Van Zyl, D.J.A., Eds.; Society for Mining, Metallurgy, and Exploration: Golden, CO, USA, 2000.
10. Hustrulid, W.A.; McCarter, M.K.; Van Zyl, D.J.A. *Slope Stability in Surface Mining*; Society for Mining, Metallurgy, and Exploration: Littleton, CO, USA, 2000; pp. 81–88.
11. Read, J.; Stacey, P. *Guidelines for Open Pit Slope Design*, 1st ed.; CSIRO: Canberra, Australia, 2009.
12. Eberhardt, E. *Rock Slope Stability Analysis—Utilization of Advanced Numerical Techniques*; Earth and Ocean Sciences at UBC Report; University of British Columbia (UBC): Vancouver, BC, Canada, 2003.
13. Stacey, T.R.; Swart, A.H. *Booklet Practical Rock Engineering Practice for Shallow and Opencast Mines*; The Safety in Mines Research Advisory Committee (SIMRAC): Johannesburg, South Africa, 2001.
14. Suman, S. *Slope Stability Analysis Using Numerical Modelling*; B.Tech Report; National Institute of Technology: Rourkela, India, 2015.
15. Abramson, L. *Slope Stability and Stabilization Methods*, 2nd ed.; Wiley: New York, NY, USA, 2002.
16. Eberhardt, E.; Stead, D.; Coggan, J.S. Numerical analysis of initiation and progressive failure in natural rock slopes. *Int. J. Rock Mech. Min. Sci.* **2004**, *41*, 69–87. [CrossRef]
17. Johansson, J. *Impact of Water-Level Variations on Slope Stability*; Luleå University of Technology: Luleå, Sweden, 2014.
18. Zhu, J.; Xue, Y.; Tao, Y.; Zhang, K.; Li, Z.; Zhang, X.; Yang, Y. The Numerical Simulation on the Stability of Steep Rock Slope by DDA. Bydgoszcz, Poland. *AIP Conf. Proc.* **2017**, *1839*, 020134.
19. Hoek, E.; Read, J.; Karzulovic, A.; Chen, Z.Y. Rock slopes in Civil and Mining Engineering. In Proceedings of the International Conference on Geotechnical and Geological Engineering, GeoEng, Melbourne, Australia, 19–24 November 2000.
20. Li, A.J.; Cassidy, M.J.; Wanga, Y.; Merifield, R.S.; Lyamin, A.V. Parametric Monte Carlo studies of rock slopes based on the Hoek–Brown failure criterion. *Comput. Geotech.* **2012**, *45*, 11–18. [CrossRef]
21. Duncan, J.; Wright, S.; Brandon, T. *Soil Strength and Slope Stability*, 2nd ed.; John Wiley & Sons: Hoboken, NJ, USA, 2005.
22. Fleurisson, J.-A.; Cojean, R. *Error Reduction in Slope Stability Assessment*; MINES ParisTech: Paris, France, 2014.
23. Stacey, T.R. *Mechanics and Design of Major Rock Slopes, Lecture Note*; The University of Witwatersrand: Johannesburg, South Africa, 2018; *unpublished*.
24. Gundewar, C. *Application of Rock Mechanics In surface and Underground Mining*; Indian Bureau of Mines: Nagpur, Maharashtra, India, 2014; pp. 62–66.
25. Valdivia, C.; Lorig, L. Slope stability at Escondida mine. In *Slope Stability in Surface Mining*; Hustrulid, M., Zyl, V., Eds.; SME: Golden, CO, USA, 2000; pp. 153–162.
26. Ceryan, N.; Kesimal, A.; Ceryan, S. Probabilistic Analysis Applied to Rock Slope Stability: A Case Study from Northeast Turkey. In *Integrating Disaster Science and Management*; Samui, P., Kim, D., Ghosh, C., Eds.; Elsevier: Amsterdam, The Netherlands, 2018; pp. 221–261.

27. Corkum, A.; Martin, C. Analysis of a rockslide stabilized with a toe-berm: A case study in British Columbia, Canada. *Int. J. Rock Mech. Min. Sci.* **2004**, *41*, 1109–1121. [CrossRef]
28. Chiwaye, H.; Stacey, T. A comparison of limit equilibrium and numerical modelling approaches to risk analysis for open pit mining. *S. Afr. Inst. Min. Metall.* **2010**, *110*, 571–580.
29. Stead, D.; Eberhardt, E.; Coggan, J. Developments in the characterization of complex rock slope deformation and failure using numerical modelling techniques. *Eng. Geol.* **2006**, *83*, 217–235. [CrossRef]
30. Fredj, M.; Hafsaoui, A.; Boukarm, R.; Nakache, R.; Saardoum, A. Numerical Modelling of Slope Stability in Open Pit Phosphate. Prague, Czech, World Multidisciplinary Earth Sciences Symposium (WMESS). *IOP Conf. Ser. Earth Environ. Sci.* **2018**, *221*, 012020. [CrossRef]
31. Griffiths, D.; Lane, P. Slope stability analysis by finite elements. *Geotechnique* **1999**, *49*, 387–403. [CrossRef]
32. Hammah, R.; Yacoub, T.; Corkum, B. *The Shear Strength Reduction Method for the Generalized Hoek-Brown Criterion*; American Rock Mechanics Association (ARMA): Anchorage, AK, USA, 2005.
33. Alzo'ubi, A. Rock slopes processes and recommended methods for analysis. *Int. J. Geomate* **2016**, *11*, 2520–2527. [CrossRef]
34. Tang, X.-T.; Li, D.-Q.; Rong, G.; Phoon, K.-K. Zhou Impact of copula selection on geotechnical reliability under incomplete probability information. *Comput. Geotech.* **2013**, *49*, 264–278. [CrossRef]
35. Lei, Q.; Latham, J.-P.; Tsang, C.-F. The use of discrete fracture networks for modelling coupled geomechanical and hydrological behaviour of fractured rocks. *Comput. Geotech.* **2017**, *85*, 151–176. [CrossRef]
36. Ahmad, F.; Tang, X.-W.-N.; Wróblewski, P.; Ahmad, M. Prediction of slope stability using Tree ugmented Naive-Bayes classifier: Modeling and performance evaluation. *Math. Biosci. Eng.* **2022**, *19*, 4526–4546. [CrossRef]
37. Haghshenas, S.S.; Haghshenas, S.S.; Geem, Z.W.; Kim, T.H.; Mikaeil, R.; Pugliese, L.; Troncone, A. Application of Harmony Search Algorithm to Slope Stability Analysis. *Land* **2021**, *10*, 1250. [CrossRef]
38. Huang, L.; Liu, F.; Dong, V. Geotechnical Slope Protection Model Based on Genetic Algorithm. *Nat. Environ. Pollut. Technol.* **2019**, *18*, 1509–1513.
39. Figueiredo, L.; Barros, M.; Hammah, R.; Santos, Y.; Souza, T.; Nogueira, C. *Case Study: Open Pit Three-Dimensional Slope Stability Back-Analysis for an Anisotropic Iron Ore Rock Mass*; Australian Centre for Geomechanics: Perth, Australia, 2021.
40. Tugelbayeva, G.K. Mathematical models for numerical solution of nonstationary problems of geomechanics. *AIP Conf. Proc.* **2020**, *2312*, 050024.
41. Nemirovisky, Y.V.; Tyrymov, A.A. Mathematical modeling in problems of geomechanics of layered rock massifs. *AIP Conf. Proc.* **2021**, *2448*, 020025.
42. Pirklbauer, A. The influence of land use on the stability of slopes with examples from the European Alps. *Mitt. Naturwiss. Ver. Steiermark* **2002**, *132*, 43–62.
43. Wyllie, D.C.; Mah, C.W. *Rock Slope Engineering: Civil and Mining*, 4th ed.; Spon Press: New York, NY, USA, 2004.
44. Babikar, A.; Smith, C.; Gilbert, M.; Ashbly, J. Non-associative limit analysis of the toppling-sliding failure. *Int. J. Rock Mech. Min. Sci.* **2014**, *71*, 1–11. [CrossRef]
45. Kliche, C. *Rock Slope Stability*; Society for Mining, Metallurgy and Exploration: Littleton, CO, USA, 1999.
46. Hoek, E.; Bray, J. *Rock Slope Engineering*, 3rd ed.; CRC: Boca Raton, FL, USA, 1981.
47. Alabideen, K.; Helal, M. Determination of the Safe Orientation and Dip of a Rock Slope in an Open Pit Mine in Syria Using Kinematic Analysis. *Al-Nahrain Univ. Coll. Eng. J. (NUCEJ)* **2016**, *19*, 33–45.
48. Goodman, R.E.; Kieffer, D.S. Behaviour of rock in slopes. *J. Geotech. Geoenviron. Eng.* **2000**, *126*, 675–684. [CrossRef]
49. Hudson, J.A.; Harrison, J.P. *Engineering Rock Mechanics*; Elsevier: Oxford, UK, 1997; p. 444.
50. Low, B.K. Reliability analysis of rock wedges. *J. Geotech. Geoenviron. Eng.* **1997**, *123*, 498–505. [CrossRef]
51. Coates, D.F. *Rock Mechanics Principles*; Canadian Centre for Mineral and Energy Technology (CANMET): Energy, Mines, and Resources Canada in Ottawa: Ottawa, ON, Canada, 1981; p. 410.
52. Dahner-Lindqvist, C. Liggvagga Stabiliteten I Kiirunavaara. In Proceedings of the Bergmekanikdagen, Papers presented at Rock Mechanics Meeting, Stockholm, Sweden, 19 March 1992; pp. 37–52.
53. Prakash, B. *Design of Stable Slope for Open Cast Mines*; B.Tech Report; National Institute of Technology: Rourkela, India, 2009.
54. Sjöberg, J. Analysis of Large-Scale Rock Slopes. Ph.D. Thesis, University of Technology, Luleå, Sweden, 1999.
55. Sjöberg, J. *Failure Mechanisms for High Slopes in Hard Rock. Colorado, Denver, USA, Slope Stability in Surface Mining*; Hustrulid, M., Zyl, V., Eds.; Society for Mining, Metallurgy & Exploration: Englewood, CO, USA, 2000.
56. Adeyeri, J. *Technology and Practice in Geotechnical Engineering: Stability of Earth Masses and Slopes*; IGI Global: Pennsylvania, PA, USA, 2014; pp. 528–588.
57. Guo, S.; Qi, S.; Yang, G.; Zhang, S.; Saroglou, C. An Analytical Solution for Block Toppling Failure of Rock Slopes during an Earthquake. *Appl. Sci.* **2017**, *7*, 1008. [CrossRef]
58. Chaulya, S.; Prasad, G. Slope Failure Mechanism and Monitoring Techniques. In *Sensing and Monitoring Technologies for Mines and Hazardous Areas: Monitoring and Prediction Technologies*; Elsevier: Amsterdam, The Netherlands, 2016; pp. 1–86.
59. Chakraborty, A.; Goswami, D. State of the art: Three-Dimensional (3D) Slope-Stability Analysis. *Int. J. Geotech. Eng.* **2016**, *10*, 493–498. [CrossRef]
60. Wines, D. A comparison of slope stability analyses in two and three dimensions. *J. South. Afr. Inst. Min. Metall.* **2016**, *116*, 399–406. [CrossRef]

61. Totonchi, A.; Askari, F.; Farzaneh, O. 3D stability analysis of concave slopes in plan view using linear finite element and lower bound method. *Iran. J. Sci. Technol. Trans. Civ. Eng.* **2012**, *36*, 181–194.
62. Cala, M. Convex and concave slope stability analyses with numerical methods. *Arch. Min. Sci.* **2007**, *52*, 75–89.
63. Zhang, Y.; Chen, G.; Zheng, L.; Li, Y.; Zhuang, X. Effects of geometries on three-dimensional slope stability. *Can. Geotech. J.* **2013**, *50*, 233–249. [CrossRef]
64. Dana, H.; Kakaie, R.; Rafiee, R.; Bafghi, A. Effects of geometrical and geomechanical properties on slope stability of open-pit mines using 2D and 3D finite difference methods. *J. Min. Environ.* **2018**, *9*, 941–957.
65. Keykha, H.; Huat, B.; Asadi, A.; Moayedi, H. The effect of discontinuities on stability of rock blocks in tunnel. *Int. J. Phys. Sci.* **2011**, *6*, 7132–7138.
66. Liu, S.; Huang, X.; Zhou, X.; Hu, J.; Wang, W. Soil-Rock Slope Stability Analysis by Considering the Nonuniformity of Rocks. *Math. Probl. Eng.* **2018**, *2018*, 3121604. [CrossRef]
67. Kolapo, P.; Munemo, P. Investigating the correlations between point load strength index, uniaxial compressive strength and Brazilian tensile strength of sandstones. A case study of QwaQwa sandstone deposit. *Int. J. Min. Miner. Eng.* **2021**, *12*, 67–83. [CrossRef]
68. Brian, D.; Greg, M. Rockfall triggering by cyclic thermal stressing of exfoliation fractures. *Nat. Geosci.* **2016**, *9*, 395–400.
69. Cravero, M.; Labichino, G. Analysis of the flexural failure of an overhanging rock slab. *Int. J. Rock Mechanics Min. Sci.* **2004**, *41* (Suppl. S1), 605–610. [CrossRef]
70. Borrelli, L.; Greco, R.; Gulla, G. Weathering grade of rock masses as a predisposing factor to slope instabilities: Reconnaissance and control procedures. *Geomorphology* **2007**, *87*, 158–175. [CrossRef]
71. Wieczorek, G.; Jager, S. Triggering mechanisms and depositional rates of postglacial slope-movement processes in the Yosemite Valley, California. *Geomorphology* **1996**, *15*, 17–31. [CrossRef]
72. Shang, J.; West, L.J.; Hencher, S.R.; Zhao, Z. Geological discontinuity persistence: Implication and quantification. *Eng. Geol.* **2018**, *241*, 41–54. [CrossRef]
73. Shang, J.; Zhao, Z.; Hu, J.; Handley, K. 3D particle-based DEM investigation into the shear behaviour of incipient rock joints with various geometries of rock bridges. *Rock Mech. Rock Eng.* **2018**, *51*, 3563–3584. [CrossRef]
74. Kalenchuk, K.; Diederichs, M.; Mckinnon, S. Characterizing block geometry in jointed rock masses. *Int. J. Rock Mechanics Min. Sci.* **2006**, *43*, 1212–1225. [CrossRef]
75. Higuchi, K.; Fujisawa, K.; Asai, K.; Pasuto, A.; Marcato, G. Development of slope failure displacement detection sensor using optical fiber in the OTDR method. In Proceedings of the 44th Colloquium of Japan Slope Failure Society, Sasebo, Japan, 2015; pp. 315–318.
76. Ganapathy, G.; Mahendran, K.; Sekar, S. Need and urgency of slope failure risk planning for Nilgiris district, Tamil Nadu State, India. *Intern. J. Geosci.* **2012**, *1*, 29–40.
77. Central Geological Survey. Taiwan Geoscience Portal. 2014. Available online: https://twgeoref.moeacgs.gov.tw/GipOpenWeb/wSite/mp (accessed on 10 June 2021).
78. Tiwari, B.; Beena, A.; Mohammed, K.; Samin, D.; Rosalie, C. Developments in Earth Retention, Support Systems, and Tunneling. In Proceedings of the International Foundations Congress & Equipment Expo, Orlando, FL, USA, 5–6 March 2018.
79. Tiwari, B.; Caballero, S. Experimental Model of Rainfall Induced Slope Failure in Compacted Clays. *Geotech. Spec. Publ.* **2015**, *256*, 1217–1226.
80. Aladejare, A.; Akeju, V. Design and Sensitivity Analysis of Rock Slope Using Monte Carlo Simulation. *Geotech. Geol. Eng.* **2020**, *38*, 573–585. [CrossRef]
81. Liu, Q.; Li, J. *Effects of Water Seepage on the Stability of Soil-Slopes*; Procedia IUTAM: Beijing, China, 2015.
82. Zhu, Y.; Xiao, Y. Slope Stability from a Hydrological Perspective: Taking Typical Soil Slope as an Example. *Adv. Civ. Eng.* **2020**, *17*, 2020. [CrossRef]
83. Tandon, R.; Gupta, V. The control of mineral constituents and textural characteristics on the petrophysical & mechanical (PM) properties of different rocks of the Himalaya. *Eng. Geol.* **2013**, *153*, 125–143.
84. McNeilly, M.; Funkhouser, M.; Lupo, J. Design and Evaluation of Ground Support for the Exchange Place Station Improvements Project. In Proceedings of the North American Tunneling Conference, Atlanta, GA, USA, 17–22 April 2004.
85. Yasir, S.; Awang, H.; Ayub, M. The Relationship of Sandstone's Strength with Mineral Content and Petrographic Characteristics in Sungai Tekai, Jerantut, Pahang. In Proceedings of the AIP Conference Proceedings, Depok, Indonesia, 6–8 November 2021.
86. Villeneuve, M.; Heap, M. Calculating the cohesion and internal friction angle of volcanic rocks and rock masses. *Volcanica* **2022**, *4*, 279–293. [CrossRef]
87. Sullivan, R.; Anderson, R.; Biesiadecki, J.; Bond Stewart, H. Cohesions, friction angles, and other physical properties of Martian regolith from Mars Exploration Rover wheel trenches and wheel scuffs. *J. Geophys. Res.* **2011**, *116*, 1–38. [CrossRef]
88. Zhou, J.-W.; Shi, C.; Xu, F.-G. Geotechnical Characteristics and Stability Analysis of Rock-Soil Aggregate Slope at the Gushui Hydropower Station, Southwest China. *Sci. World J.* **2013**, *2013*, 540636. [CrossRef] [PubMed]
89. Harabinová, S.; Panulinová, E. Impact of Shear Strength Parameters on Slope Stability. In Proceedings of the MATEC Web of Conferences, Tleń/Osie, Poland, 8–11 September 2020.
90. Yin, Z.-Q.; Hu, Z.-X.; Wei, Z.-D.; Zhao, G.-M.; Hai-Feng, M.; Zhang, Z.; Feng, R.-M. Assessment of Blasting-Induced Ground Vibration in an Open-Pit Mine under Different Rock Properties. *Adv. Civ. Eng.* **2018**, *2018*, 4603687. [CrossRef]

91. Sutejo, Y.; Gofar, N. Effect of Area Development on the stability of cut slopes. In Proceedings of the 5th International Conference of Euro Asia Civil Engineering Forum (EACEF-5), Surabaya, East Java, Indonesia, 15–18 September 2015.
92. Zou, D.; Kaiser, P.K. Determination of in situ stresses from excavation-induced stress changes. *Rock Mech. Rock Eng.* **1990**, *23*, 167–218. [CrossRef]
93. Stacey, T.; Ndlovu, X.; Ortlepp, W. *Dynamic Rock Failures Due to "High" Stress at Shallow Depth*; Deep Mining: Australia Centre for Geomechnaics: Perth, Australia, 2007.
94. Stacey, T.; Wesseloo, J. In situ stresses in mining areas in South Africa. *J. South Afr. Inst. Min. Metall.* **1998**, *98*, 365–368.
95. Vazaios, I.; Farahmand, K.; Vlachopoulos, N.; Dederichs, M. Effects of confinement on rock mass modulus: A synthetic rock mass modelling (SRM) study. *J. Rock Mech. Geotech. Eng.* **2018**, *10*, 436–456. [CrossRef]
96. Peng, C.; Guo, Q.; Yan, Z.; Wang, M.; Pan, J. Investigating the Failure Mechanism of Jointed Rock Slopes Based on Discrete Element Method. *Adv. Civ. Eng.* **2020**, *2020*, 8820158.
97. Ryan, T. Shear Strength of Closely Jointed Porphyry Rock Masses. In Proceedings of the 40th U.S. Symposium on Rock Mechanics (USRMS): Rock Mechanics for Energy, Mineral and Infrastructure Development in the Northern Regions, Anchorage, AK, USA, 25–29 June 2005.
98. Pariseau, W.; Puri, S.; Schmelter, S. A new model for effects of impersistent joint sets on rock slope stability. *Int. J. Rock Mech. Min. Sci.* **2008**, *45*, 122–131. [CrossRef]
99. Cundall, P.; Damjanac, B. Considerations on Slope Stability in a Jointed Rock Mass. In *50th U.S. Rock Mechanics/Geomechanics Symposium*; America Rock Mechanics Association: Houston, TX, USA, 2016.
100. Carlàab, T.; Farinab, P.; Intrieri, E.; Botsialas, K.; Casaglib, N. On the monitoring and early-warning of brittle slope failures in hard rock masses: Examples from an open-pit mine. *Eng. Geol.* **2017**, *228*, 71–81. [CrossRef]
101. Simmons, J.; Simpson, P. Composite failure mechanisms in coal measures rock masses-myths and reality. *Journal South. Afr. Inst. Min. Metall.* **2006**, *106*, 459–469.
102. Wesseloo, J.; Dight, P. Rock mass damage in hard rock open pit mine slopes. In Proceedings of the 2009 International Symposium on Rock Slope Stability in Open Pit and Civil Engineering, Santiago, Chile, 10 September 2009; Read, J.R., Ed.; University de Los Andes: Los Andes, Chile, 2009; Volume CD ROM, p. 7.
103. Hoek, E. Fundamentals of slope design. In Proceedings of the Keynote address at Slope Stability 2009, Santiago, Chile, 9–11 November 2009.
104. Hu, Y.; Ren, F.; Ding, H.; Fu, Y.; Tan, B. Study on the Process and Mechanism of Slope Failure Induced by Mining under Open Pit Slope: A Case Study from Yanqianshan Iron Mine, China. *Adv. Civ. Eng.* **2019**, 1–27. [CrossRef]
105. Zevgolis, I.; Deliveris, A.; Koukouzas, N. Slope failure incidents and other stability concerns in surface lignite mines in Greece. *J. Sustain. Min.* **2019**, *18*, 182–197. [CrossRef]
106. DTE. Down to Earth (DTE). Available online: https://www.downtoearth-indonesia.org/story/protests-over-fatal-collapse-freeportrio-tinto-west-papua-mine (accessed on 19 July 2021).
107. Graham, R. The Northern Miner: Slippage Causes Deaths at Grasberg. 2003. Available online: https://www.northernminer.com/news/slippage-causes-deaths-at-grasberg/1000143698 (accessed on 25 July 2021).
108. Reuters. Reuters: Vedanta Zinc suspends mining at South Africa's Gamsberg mine after accident. 2020. Available online: https://www.reuters.com/article/us-vedanta-zinc-accident-idINKBN27Y0VO (accessed on 4 April 2021).
109. Petley, D. Advance Earth Science. 2020. Available online: https://blogs.agu.org/landslideblog/2020/12/15/the-large-landslide-at-gamsberg-mine-high-resolution-satellite-images/#:~{}:text=used%20with%20permission.-,The%20large%20landslide%20at%20Gamsberg%20mine%3A%20high%20resolution%20satellite%20images,been%20anti (accessed on 9 April 2021).
110. Voight, B.; Kennedy, B.A. Slope failure of 1967–1969, Chiquicamata Mine Chile. In *Rockslides & Avalanches*; Voight, B., Ed.; Developments in Geotechnical Engineering, 14b; Elsevier: Amsterdam, The Netherlands, 1979; pp. 595–632.
111. Serna, M.; Williams, D.; Ruest, M. Analysis of Kennecott Utah copper's Bingham Canyon mine pit wall slides. In Proceedings of the Tailing and Mine Waste, Denver, CO, USA, 2–5 October 2016.
112. Pankow, K.L.; Moore, J.R.; Hale, J.M.; Koper, K.D.; Kubacki, T.; Whidden, K.M.; McCarter, M.K. Massive landslide at Utah copper mine generates wealth of geophysical data. *Geol. Soc. Am.* **2014**, *24*, 4–9. [CrossRef]
113. Hibert, C.; Ekström, G.; Stark, C. Dynamics of the Bingham Canyon Mine landslides from seismic signal analysis. *Geophys. Res. Lett.* **2014**, *41*, 4535–4541. [CrossRef]
114. Blackwell, G.H.; Calder, P.N. Practical aspects of wall stability at Brenda Mines Ltd., Peachland, B.C. In Proceedings of the Third International Conference on Stability in Surface Mining, Vancouver, BC, Canada, 1–3 June 1981; Brawner, C.O., Ed.; Society of Mining Engineers, AIME: New York, NY, USA, 1982; pp. 573–605.
115. Blake, W. Finite Element Model Study of Slope Modification at the Kimbley Pit. *Trans. Soc. Min. Eng.* **1968**, *241*, 525–532.
116. Brawner, C.O. Chapter 1, Introduction. In Proceedings of the First International Conference on Stability in Open Pit Mining, Vancouver, BC, Canada, 23–25 November 1971; Brawner, C.O., Milligan, V., Eds.; Society of Mining Engineering: New York, NY, USA, 1971; pp. 1–3.
117. Brummer, R.; Li, H.; Moss, A. The transition from open pit to underground mining: An unusual slope failure mechanism at Palabora. In Proceedings of the International Conference S44 Stability of Rock Slopes in Open Pit Mining and Civil Engineering, Cape Town, South Africa, 3–6 April 2006.

118. Bye, A. Stability assessment and slope design at Sandsloot open pit, South Africa. *Int. J. Rock Mech. Min. Sci.* **2001**, *38*, 449–466. [CrossRef]
119. Call, R.; Ryan, T.; Barkely, R. Geotechnical analysis for open pit mining in areas of large-scale slope instability. In Proceedings of the Innovative Mine Design for the 21st Century, Kingston, ON, Canada, 23–26 August 1993; A.A. Balkema: Rotterdam, The Netherlands; pp. 45–56.
120. Golder Associate (UK) Ltd. *Slope Stability Review East Football, Aznalcollar Mine*; Draft report to Boliden Aspira: Seville, Spain, 1995.
121. Jian, W.X.; Xu, Q.; Yang, H.F.; Wang, F.W. Mechanism and failure process of Qianjiang Ping landslide in the Three Gorges Reservoir, China. *Environ. Earth Sci.* **2014**, *72*, 2999–3013. [CrossRef]
122. Kayesa, G. Prediction of Slope Failure at Letlhakane Mine with the Geomos Slope Monitoring System. In *International Symposium on Stability of Rock Slopes in Open Pit Mining and Civil Engineering*; The South African Institute of Mining and Metallurgy: Cape Town, South Africa, 2006.
123. Martin, D.C. Deformation of Open Pit Mines Slopes by Deep Seated Toppling. *Int. J. Surf. Min. Reclam.* **1990**, *4*, 153–164. [CrossRef]
124. Martin, D.C.; Mehr, E.F. Assesment of Slope Deformation and Deep Seated Instability in the CAssiar Open Pit. *CIM Bull.* **1993**, *86*, 58–67.
125. Marthis, J.I.; Robertson, A.M.; Robertson, R.B. Mining Below Sea Level: Design and Performance of the South Slope Expansion at Island Copper. In Proceedings of the Innovative Mine for the 21st Century, Kingston, ON, Canada, 23–26 August 1993; Balkema: Rotterdam, The Netherlands, 1993; pp. 57–65.
126. Miller, V. The northeast tripp slide—A 11.7 million cubic Meter Wedge Failure at Kennecott's Nevada Mine Division. In Proceedings of the 3rd International Conference on Stability in Surface Mining, Vancouver, BC, Canada, 1 December 1982; Society of Mining Engineers, A.I.M.E.: New York, NY, USA; pp. 743–752.
127. Nilsen, B.; Hagen, R. Stabilitwtsproblemer og forslag till ukonvensjonell Sikring ved Tellness dagbrudd. *Fjellsprengningsteknikk Bergmek./Geotek.* **1990**, *23*, 1–20. (In Norwegian)
128. Newcomen, H.; Martin, D. Geotechnical Assessment of the Southeast wall slope failre at Highmont Mine, British Columbia. *CIM Bull.* **1988**, *81*, 71–76.
129. Rapiman, M. Slope stability and rock mechanics analyses Chiquicamata mine, Codelco, Chile. In Proceedings of the Innovative Mine Design for the 21st Century, Kingston, ON, Canada, 23–26 August 1993; A.A. Balkema: Rotterdam, The Netherlands, 1993; pp. 35–44.
130. Pritchard, M.; Savigny, K. Numerical mobility of toppling. *Can. Geotech. J.* **1990**, *27*, 823–834. [CrossRef]
131. Reid, G.; Stewart, D. A large-Scale Toppling Failure at Afton. In Proceedings of the International Symposium on Geotechnical Stability in Surface Mining, Calgary, AL, Canada, 6–7 November 1986; A.A Balkema: Rotterdam, The Netherlands, 1996; pp. 215–223.
132. Ren, S.-L.; He, M.-C.; Pang, S.-H.; Li, M.-N.; Xu, H.-T. Stability analysis of open-pit gold mine slopes and optimization of mining scheme in Inner Mongolia, China. *J. Mt. Sci.* **2020**, *17*, 2997–3011. [CrossRef]
133. Stacey, P.F. Pit Slope Designs for the 21st Century. In Proceedings of the Innovative Mine Design for the 21st Century, Kingston, ON, Canada, 23–26 August 1993; A.A. Balkema: Rotterdam, The Netherlands, 1993; pp. 3–11.
134. Zavodni, Z.M.; McCarter, M.K. Main Hill Slide Zone, Utah Copper Division. In *The 17th US Symposium on Rock Mechanics (USRMS)*; Monograph on Rock Mechanics Applications in Mining, AIME: New York, NY, USA, 1977; pp. 84–92.
135. Niroumand, H.; Kassim, K.A.; Ghafooripour, A.; Nazir, R.; Far, S.Y.Z. Investigation of Slope Failures in Soil Mechanics. *Electron. J. Geotech. Eng.* **2012**, *17*, 2703–2719.
136. Roux, M.; Terbrugge, M.; Badenhorst, M. Slope management at Navachab Gold Mine, Namibia. In Proceedings of the South Africa Institute of Mining and Metallurgy: International Symposium on Stability of Rock Slopes in Open Pit Mining and Civil Engineering, Cape Town, South Africa, 12–14 October 2015.
137. Singh, P.; Kainthola, A.; Prasad, S.; Singh, T. Protection Measures on the Failed Cut-slope along the Free Expressway, Chembur, Mumbai, India. *J. Geol. Soc. India* **2015**, *86*, 687–695. [CrossRef]
138. Normaniza, O.; Faisal, H.; Barakbah, S. Engineering properties of Leucaena leucocephala for prevention of slope failure. *Ecol. Eng.* **2008**, *31*, 215–221. [CrossRef]
139. Karam, K.; He, M.; Sousa, L. Slope stability risk management in open pit mines. In Proceedings of the 7th GiT4NDM and 5th EOGL International Conference, Abu Dhabi, United Arab Emirates, 14–16 March 2015; UAE University: Al-Ain, United Arab Emirates, 2015.
140. Brawner, C. Control of groundwater in surface mining. *Int. J. Mine Water* **1982**, *1*, 1–16. [CrossRef]
141. Kecojevic, V.; Mrugala, M.; Simic, R.; Stankovic, R. A computer-aided system for design of drainage facilities in surface mining. *Mine Water Environ.* **2003**, *22*, 149–154. [CrossRef]
142. Cawood, F.; Stacey, T. Survey and geotechnical slope monitoring considerations. *J. S. Afr. Inst. Min. Metall.* **2006**, *106*, 495–502.
143. Simonetta, C.; Alberto, B.; Corrado, P.; Stefano, F.; Giovanni, S.; Giulia, T. Reinforcement works for the slope stabilization: Standard and new approaches for the use of micropiles and anchors. In Proceedings of the 11th International Workshop on Micropiles, ISM, Milano, Italy, 3–6 October 2012.
144. Hua, Z.; Yang, L.; Qiang, C. Numerical Simulation of Reinforcement for Rock Slope with Rock Bolt (Anchor Cable) Frame Beam. *J. Highw. Transp. Res. Dev.* **2008**, *3*, 65–71.

145. Sazzad, M.; Hie, A.; Hossain, M.S. Stability analysis of reinforcement slope using FEM. *Int. J. Adv. Struct. Geotech. Eng.* **2016**, *5*, 83–88.
146. Bui, X.-N.; Nguyen, H.; Choi, Y.; Nguyen-Thoi, T.; Zhou, J.; Dou, J. Prediction of slope failure in open-pit mines using a novel hybrid artificial intelligence model based on decision tree and evolution algorithm. *Sci. Rep.* **2020**, *10*, 1–17. [CrossRef] [PubMed]
147. Luo, Z.; Bui, X.N.; Nguyen, H. A novel artificial intelligence technique for analyzing slope stability using PSO-CA model. *Eng. Comput.* **2021**, *37*, 533–544. [CrossRef]
148. Kothari, U.C.; Momayez, M. Machine Learning: A Novel Approach to Predicting Slope Instabilities. *Int. J. Geophys.* **2018**, 1–10. [CrossRef]
149. Chebrolu, A.; Sasmal, S.K.; Behera, R.N.; Das, S.K. Prediction of Factor of Safety for Slope Stability Using Advanced Artificial Intelligence Techniques. In *Advances in Intelligent Systems and Computing*; Springer: Singapore, 2019; pp. 173–181.
150. Du, S.; Feng, G.; Wang, J.; Feng, S.; Malekian, R.; Li, Z. A New Machine-Learning Prediction Model for Slope Deformation of an Open-Pit Mine: An Evaluation of Field Data. *Energies* **2019**, *12*, 1288. [CrossRef]
151. Ferentinou, M.; Fakir, M. Integrating rock engineering systems device and artificial neural networks to predict stability conditions in an open pit. *Eng. Geol.* **2018**, *246*, 293–309. [CrossRef]

mining MDPI

Article

Time-Dependent Model for Brittle Rocks Considering the Long-Term Strength Determined from Lab Data

James Innocente [1], Chrysothemis Paraskevopoulou [2,*] and Mark S. Diederichs [1]

[1] Department of Geological Sciences and Geological Engineering, Queen's University, Kingston, ON K7L 3N6, Canada; j.innocente@queensu.ca (J.I.); diederim@queensu.ca (M.S.D.)
[2] School of Earth and Environment, University of Leeds, Leeds LS2 9JT, UK
* Correspondence: c.paraskevopoulou@leeds.ac.uk

Abstract: The excavation of tunnels in brittle rocks with high in-situ strengths under large deviatoric stresses has been shown to exhibit brittle failure at the periphery of tunnels parallel to the maximum in-situ stress. This failure can either occur instantaneously or after several hours due to the strength degradation that is implicitly and indirectly considered in typical brittle constitutive models. While these models are powerful tools for engineering analyses, they cannot predict the time at which brittle rupture occurs, but rather, they show a possible failure pattern occurring instantaneously. In this paper, a model referred to as the long-term strength (LTS) model is introduced and implemented into FLAC2D. The model is built as a modified version of the CVISC model, introduced by Itasca, by adding a strength decay function. This function is developed from lab-scale time-to-failure (TTF) data. The LTS model is verified against its corresponding analytical solution using a constant stress creep lab test and implemented into a tunnel-scale model using the geometry, stress, and geologic conditions from the Atomic Energy of Canada Limited Underground Research Laboratory (AECL URL). The results of the LTS tunnel model are then compared to an identical model using the Cohesion Weakening Friction Strengthening (CWFS) approach.

Keywords: long-term strength and time-to-failure; time-dependency and creep; non-Newtonian viscosity

1. Introduction

The ongoing development and calibration of models for rock strength and deformation around tunnel peripheries is important in rock mechanics. The need for constitutive models in the field of rock mechanics allows engineers and scientists to obtain estimates of yield zones, displacements, and ground settlements so that support requirements and excavation methodologies can be optimized [1,2]. Classical rock strength models, such as the Mohr–Coulomb criterion [3] and the Hoek–Brown criterion [4–6] have been developed for use in moderately-jointed rock masses where failure is the result of block rotation, or where 30 < GSI < 65 [7]. In more massive rock masses under low-to-moderate confinement, failure becomes the result of extensile processes rather than shear based [8,9] which the classical failure criteria do not consider.

Early attempts to capture brittle behaviour in rock masses have used an iterative elastic approach as shown in [10,11]. It has been shown that when the intact strength of rock masses initially has a near-zero frictional strength, the failure behaviour at the tunnel periphery matches those observed in the AECL URL [11]. The mechanics of near-zero frictional strength show that when intact, the strength that is controlled by cohesion and friction is only mobilized when rupture occurs. Following this work, the CWFS approach was introduced by [12] and the DISL approach by [8].

The strength of brittle rocks and rock masses is also considered time-dependent, e.g., [8,13–18] where brittle failure at the excavation scale can manifest immediately or after some time due to subdued crack growth and interaction. The lower limit at which these cracks can develop and grow with time is the crack-initiation stress threshold [8]. The authors of [19] compiled several long-term strength tests conducted on brittle rocks from published sources and showed that as the time-to-failure increases, the lower limit of stress at which the rocks fail approaches the crack-initiation threshold.

The CWFS and DISL methods have been shown to capture the brittle behaviour of rocks at the excavation scale but fail to capture the time-to-failure and time-dependent deformations in these models which can be an important aspect to consider in an engineering design. An alternative approach to modelling brittle failure with time-dependent deformations based on time-to-failure lab tests is proposed and explored, and the subsequent results are validated against an equivalent CWFS analysis. The primary aim of this paper is to develop a time-dependent model, the long-term strength (LTS) model, that can adequately capture creep and strength degradation leading to progressive failure and rupture in brittle rocks. This paper provides an overview of the existing CWFS model and its applicability when modelling underground excavations in brittle rock. The CWFS model is used as a verification tool for the proposed LTS model to show that both the magnitude and the geometry of failure around a circular tunnel in brittle rock are equivalent.

2. Background

This section provides an overview of the different types of time-dependent behaviours observed at the lab or at the excavation scale as well as the failure modes and mechanisms of brittle rocks around and away from tunnel peripheries.

2.1. Failure in Brittle Rocks

The proper classification of rock masses and their associated in-situ strength at the excavation scale has been a significant research topic for many researchers, e.g., [4,5,7,8,20–28]. The use of engineering design equations, such as the GSI system, to determine rock strength parameters were developed and calibrated for use in moderately blocky rock masses (30 < GSI < 65), wherein the failure process is associated to block rotation. The author of [7] demonstrated that for more intact rocks (GSI > 65), the GSI equations are not valid because blocks cannot form without failure through intact rock first. Such failure typically occurs due to the formation of axial cracks parallel to the direction of the maximum applied stress, which is related to the tensile strength of the rock [7,29].

In recognition that classical failure criteria do not consider the effect of axial splitting, but rather shear fracturing, other approaches to modelling brittle failure, namely in continuum models, have been developed. Such models include the cohesion weakening, the friction strengthening (CWFS) model [11], and the damage initiation spalling limit (DISL) model [7]. These models are based on the respective crack-initiation (CI) and crack-damage (CD) thresholds as defined by [25]. When in-situ stresses are above the CI threshold, new fractures initiate and propagate with time, ultimately controlling the long-term strength (LTS) of the rock in low-to-moderate confinement. At higher confinement levels, the initiation of fractures becomes inhibited, changing the failure modes from tensile to shear rupturing [30,31], as shown in Figure 1.

At low-to-moderate confinements, the failure behaviour as governed by the DISL and CWFS approaches correspond well with empirical observations made in massive, brittle rock with stresses around the periphery of excavations at or exceeding the CI threshold, as performed in the following studies [7,8,32,33], among which a review of the CI and CD thresholds in various rocks found that the average CI to UCS ratio is between 0.4 and 0.55 whereas the average CD to UCS ratio is between 0.75 and 0.9. When rocks in low confinement are loaded to the CI threshold, they will experience continued fracture growth with time, leading to failure, whereas if they are loaded to the CD threshold, they will experience rapid crack growth and interaction, leading to sudden rupture (spalling), e.g., [22,34–36].

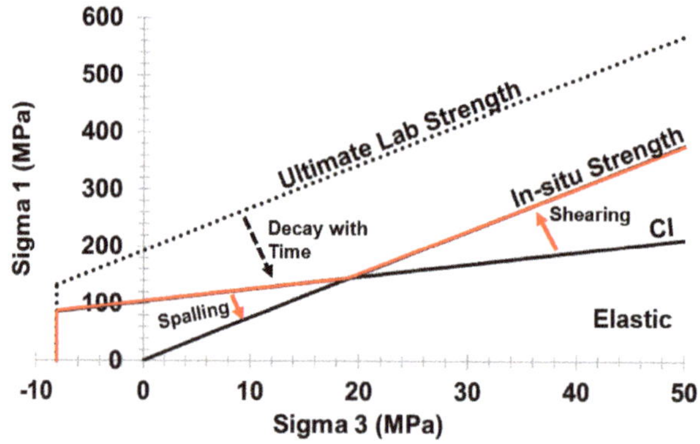

Figure 1. Strength envelope of the LdB granite at the AECL URL (CWFS) used to describe in−situ strength and development of brittle failure. Modified from [8,25].

The author of [8] provides an outline of which conditions using the DISL and CWFS approaches are most appropriate based on the ratio of UCS to tensile strength and geological strength index (GSI) from [24] shown in Table 1. The author of [37] later provides a set of guidelines for determining equivalent CWFS parameters based on the commonly used Mohr–Coulomb criterion. Using the strength data for the LdB granite at the AECL URL provided in [37] modified from [8], a typical CWFS analysis can be used as a baseline for comparison to the LTS model.

Table 1. Constitutive model selection based on rock strength and GSI, modified from [8]. Note that UCS and T indicate intact unconfined compressive strength and tensile strength, respectively.

Strength Ratio	GSI < 55	GSI = 55–65	GSI = 65–80	GSI > 80
UCS/T < 9	GSI	GSI	GSI	GSI
UCS/T = 9–15	GSI	GSI	GSI	GSI or CWFS *
UCS/T = 15–20	GSI	GSI or CWFS *	CWFS or GSI *	CWFS
UCS/T > 20	GSI	GSI or CWFS *	CWFS	CWFS

* indicates most appropriate analysis first.

2.2. Creep in Rock Mechanics

Creep in rock mechanics is defined as the accumulation of shear strain under constant stress without change in volume. Creep behaviour in solids has long been studied since the early 20th century by many researchers, e.g., [13,17,38–49]. Among the early experimental studies on creep [39,50] were performed on steel while [13] was the first researcher to apply the study of creep on geomaterials, namely on talc, shale, and crystals of halite and calcite at various levels of confinement. In [38], the author introduced the idea of three distinct stages of creep while studying creep in metals, as shown in Figure 2.

When a load is applied to a creeping material, its instantaneous deviatoric and volumetric behaviour is described by Hooke's Law [51]. After time is considered, the accumulated deviatoric strains increase at a decreasing rate (primary stage creep). If the load is held constant thereafter, the deviatoric strains increase but at a constant rate (secondary stage creep), after which the material may or may not enter the tertiary stage (yield). In theory, the volumetric strains do not change throughout the creep process; however, this assumption is only valid if the deviatoric strains develop from true creep processes such as solid diffusion, dislocation creep, or solution transfer, e.g., [13,41,52,53]. In practice, strain accumulation due to creep processes are only observed at the relatively short timescale in rock salt, potash, steel, and other ductile materials, e.g., [39,43,50]; however, true creep processes may also be observed in strong, brittle rocks given appropriate environmental conditions and longer time spans.

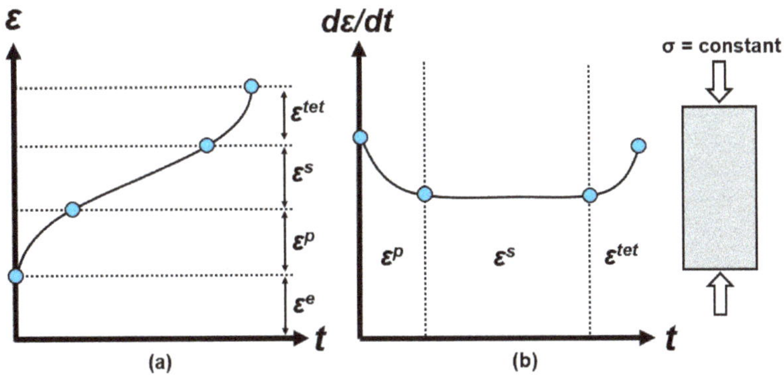

Figure 2. General creep curve in axial strain-time space for a specimen of rock under an applied constant load showing (**a**) the three stages of creep (primary, secondary, and tertiary) and (**b**) associated strain-rate time curve. The strain-rate curve is also referred to as the "bathtub" curve. Note that superscripts *e*, *p*, *s*, and *tet* denote elastic, primary, secondary, and tertiary, respectively.

At the tunnel scale, creep is often observed as the decrease in tunnel radius with time, e.g., [49,54]. This behaviour can also be associated with squeezing and swelling, which are attributed to weak and soft rocks [55,56]. Squeezing is defined as the advance of rock into a tunnel without a perceptible volume change due to the presence of micaceous minerals with low swelling capacity. Swelling is defined as the expansion of rock limited to rocks which contain clay minerals such as montmorillonite or other minerals with high swelling capacity [57].

Rheological Models

The term "rheology" refers to the branch of study related to the flow of liquids and solids under an applied force in which they deform plastically rather than elastically [58]. Rheological creep models are models that have been built up from simple mechanical analogues such as springs (Hookean elements), dashpots (Newtonian elements), and plastic sliders (St. Venant elements) as described by some constitutive model, typically the Mohr–Coulomb criterion. These elements can then be combined in series or parallel in many ways to describe the strain-time behaviour of rocks at the lab scale or in-situ. The author of [59] provides various examples of ways in which these elements can be arranged to describe different observed behaviours. The Hookean element is described by Hooke's Law [51] in which the displacement of a spring is linearly proportional to the stress acting on the spring and the stiffness of the spring.

The time-dependent aspect of rock deformation in rheology is represented by the Newtonian element which follows Newton's law of viscosity [60]. Newton's law of viscosity states that a material or fluid under applied constant stress will exhibit a constant rate of deformation with time. The material's resistance to this deformation is referred to as its viscosity, which is a material property and does not change with stress, time, or accumulated deformation. Conversely, non-Newtonian fluids are materials that do not obey Newton's law of viscosity because their respective viscosity is not constant at either given stress, strain-rate, or deformation level. One such material is Bingham plastic which behaves as a solid at low stress (does not flow) but flows as a viscous fluid at high stresses [61]. The Newtonian element can adequately capture the time-dependent aspect of rock deformation; however, it cannot capture the instantaneous response like the Hookean element. In addition, the mechanics of the viscous element allow strains accumulated with time to be fully recovered with reversal in boundary conditions regardless of the magnitude of accumulated strains, meaning the element lacks plasticity as well. A rheological model is a model that incorporates the Hookean, Newtonian, and St. Venant elements in some combination. These models can be visco-elastic, visco-elastic-plastic, or elasto-visco-plastic. Table 2 highlights some common rheological models used in the study and the modelling of creeping materials, e.g., [49,59,62–64].

Table 2. Visco-elastic-plastic rheological models with their associated mechanical analogues; analytical solutions; and stress, strain–time behaviour, modified from [18]. Note that G, K, G_K, η_K, and η_M denote the shear modulus, bulk modulus, Kelvin viscosity, and Maxwell viscosity, respectively, and p, q, t, and ε denote mean stress, deviator stress, time, and strain, respectively.

Model	Mechanical Analogue	Stress, Strain-Time Behaviour
Maxwell	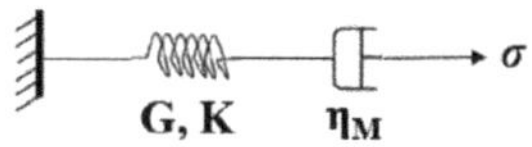 $\varepsilon = \frac{p}{3K} + \frac{q}{3G} + \frac{qt}{3\eta_M}$	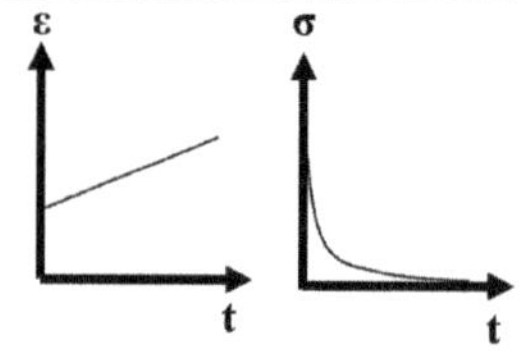
Kelvin -Voigt	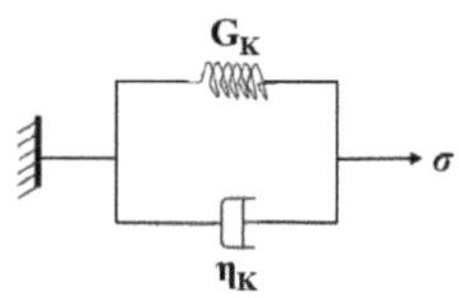 $\varepsilon = \frac{q}{3G_K}\left[1 - \exp\left(-\frac{G_K}{\eta_K}t\right)\right]$	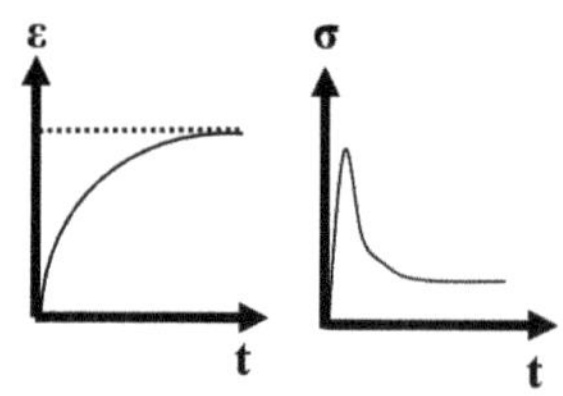
Zener	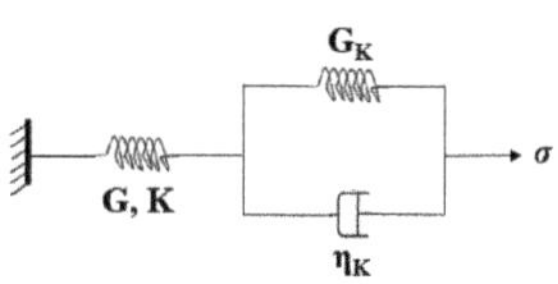 $\varepsilon = \frac{p}{3K} + \frac{q}{3G} + \frac{q}{3G_K}\left[1 - \exp\left(-\frac{G_K}{\eta_K}t\right)\right]$	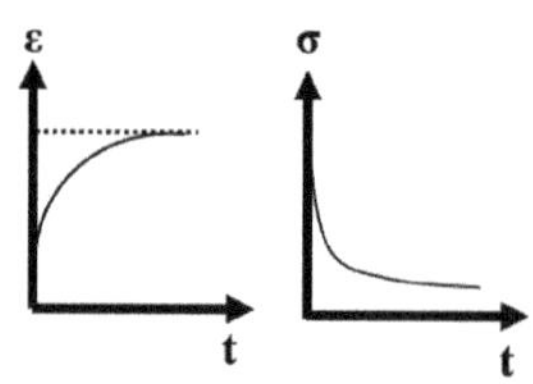
Burgers	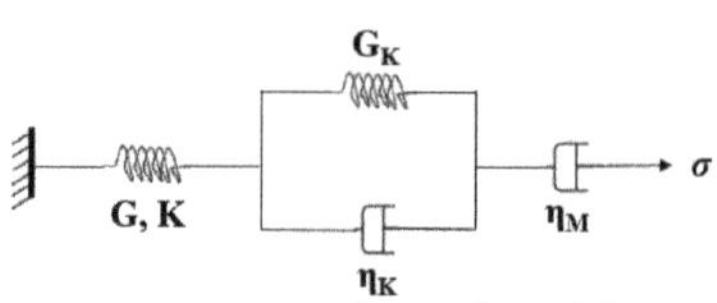 $\varepsilon = \frac{p}{3K} + \frac{q}{3G} + \frac{q}{3G_K}\left[1 - \exp\left(-\frac{G_K}{\eta_K}t\right)\right] + \frac{qt}{3\eta_M}$	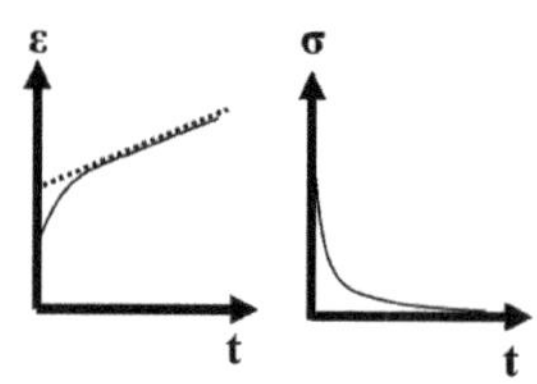
CVISC	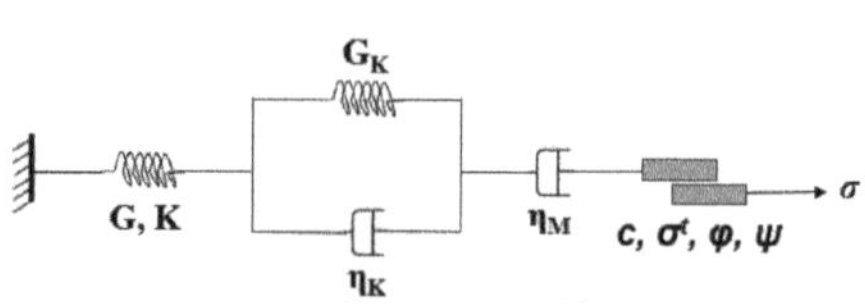 $\varepsilon = \frac{p}{3K} + \frac{q}{3G} + \frac{q}{3G_K}\left[1 - \exp\left(-\frac{G_K}{\eta_K}t\right)\right] + \frac{qt}{3\eta_M}$	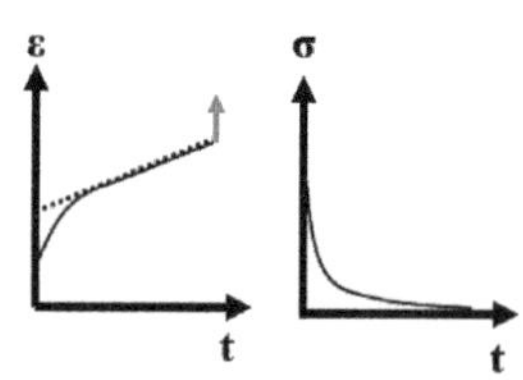
Simplified Cividini and Gioda	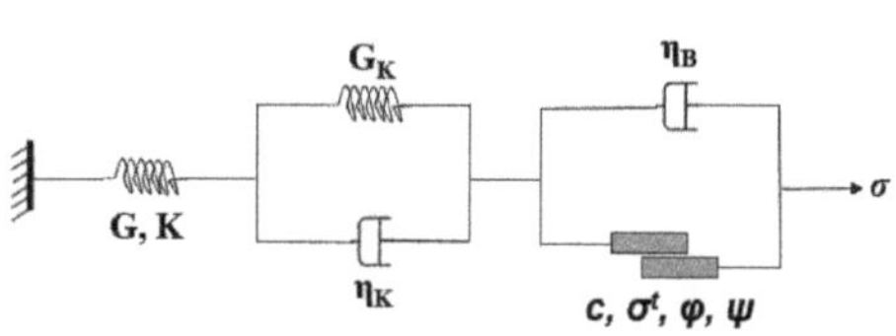	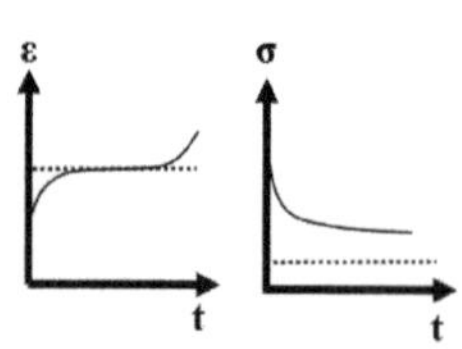

The Maxwell, Kelvin–Voigt, Zener, and Burgers models are referred to as visco-elastic models. The CVISC model is a visco-elastic-plastic model introduced by [65] and the simplified Cividini and Gioda is an elastic-visco-plastic model introduced by [66]. It is important to note the distinction between 'elasto-plastic' and 'visco-plastic', denoting no connection between creep and plasticity and a direct connection, respectively. The connection between creep and plasticity is complex and often difficult to determine. The Burgers and CVISC models are preferable for practical applications [67]; however, there are limitations to the model as described in [68–70] and in this paper.

Other models to examine creep exist, such as empirical or phenomenological models, as well as general theories [64]. Empirical models are models built purely from curve-fitting of lab data from constant-stress or stress-relaxation tests and are generally given as closed form or differential solutions. General theories are the most advanced aspects of numerical modelling and are generally very robust in their use case. Perzyna's overstress theory is one such example of a general theory [71]. The further analysis and application of empirical models and general theories is out of the scope of this paper.

2.3. Interpreting Time-to-Failure Lab Results in Brittle Rocks

The most common method for determining the long-term strength of brittle rocks is by conducting a series of uniaxial compressive strength (UCS) tests to determine the average strength of the rock. This suite of tests should follow the methodology as outlined in [72–74]. The long-term strength tests comprise loading a standard cylinder of core to some stress that is less than its UCS but more than its crack initiation (CI) threshold. Once the desired stress is reached, it is held and the time to failure (TTF) is recorded and compared to the applied driving stress ratio (**DSR**) which is historically presented in Equation (1) as:

$$\mathbf{DSR}_{\sigma_3=0} = \frac{\sigma_1}{\mathbf{UCS}} \tag{1}$$

where:

- σ_1 is the applied axial stress.

Ref. [19] reviews the current state of practice for analysing the long-term strength of brittle rocks and proposes that the model shown in Figure 3 be used to calculate the TTF for igneous rocks when under uniaxial stress conditions. Based on an earlier formulation from [75], The author of [19] developed a set of Equations (2) and (3)

$$t_f = \left(-\frac{C - \ln(100DSR)}{A}\right)^{-\frac{1}{B}} \text{ for } t_f > 10 \text{ s and DSR} > \frac{\exp(C)}{100} \tag{2}$$

$$C = \ln\left(\frac{CI}{UCS} \times 100\%\right) \tag{3}$$

where:

- t_f is the time-to-failure;
- C is an asymptote control parameter;
- A and B are curve-fitting constants that are determined empirically.

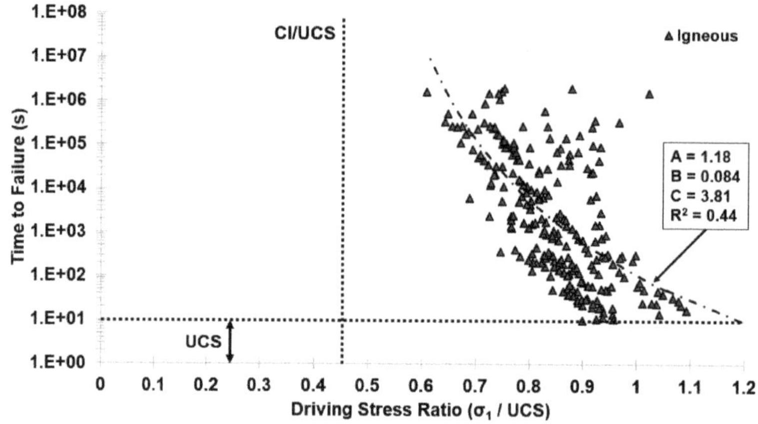

Figure 3. Long-term strength data for various rock types and their respective exponential model fits as presented in [19]. The average CI threshold [76] is added to represent the lower bound of long-term strength.

Note that time-to-failure is a function of intact material properties and does not change with time; therefore, it can be considered as a material property that is a function of applied stress. To be able to use the TTF equations as shown in Equations (2) and (3), the effect of confinement on strength must be considered; however, very few long-term strength tests have been conducted under confined conditions apart from the suite of tests conducted by [77,78] in which it is shown that the effect of confinement clearly affects the absolute TTF under constant applied stress. The TTF as shown in Equations (2) and (3) is a function of unconfined conditions only. To account for this, the DSR must be modified as **DSR*** shown by [79] in Equation (4):

$$\mathbf{DSR}^* = \frac{\sigma_1 - \sigma_3}{\sigma_1^P - \sigma_3} = \frac{q}{\sigma_1^P - \sigma_3} \tag{4}$$

where:

- σ_3 is the confinement, or minimum principal stress;
- σ_1^P is the peak strength of the rock at a given confinement level;
- q is the deviator stress.
- **DSR*** is the modified DSR

The justification for the modified DSR equation is shown in Figure 4. In unconfined conditions, the DSR equation simplifies to that as shown in Equation (1). It is assumed that the same DSR under both unconfined and confined conditions will lead to the same TTF as shown in Equation (2). Equation (3) does not need further modification as the ratio of CI to UCS is also assumed constant with relatively small increases in confinement; however, this assumption is only valid within the spalling limit of the material, which is within the range of $10 \leq \sigma_1/\sigma_3 \leq 20$ [80,81]. It should also be noted that the DSR and TTF formulations are for two-dimensional problems, but they can be modified for use in three-dimensional problems.

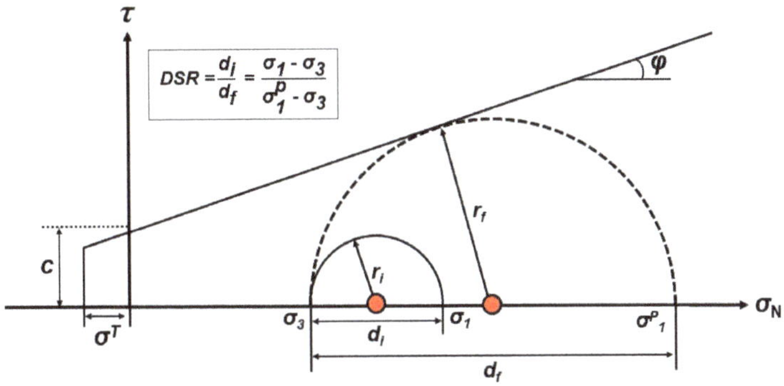

Figure 4. Schematic of variables used within the DSR equation for confined conditions using the Mohr–Coulomb criterion, modified from [79].

3. Proposed Long-Term Strength (LTS) Model

This section describes the long-term strength (LTS) model that explicitly considers time in two dimensions. The proposed LTS model is built on the existing CVISC creep model by modifying the attached Mohr–Coulomb plastic slider, as shown in Table 2. The modification includes degrading the strength of the criterion as per the TTF equations in Section 2.3. The process for degrading strength with time is outlined in this Section.

In two dimensions, the Mohr–Coulomb failure criterion is expressed in Equations (5) and (6) as:

$$\sigma_1^P = UCS + s\sigma_3 \tag{5}$$

$$s = \frac{1 + \sin\varphi}{1 - \sin\varphi} \tag{6}$$

where:

- φ is the friction angle as shown in Figure 4.

Using Equation (5), the strength of the rock at the excavation scale can be determined throughout the FLAC2D [65] grid. With the data from Figure 3, the theoretical time-to-failure can be determined throughout the FLAC2D grid using the equations described below. A key assumption in the development of this model is that the failure resulting from stress corrosion is the result of a linear loss in cohesion with time that is a function of the in-situ stress conditions. This assumption is made based on the log-linear relation that laboratory data shows when brittle rock materials are subjected to constant load as [19] thoroughly analysed and discussed lab datasets from a range of rock materials. In the finite difference model, this can be achieved by introducing a damage variable (R), which is calculated as shown in Equation (7).

$$\frac{dR}{dt} = (1 - DSR)\left(\frac{1}{t_f}\right) \tag{7}$$

The damage is then iterated through time using the creep plugin available for FLAC2D. It should be noted that damage can be calculated directly with time; however, this limits the applicability of the equation to monotonic loading conditions only, whereas iterating the variable allows it to apply to more complex loading conditions such as those encountered in underground mines and other excavations. The damage function is then used to decrease the ultimate strength (σ_1^P) of each time-step and tensile strength (σ^T), as shown in Equations (8) and (10). The damage to ultimate strength is used to calculate the new cohesion, as per Equation (9).

$$UCS^* = R\sigma_1^P - s\sigma_3 \tag{8}$$

$$c^* = \frac{UCS^*(1 - \sin(\varphi))}{2\cos(\varphi)} \quad (9)$$

$$\sigma^{T*} = R\sigma_i^T \quad (10)$$

where:

- UCS^* is the new UCS value after weakening;
- c^* is the new cohesion after weakening;
- σ_i^T is the initial intact tensile strength;
- σ^{T*} is the new tensile strength after weakening.

Equations (5)–(10) provide the basis needed to begin verifying the model at the lab scale and applying it at the excavation scale. Figure 5 provides a schematic of the weakening behaviour as described by the LTS model and Figure 6 provides a schematic workflow for the overall strength degradation model when implemented into a finite-difference modeller such as FLAC2D.

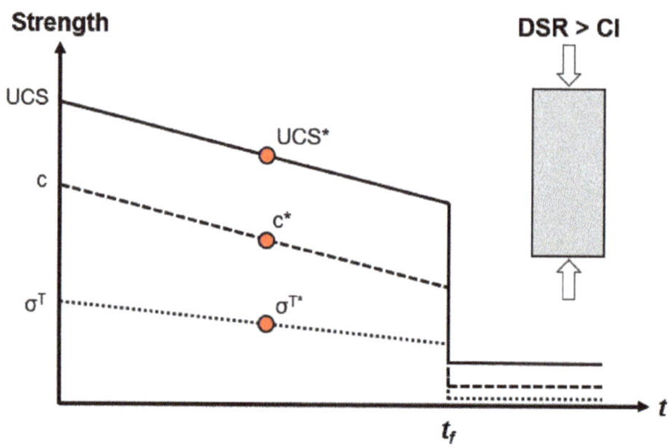

Figure 5. Schematic representation of the LTS model and the effect on the respective strength parameters when a sample is subjected to an applied constant load greater than its CI threshold. Note that friction angle has been omitted as it is assumed constant until failure. Values after tf are residual (post-yield), * denotes the change over time.

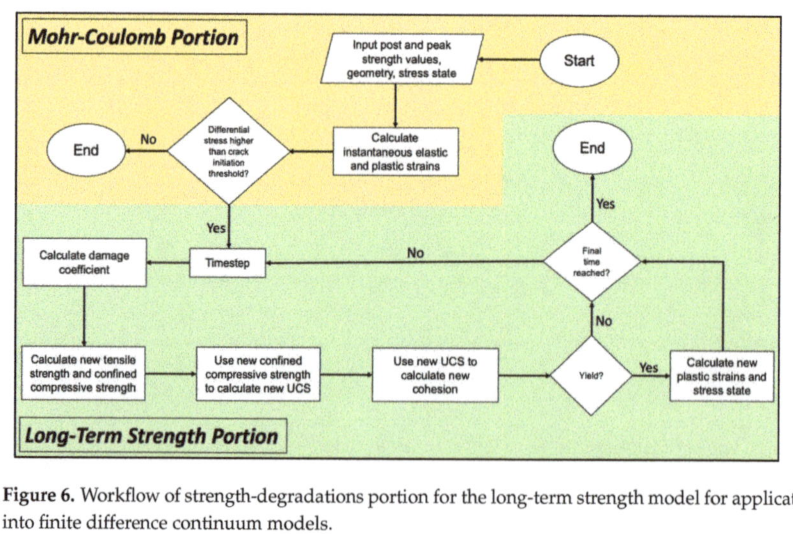

Figure 6. Workflow of strength-degradations portion for the long-term strength model for application into finite difference continuum models.

3.1. Effect on Applied Stress and Confinement on Secondary Strain-Rates and Viscosities

It has long been shown that brittle rocks do experience time-dependent strains when under applied stress, e.g., [36,48,82–89]. It is generally theorized that the strains associated to creep in brittle rocks are not due to true creep mechanics (continuum processes), but rather, crack initiation and propagation with time which are associated to brittle failure mechanisms. Continuum models (such as those created in FLAC2D) cannot explicitly capture brittle behaviour due to the discontinuum nature of brittle failure. Therefore, to capture these strains, the Burgers model for creep can be implemented in a model to account for brittle deviatoric strains (recall that Burger's creep model is deviatoric only). It is important that the time-dependent aspect of brittle failure is well understood as creep at the tunnel scale allows for stress relaxation [70].

The author of [39] studied the effect of creep in steel and showed that the secondary strain rate can be expressed as shown in Equation (11).

$$\dot{\varepsilon}^{s} = \alpha\overline{\sigma}^{\beta} \tag{11}$$

$$\overline{\sigma} = \sqrt{3J_2} = \left(0.5[(\sigma_{11} - \sigma_{22}) + (\sigma_{22} - \sigma_{33})^2 + (\sigma_{33} - \sigma_{11})^2 + 6(\sigma_{12}^2 + \sigma_{23}^2 + \sigma_{31}^2)]\right)^{0.5} \tag{12}$$

where:

- $\dot{\varepsilon}^{s}$ is the secondary strain rate;
- $\overline{\sigma}$ is the equivalent Von–Mises stress;
- J_2 is the second invariant of the deviatoric stress tensor;
- σ_{ii} and σ_{ij} are components of the Cauchy stress tensor;
- α and β are curve-fitting constants.

Conversely, the secondary strain rate as described by the Maxwell and Burgers equations is shown in Equation (13).

$$\dot{\varepsilon}^{s} = \frac{q}{3\eta_M} \tag{13}$$

From Equation (11), it is clear that with a change in deviatoric stress, the secondary strain rate changes linearly, assuming that the secondary viscosity term (η_M) is constant whereas the secondary strain rate as described by Equation (12) changes exponentially with changes in stress, as described in [70]. This means that the behaviour as described by the Power Law is for non-Newtonian fluids, whereas the behaviour as described by the rheological models is for Newtonian fluids. The key behaviour of Newtonian fluids is that the viscosity of the material does not change with stress, strain-rate, or deformation, but a non-Newtonian fluid's viscosity is not constant [60]. In practice, it is typical that one average value is used for the secondary viscosity in the Burgers/CVISC model [31,88–90]; however, as shown in Figures 7 and 8, this is only applicable when the expected stresses are monotonic and unchanging from the lab-calibrated values.

A general equation for the secondary viscosity, or Maxwell viscosity (η_M), as shown in Figure 7 can be written in Equation (14) as:

$$\eta_M = \chi\exp(\kappa q) \tag{14}$$

where:

- χ is some function of confinement;
- κ is the rate of change in secondary viscosity with changing deviator stress;
- q is the deviator stress.

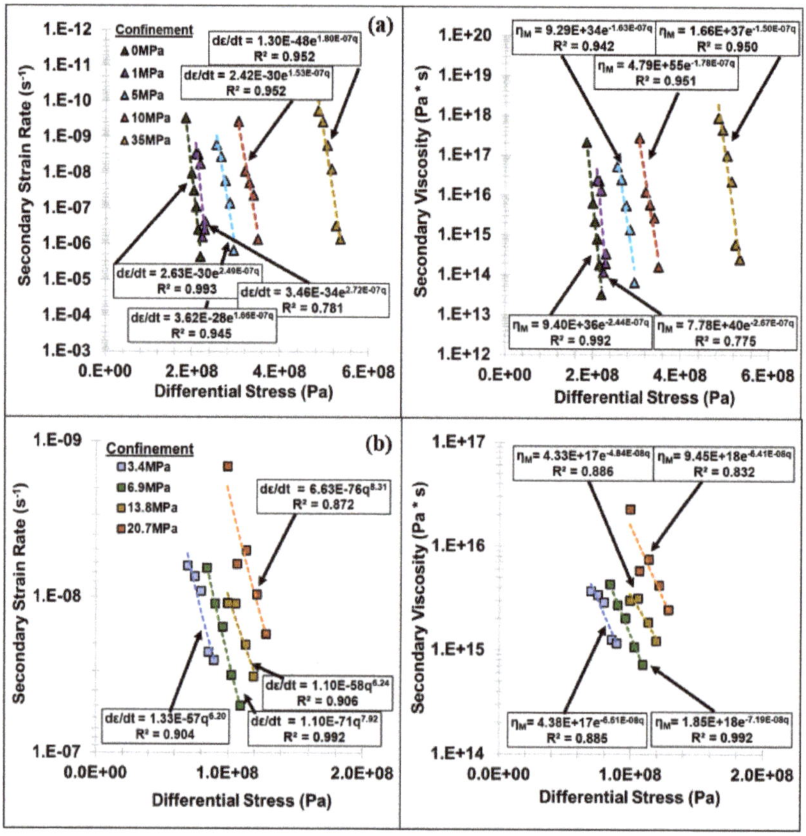

Figure 7. Secondary strain rates and associated Maxwell viscosities at varying levels of confinement for (**a**) LdB granite from [77], (**b**) Gyda sandstone from [91]. Note that η_M denotes Maxwell viscosity, q is the deviatoric stress, and $d\varepsilon/dt$ is strain rate.

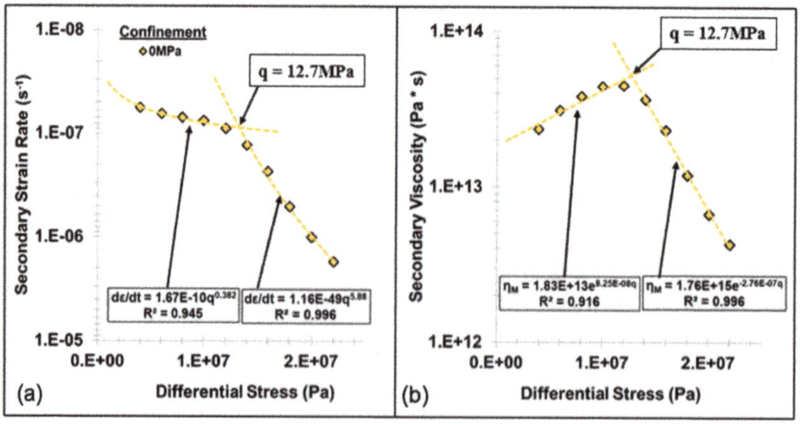

Figure 8. (**a**) Secondary strain rate and (**b**) Maxwell viscosity versus applied stress for rock salt from [92].

As shown in Figure 7, viscosity is also dependent on confinement; however, the rate of change (κ) in viscosity at any confinement level is constant. To effectively capture this behaviour of confinement dependency, the behaviour of the variable 'χ' must be determined. Figure 9 shows the change in 'χ' with confinement assuming an average value of -2.004×10^{-7} Pa^{-1} for 'κ' for LdB granite. From Figure 9, there is a clear pattern in 'χ' with confinement, with it increasing exponentially with increased confinement. Plugging in the Equation shown in Figure 9 into Equation (14), the Maxwell viscosity for LdB granite is fully expressed in Equations (15) and (16) as:

$$\eta_M = [4.28E + 34\exp(1.77E - 6(\sigma_3))]\exp(-2.004E - 7(q)) \quad (15)$$

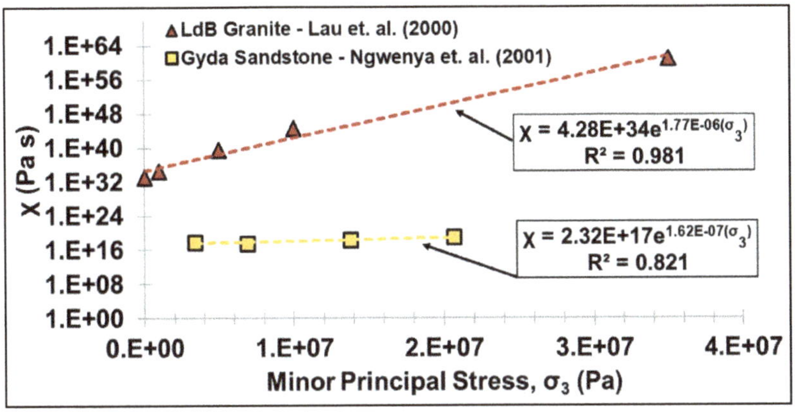

Figure 9. Plot of variable χ versus confinement for LdB granite from [77] and Gyda sandstone from [91].

Which simplifies to:

$$\eta_M = 4.28E + 34\exp[1.77E - 6(\sigma_3) - 2.004E - 7(q)] \quad (16)$$

where:

- η_M is the Maxwell (secondary) viscosity;
- q is the deviator stress $(\sigma_1 - \sigma_3)$;
- σ_3 is the minor principal stress.

The resulting viscosities for LdB granite in the AECL URL tunnel are shown in Figure 10. Note that this is an empirical approach to accounting for effects of confinement in brittle rock creep and that none of the constants in Equation (16) have any significance to real-world mechanisms. Additionally, it has been shown that a Maxwell material most likely behaves as a non-Newtonian fluid, rather than as a Newtonian fluid, and the same can likely be said for a Kelvin type material; however, this distinction is out of the scope of this analysis. In simple loading conditions, such as those shown in the following numerical models, strains are more sensitive to secondary viscosities over long periods of time rather than primary viscosities; therefore, the distinction between Newtonian and non-Newtonian viscosity for primary creep is insignificant.

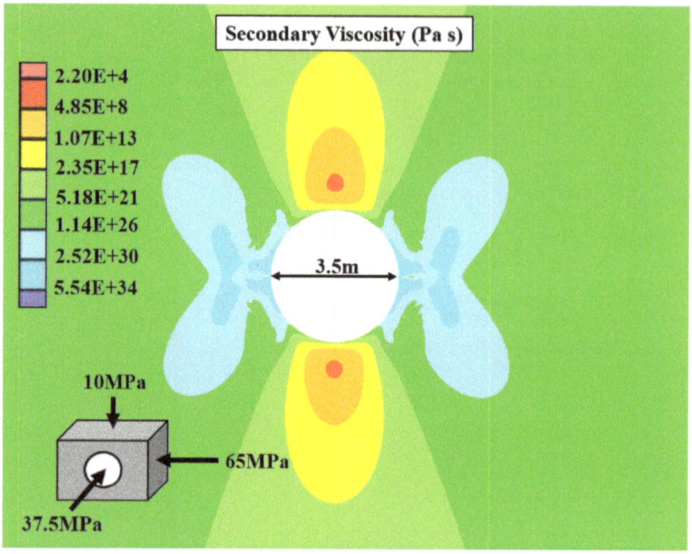

Figure 10. Instantaneous secondary viscosity contours around the 3.5 m diameter AECL URL tunnel in LdB granite.

3.2. Lab Scale Verification of Model

The long-term strength and modified CVISC model are implemented into FLAC2D [65], which is a finite-difference continuum numerical modelling software. This section will provide a lab-scale verification of the model as well as a tunnel-scale analysis using the AECL URL geometry in LdB granite. The material properties used are shown in Table 3, which are average values for LdB granite. The strength values are from [8], and the creep properties are from [19], which were calculated using lab data from [77]. The goal of the lab-scale verification model is to show that under varying confinement conditions, the proposed long-term strength model accurately captures the predicted analytical time-to-failure.

Table 3. Intact strength, long-term strength, and creep parameters for LdB granite. Peak and residual strength values calibrated from [8] and time-dependent values from [19,77]. Variables K, G, c, φ, σ^T, and η denote bulk modulus, shear modulus, cohesion, friction angle, tensile strength, and viscosity, respectively. Additionally, subscripts p, r, M, and K denote peak, residual, Maxwell, and Kelvin, respectively. Variables A, B, and C are shown in Equation (2). Note that 'E' is scientific notation (i.e., 4.08E + 14 = 4.08 × 10^{14}) and 'exp' denotes and exponential of 'e' (i.e., exp(x) = e^x).

Parameter	Value
K (GPa)	58
G (GPa)	25
c_p (MPa)	40
c_r (MPa)	0.1
φ_p (deg)	50
φ_r (deg)	22
$\sigma^T{}_p$ (MPa)	8
$\sigma^T{}_r$ (MPa)	0
η_M (Pa s)	$4.28\text{E} + 34\exp[1.77\text{E} - 6(\sigma_3) - 2.004\text{E} - 7(\text{q})]$
η_K (Pa s)	4.08E + 14
G_K (GPa)	107
A	1.18
B	0.084
C	3.81

The geometries used for both the lab-scale verification and excavation scale models are shown in Figure 11 with the FLAC grid. The grid for the lab verification is relatively coarse for computing efficiency and should not affect the TTF in any significant way. Additionally, the lab-scale model uses a DSR of 0.75 for each confinement level, which is achieved by changing the applied stress on the top and bottom of the sample. This is performed in such a way that the TTF is the same for each run and making comparisons between them is easier. The excavation scale model uses stresses modified from the AECL URL tunnel as provided in [8]. The most notable change is the change in out-of-plane stress which was changed from the published 43 MPa to the 37.5 MPa (average of in-plane stresses such that out-of-plane stress is effectively ignored). The grid in the tunnel model is radial with a very dense mesh around the excavation, becoming gradually coarser towards the model boundaries, which are 17.5 m from the tunnel centre.

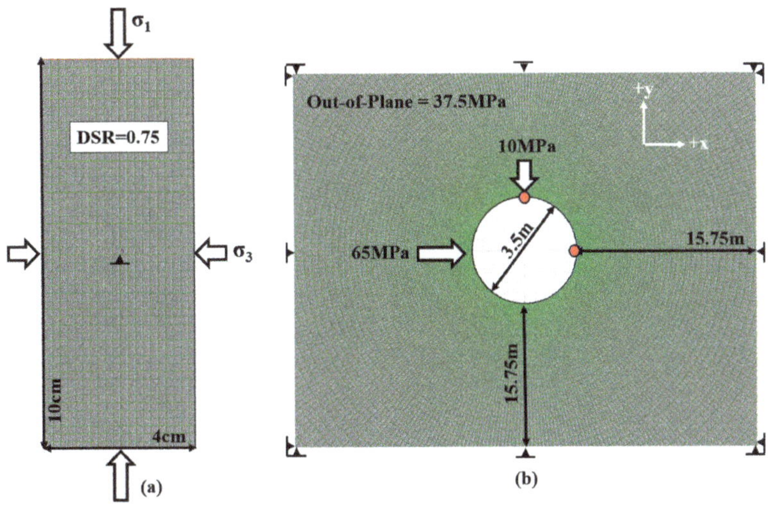

Figure 11. Model geometries from FLAC2D for (**a**) the lab-scale verification and (**b**) the tunnel-scale model modified from the AECL URL tunnel. Note that model dimensions and boundary conditions (pins) are shown. Red dots represent measurement points for displacement.

The results of the lab-scale verification can be seen in Figure 12, which plots the axial strain and cohesion of the sample versus time with the analytical TTF overlain as a dotted red line. The effective secondary viscosity is also shown and is calculated from Equation (16); however, each of the values is high enough that it is insignificant in the time span used. Failure in each of the models is represented by the marked rapid decrease in cohesion to its residual state (0.1 MPa) followed by the rapid increase in strain rate. The intact cohesion value for each confinement level at failure is shown to decrease with increasing confinement as expected. The lag between cohesion loss and strain increase can be attributed to the increase in unbalanced forces resulting from failure. From each of the plots in Figure 12, it is presented that the loss in cohesion correlates with the analytical TTF, showing that, as formulated, the long-term strength model is adequate and can be carried over to a tunnel-scale model.

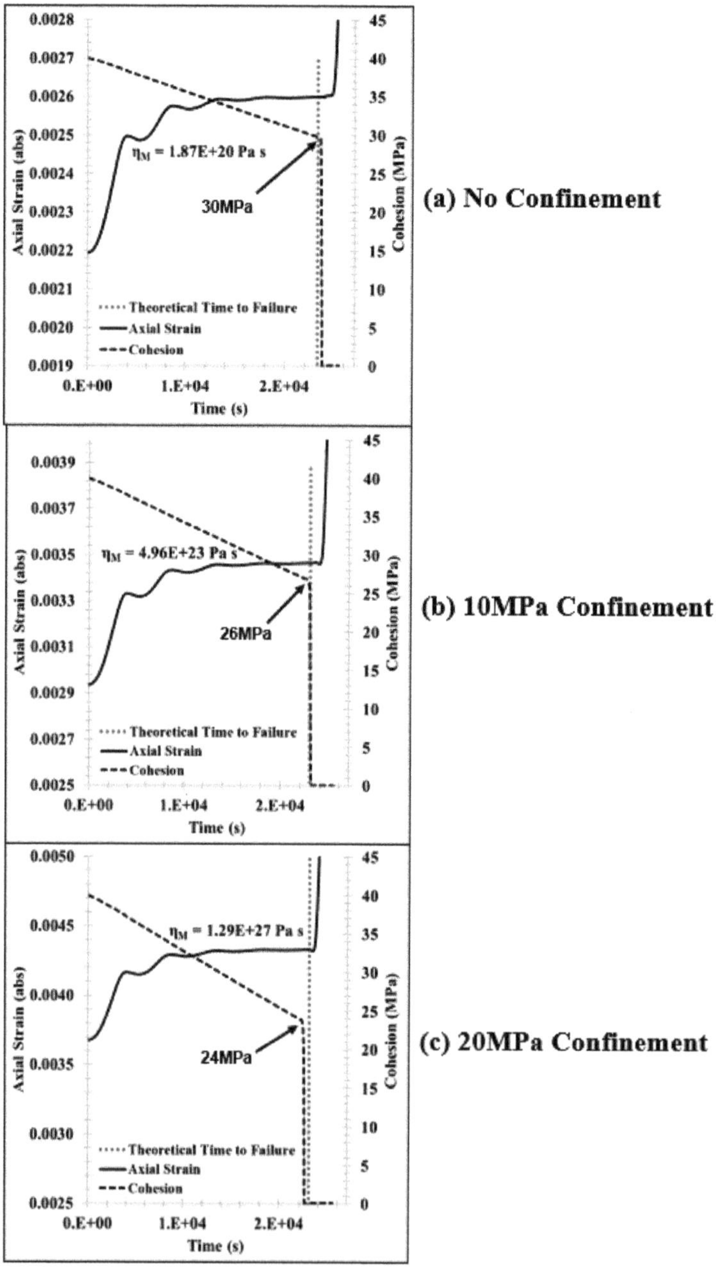

Figure 12. Axial strain and cohesion with time under a constant applied stress for (**a**) no confinement, (**b**) 10 MPa confinement and (c) 20 MPa confinement. All models are run with a DSR of 0.75 and failure is marked by the rapid decrease in cohesion with time. The cohesion value at failure is shown for each run.

3.3. Tunnel Scale Model

The tunnel-scale model for LTS is developed using the geometry shown in Figure 11. This geometry is similar to that shown in [8] for the AECL URL. In the model, the tunnel is developed "instantaneously" in a Mohr–Coulomb strain-softening medium using the values as shown in Table 3. At the tunnel scale, the goal of the LTS model is to accurately capture the typical tension cracks that form in the walls of the tunnel parallel to the minimum principal stress as well as the brittle overbreak "notch", described by [7,8,10,76], that forms in the periphery of the tunnel in the direction of the maximum applied stress. To compare and validate the applicability of the LTS model, it will be compared to a CWFS model run in the exact same conditions using the values shown in Table 4.

Table 4. Values used for the CWFS validation modified from [8]. Note that ε_c^P and ε_φ^P are plastic strain values for residual state to be reached for cohesion and friction angle, respectively.

Parameter	Value
K (GPa)	58
G (GPa)	25
c_P (MPa)	40
c_r (MPa)	0.1
φ_P (deg)	20
φ_r (deg)	50
$\sigma^T{}_P$ (MPa)	8
$\sigma^T{}_r$ (MPa)	0
ε_c^P (%)	0.3
ε_φ^P (%)	0.3

The in-situ stresses and associated strains from the CWFS model are shown in Figure 13 and the instantaneous response from the LTS model is shown in Figure 14. From comparing the results in Figures 13 and 14 (left), it can be seen that the typical Mohr–Coulomb-based analysis is not adequate for capturing the brittle overbreak that is observed in the CWFS analysis as no failure occurs instantaneously according to the LTS model. Figure 14 (right) then shows the in-situ stresses and strains from the LTS model after 7 h, the time of yield in the floor and roof of the tunnel. Here, the stresses have relaxed around the roof and the floor and redistributed accordingly, indicating rupture. The shear and volumetric strains do not, however, match the pattern as shown in the CWFS analysis and can likely be attributed to numerical noise.

To validate the results of the LTS model, the failure geometries must be compared. From the CWFS model, a typical "notch" forms in the roof and the floor to a depth of 0.75 m to 0.80 m as well as tension cracks forming in the walls. From the LTS model, the tension cracks in the wall form instantaneously, but the degree of failure in the walls increases at 7 h. The degree of wall displacement also matches the displacement as shown in the CWFS model. The failure in the roof and the floor of the LTS model also shows somewhat of a notch-type geometry to a depth of 0.70 m, which is marginally less than the depth as predicted by the CWFS model. When looking at the DSR in Figure 14, it can be seen that, at the depth of failure, deviatoric stresses are at about 70% of the strength of rock, indicating that further yield may occur with more time. The final displacements in the roof after failure in both models also match within marginal error.

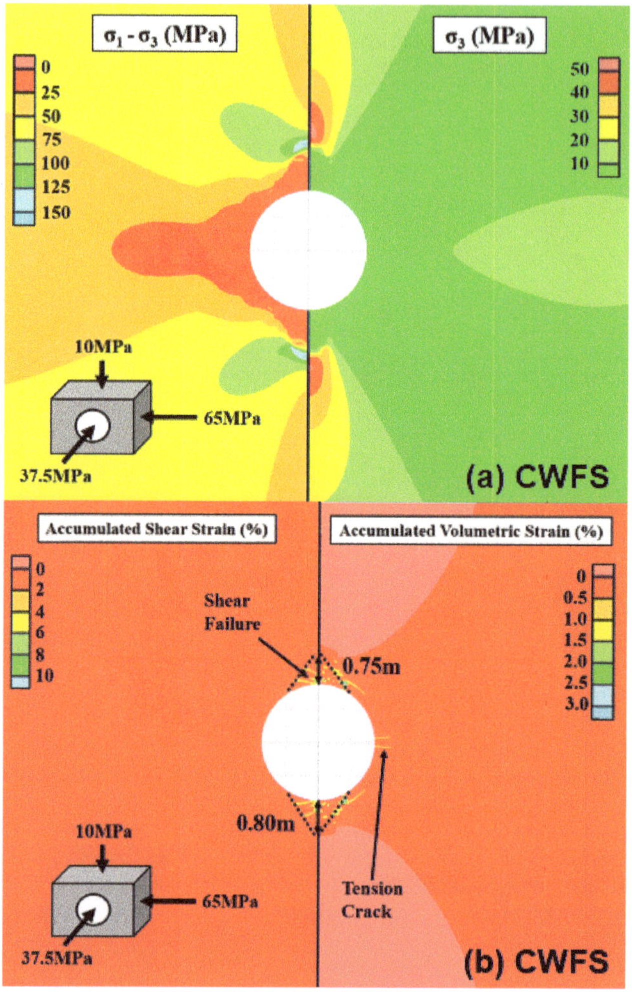

Figure 13. Instantaneous response from the CWFS analysis showing (**a**) in–situ stresses and (**b**) shear and volumetric strain with depth of yield and typical "notch"-type failure.

From Figures 13 and 14, it is clear that the failure modes between the CWFS and the LTS models differ, wherein the CWFS model, multiple shear bands develop parallel to the tunnel periphery and in the LTS model, a shear "cone" extends outwards from the tunnel periphery. The failure geometry shown by the LTS model is similar to what some researchers refer to as the "process zone" [22,23,76]. The term "process zone" refers to the small-scale buckling that occurs at the tip of the notch-type failure that gives rise to more considerable dilation, as observed in the CWFS model. Therefore, if the LTS model were to be run for even more time after the initial nucleation of the process zone, the failure geometry may begin to resemble that as shown in the CWFS model. The process zone development is described in [22] and the observed development within the AECL URL is shown in Figure 15.

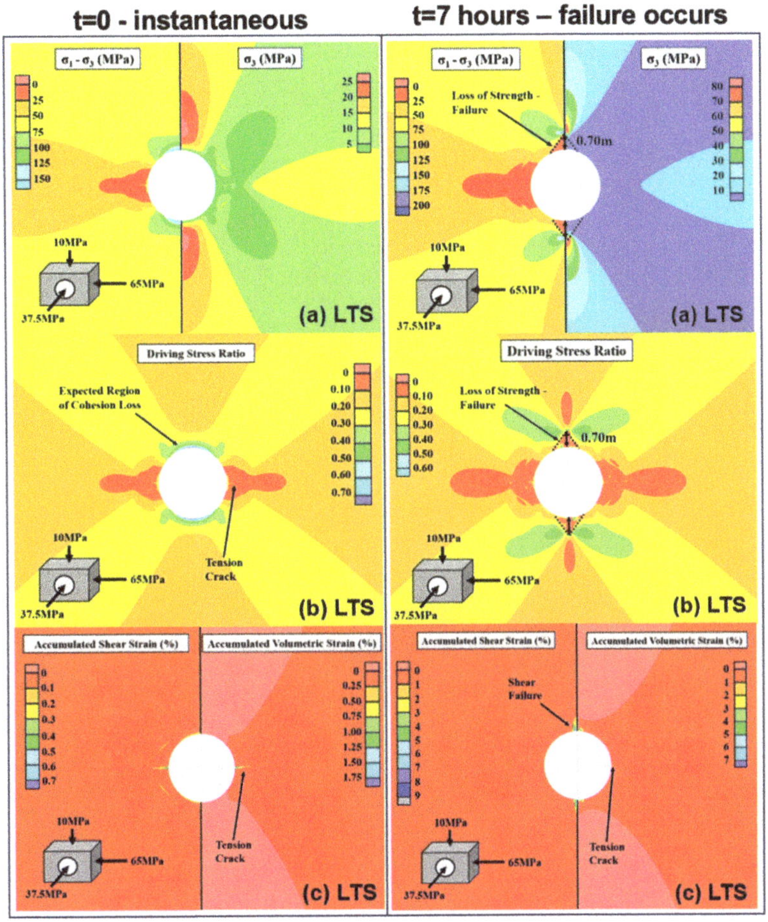

Figure 14. Instantaneous using plasticity. Left: (**a**) in–situ stresses, (**b**) driving stress ratio, and (**c**) shear and volumetric strains and after 7 h (failure time) using the LTS model. Right: (**a**) in-situ stresses, (**b**) driving stress ratio with depth of failure, and (**c**) shear and volumetric strains for the LTS model.

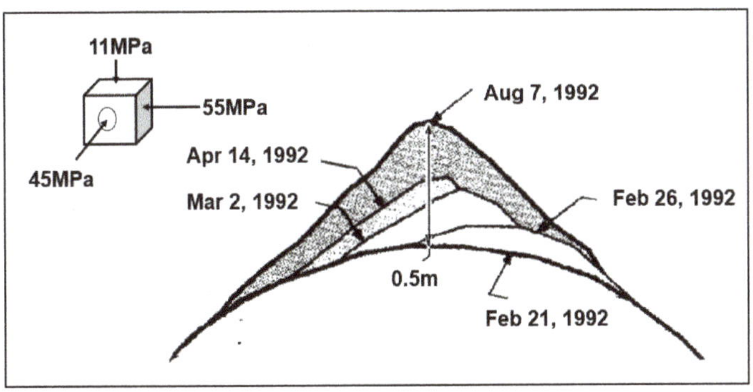

Figure 15. Observed yield profile development within the AECL URL from [22].

4. Discussion

The effect of changing deviator stress and confinement is often not considered when determining parameters for the Burgers creep model. From Figures 7 and 8, it is shown that using a single value for viscosity as shown in the Burgers model is not adequate when describing both brittle and ductile creep processes when complex loading conditions are expected. The Bailey–Norton two-component power law addresses both the exponential change in viscosity with stress as well as the transition from ductile to brittle creep at some stress. The effect of confinement can be addressed using the curve-fitting technique used in Figure 9. Using the curve-fitting technique proposed to determine secondary viscosity as a function of confinement, the viscosities throughout a tunnel model can be calculated as shown in Figure 10. The implications of using viscosities that are too low in a tunnel model, as is often the scenario when using average values from lab testing, is that tunnel convergence becomes very large compared to the observed values as well as allowing for too much stress relaxation around the tunnel periphery.

The failure process as demonstrated in Figure 13 according the LTS model assumes that up until material rupture, the rock stays intact throughout the damaging process. This contrasts with the discontinuum models demonstrated in [36,79] which showed that in discontinuum models, cracks do initiate and accumulate with time under a constant load. This highlights the advantage of using discontinuum models over continuum models when modelling brittle failure processes, which are inherently discontinuous. In contrast, [49] showed that using the Burgers creep model to model tunnel convergence along a longitudinal profile in a continuum setting in creeping ground is adequate because creep is inherently a continuum process.

The LTS model has successfully been implemented into a continuum finite-difference model to both model unconfined and heavily confined cylindrical samples of rock at the lab scale and, when modelling, the brittle overbreak encountered high deviatoric stresses at the tunnel scale. At confinement levels of 0, 10, and 20 MPa at the lab scale and DSRs of 0.75, the sample yields as denoted by a rapid increase in strain rate, similar to that of tertiary creep from Figure 2. It is important to note that the LTS model does not consider visco-plasticity unlike the simplified Cividini and Gioda model. At the tunnel scale, the LTS model must be validated to other established numerical models used to simulate brittle failure, such as the DISL and CWFS models. Using the geometry and stress conditions similar to those encountered in the AECL URL in LdB granite, both the CWFS and LTS models show similar convergence measurements but differing yield shapes after first failure as shown in Figures 13 and 14. The first failure as described by the LTS model is similar to what is referred to as the process zone, which is the preceding mechanism to spalling.

The advantage of the LTS model over other continuum-based brittle failure models is that the time it takes for yield to occur can also be calculated in addition to yield geometries. This provides further guidelines for engineering design in terms of timelines for installation support. Allowing engineers and researchers to predict TTF at the excavation scale can lead to project and support optimizations in brittle rocks, reduction of uncertainties, e.g., [93–98] in the design while avoiding tunnel failures [99] and reduction or even overcoming of cost overruns, e.g., [100–103]. This model can be modified to calculate stresses and strains in three dimensions as well as for analysis along a longitudinal displacement profile for further tunnelling optimizations.

In addition, the LTS model is relatively simple to implement and obtain parameters for, whereas more complex visco-plastic models require multiple inputs and complex lab testing to obtain parameters for and become less feasible for an engineering analysis. The workflow of the LTS model is shown in Figure 6 and the associated rheological analogue is shown in Figure 16, highlighting the necessary modifications to the existing CVISC model.

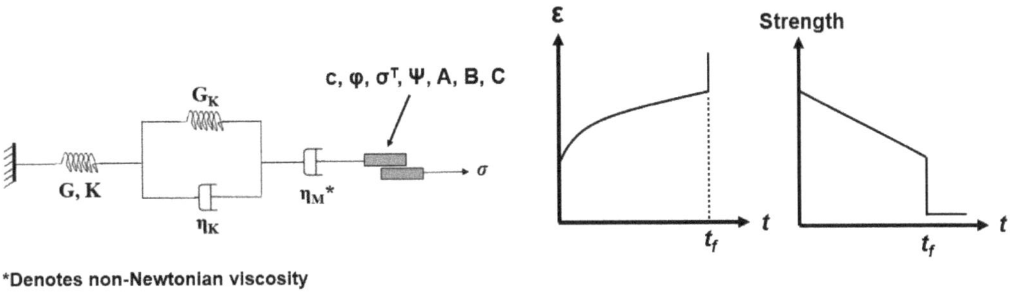

Figure 16. Conceptual schematic of the LTS model.

5. Conclusions

In this paper, a proposed long-term strength (LTS) model is developed for use in continuum models based on time-to-failure data for brittle rocks exploration at the lab scale. The advantage of this model over the conventional CWFS and DISL models is that, in addition to being able to simulate brittle breakout around tunnel peripheries, it can also simulate the time for such failure to occur, providing more information for engineering analysis and design. The model is a modified version of the CVISC creep model, where the secondary viscosity is modified to act as a non-Newtonian fluid and the Mohr–Coulomb slider experiences cohesive degradation based on the ratio of in-situ stress to strength, or driving-stress ratio, based on empirical relationships. Cohesion loss can be associated to the initiation and propagation of cracks through the material, reducing the overall effective cohesion. This opening and propagating of fractures also reduces tensile strength, which is considered as well. The friction angle is assumed constant until a residual state is reached.

The LTS model is built upon TTF lab testing wherein a cylindrical sample of rock is subject to a constant stress greater than its respective CI and less than instantaneous strength and is used to predict the ultimate stability time for any brittle rock subject to some deviator stress. The current state of practice for accounting for strength loss with time in a numerical model is to manually decrease strength parameters of the rock in stages, which results in bulk weakening; however, as shown in this analysis, only areas of rock subject to high deviator stresses weaken due to stress corrosion. The LTS model provides several advantages over classical numerical modelling techniques, including CWFS analyses, allowing engineers and scientists to weaken specific areas of rock with time using lab data as a basis. As such, a more precise analysis on the timing for support installation and excavation step sizes can be made. The model in its current state calculates TTF based solely on unconfined TTF lab tests, which likely does not reflect real-world excavation scale behaviour. Finally, it should be highlighted that to calibrate the model, further research is needed on the effects of confinement on TTF as well as field scale convergence measurements in longer time periods to ensure model validity.

Author Contributions: Conceptualization, J.I., C.P. and M.S.D.; methodology, J.I., C.P. and M.S.D.; software, J.I.; validation, J.I.; formal analysis, J.I.; investigation, J.I. and C.P.; resources, M.S.D.; data curation, J.I. and C.P.; writing—original draft preparation, J.I.; writing—review and editing, C.P. and M.S.D.; visualization, J.I., C.P. and M.S.D.; supervision, C.P. and M.S.D.; project administration, C.P.; funding acquisition, M.S.D. All authors have read and agreed to the published version of the manuscript.

Funding: This research was funded by Nuclear Waste Management Ontario—NWMO, grant number GS830 and Natural Sciences and Engineering Research Council of Canada—NSERC CRDPJ 523562-18.

Data Availability Statement: Not applicable.

Acknowledgments: The authors would like to acknowledge the contributions made by the Nuclear Waste Management Organization (NWMO) and the Natural Sciences and Engineering Research Council of Canada (NSERC) for this research would not have been possible without their continued support.

Conflicts of Interest: The authors declare no conflict of interest.

References

1. Kaiser, P.K.; McCreath, D.R.; Tennant, D.D. *Canadian Rockburst Support Handbook*; Geomechanics Research Centre/MIRARCO: Sudbury, ON, Canada, 1996.
2. Kaiser, P.K.; Amann, F.; Steiner, W. How highly stressed brittle rock failure impacts tunnel design. In *European Rock Mechanics Symposium*; Rock Mechanics and Environmental Engineering, CRC Press: Boca Raton, FL, USA, 2010; pp. 27–38.
3. Coulomb, C.A. Sur une application des regles maximis et minimis a quelques problems de statique, relatives a l'architecture. *Acad. R. Des. Sci.* **1776**, *7*, 343–382.
4. Hoek, E.; Brown, E.T. *Underground Excavations in Rock*, 1st ed.; Institution of Mining and Metallurgy: London, UK, 1980.
5. Hoek, E.; Brown, E.T. The Hoek–Brown failure criterion and GSI—2018 edition. *J. Rock Mech. Geotech. Eng.* **2019**, *11*, 445–463. [CrossRef]
6. Hoek, E.; Carranza, C.; Corkum, B. Hoek-brown failure criterion—2002 edition. In Proceedings of the 5th North American Rock Mechanics Symposium and the 17th Tunnelling Association of Canada Conference, Toronto, ON, Canada, 7–10 July 2002; pp. 267–273.
7. Diederichs, M.S. Manuel Rocha Medal Recipient Rock Fracture and Collapse under Low Confinement Conditions. *Rock Mech. Rock Eng.* **2003**, *36*, 339–381. [CrossRef]
8. Diederichs, M.S. The 2003 Canadian geotechnical colloquium: Mechanistic interpretation and practical application of damage and spalling prediction criteria for deep tunnelling. *Can. Geotech. J.* **2007**, *44*, 1082–1116. [CrossRef]
9. Hoek, E.; Brown, E.T. Practical Estimates of Rock Strength. *Int. J. Rock Mech. Min. Sci. Geomech. Abstr.* **1997**, *34*, 1165–1186. [CrossRef]
10. Martin, C.D. Seventeenth Canadian Geotechnical Colloquium: The effect of cohesion loss and stress path on brittle rock strength. *Can. Geotech. J.* **1997**, *34*, 698–725. [CrossRef]
11. Hajiabdolmajid, V.; Kaiser, P.K.; Martin, C.D. Modelling brittle failure of rock. *Int. J. Rock Mech. Min. Sci.* **2002**, *39*, 731–741. [CrossRef]
12. Cai, M.; Kaiser, P.K.; Tasaka, Y.; Maejima, T.; Morioka, H.; Minami, M. Generalized crack initiation and crack damage stress thresholds of brittle rock masses near underground excavations. *Int. J. Rock Mech. Min. Sci.* **2004**, *41*, 833–847. [CrossRef]
13. Griggs, D. Creep of Rocks. *J. Geol.* **1939**, *47*, 225–251. [CrossRef]
14. Kaiser, P.K. Time-Dependent Behaviour of Tunnels in Jointed Rock Masses. Ph.D. Thesis, University of Alberta, Edmonton, AB, Canada, 1979.
15. Lajtai, E.Z.; Scott Duncan, E.J.; Carter, B.J. The Effect of Strain Rate on Rock Strength. *Rock Mech. Rock Eng.* **1991**, *24*, 99–109. [CrossRef]
16. Pellet, F.; Hajdu, A.; Deleruyelle, F.; Besnus, F. A viscoplastic model including anisotropic damage for the time dependent behaviour of rock. *Int. J. Numer. Anal. Methods Geomech.* **2005**, *29*, 941–970. [CrossRef]
17. Paraskevopoulou, C. Time-Dependency of Rocks and Implications Associated. Ph.D. Thesis, Queen's University, Montreal, QC, Canada, 2016.
18. Paraskevopoulou, C. Time-Dependent Behavior of Rock Materials. In *Engineering Geology*; Intech Open: Leeds, UK, 2021.
19. Innocente, J.C.; Paraskevopoulou, C.; Diederichs, M.S. Estimating the long-term strength and time-to-failure of brittle rocks from laboratory testing. *Int. J. Rock Mech. Min. Sci.* **2021**, *147*, 104900. [CrossRef]
20. Barton, N. Engineering classification of rock masses for the design of tunnel support. *Rock Mech.* **1974**, *6*, 189–236. [CrossRef]
21. Bieniawski, Z.T. Geomechanics classification of rock—Masses and its application in tunnelling. In Proceedings of the Congress of the International Society for Rock Mechanics, Denver, CO, USA, 1–7 September 1974; National Academy of Sciences: Washingston, DC, USA, 1974; pp. 27–32.
22. Martin, C.D. The Strength of Massive Lac du Bonnet Granite Around Underground Openings. Ph.D. Thesis, University of Manitoba, Winnipeg, MB, Canada, 1993.
23. Diederichs, M.S. Instability of Hard Rockmasses: The Role of Tensile Damage and Relaxation. Ph.D. Thesis, University of Waterloo, Waterloo, ON, Canada, 1999.
24. Marinos, P.; Hoek, E. GSI: A Geologically Friendly Tool for Rock Mass Strength Estimation. In Proceedings of the GeoEng 2000 Conference, Melbourne, Australia, 19–24 November 2000; pp. 1422–1442.
25. Diederichs, M.S.; Martin, C.D. Measurement of spalling parameters from laboratory testing. In *ISRM International Symposium—EUROCK 2010*; Taylor & Francis Group: Lausanne, Switzerland, 2010.
26. Paraskevopoulou, C.; Perras, M.A. Investigating the long-term behaviour of brittle rocks: Visco-elastic creep parameters and time-to-failure. In Proceedings of the Progressive Rock Failure Conference, Ascona, Switzerland, 5–9 June 2017; p. 4.

27. Diederichs, M.S.; Day, J.J.; Ghazvinian, E.; Perras, M.; Paraskevopoulou, C.; Walton, G. Progressive brittle damage processes and failure in rock. In Proceedings of the PRF 2017: Progressive Failure and Long-Term Strength Degradation of Brittle Rocks ISRM Conference, Ascona, Switzerland, 5–9 June 2017.
28. Bewick, R.P.; Kaiser, P.K.; Amann, F. Strength of massive to moderately jointed hard rock masses. *J. Rock Mech. Geotech. Eng.* **2019**, *11*, 562–575. [CrossRef]
29. Tapponnier, P.; Brace, W.F. Development.of Stress-Induced Mlcrocracks m Westerly Granite. *Int. J. Rock Mech. Min. Sci. Geomech. Abstr.* **1976**, *13*, 103–112. [CrossRef]
30. Hoek, E. Brittle Failure in Rock. In *Rock Mechanics in Engineering Practice*; Stagg, K.G., Zienkiewics, O.C., Eds.; Wiley: London, UK, 1968; pp. 99–124.
31. Ottosen, N.S. Viscoelastic-viscoplastic formulas for analysis of cavities in rock salt. *Int. J. Rock Mech. Min. Sci. Geomech.* **1986**, *23*, 201–212. [CrossRef]
32. Martin, C.D.; Read, R.S.; Martino, J.B. Observation of brittle failure around a circular test tunnel. *Int. J. Rock Mech. Min. Sci.* **1997**, *34*, 1065–1073. [CrossRef]
33. Perras, M.; Diederichs, M.S.; Wannenmacher, H. Underground Excavation Behaviour of the Queenston Formation: Tunnel Back Analysis for Application to Shaft Damage Dimension Prediction. *Rock Mech. Rock Eng.* **2014**, *48*, 1647–1671. [CrossRef]
34. Lajtai, E.Z.; Schmidtke, R.H. Delayed failure in rock loaded in uniaxial compression. *Rock Mech. Rock Eng.* **1986**, *19*, 11–25. [CrossRef]
35. Diederichs, M.S.; Kaiser, P.K.; Eberhardt, E. Damage initiation and propagation in hard rock during tunnelling and the influence of near-face stress rotation. *Int. J. Rock Mech. Min. Sci.* **2004**, *41*, 785–812. [CrossRef]
36. Damjanac, B.; Fairhurst, C. Evidence for a long-term strength threshold in crystalline rock. *Rock Mech. Rock Eng.* **2010**, *43*, 513–531. [CrossRef]
37. Walton, G. Initial guidelines for the selection of input parameters for cohesion-weakening-friction-strengthening (CWFS) analysis of excavations in brittle rock. *Tunn. Undergr. Space Technol.* **2019**, *84*, 189–200. [CrossRef]
38. Da Andeade, N.C. On the Viscous Flow in Metals, and Allied Phenomena. *Proc. R. Soc. A-Math Phys. Eng. Sci.* **1910**, *84*, 1–12. Available online: http://rspa.royalsocietypublishing.org/cgi/doi/10.1098/rspa.1910.0050 (accessed on 13 January 2021).
39. Norton, F.H. *The Creep of Steel at High Temperatures*, 1st ed.; McGraw-Hill Book Company, Inc.: New York, NY, USA, 1929.
40. Griggs, D. Experimental flow of rocks under conditions favoring recrystallization. *Geol. Soc. Am. Bull.* **1940**, *51*, 1001–1022. Available online: https://pubs.geoscienceworld.org/gsabulletin/article/51/7/1001-1022/3803 (accessed on 13 January 2021). [CrossRef]
41. Le Comte, P. Creep in Rock Salt. *J. Geol.* **1965**, *73*, 469–484. [CrossRef]
42. Panet, M. Time-Dependent Deformations in Underground Works. In Proceedings of the 4th International Congress on Rock Mechanics, Montreux, Switzerland, 2–8 September 1979; p. 12.
43. Van Sambeek, L.L. *Creep of Rock Salt under Inhomogeneous Stress Conditions*; Colorado School of Mines: Golden, CO, USA, 1986.
44. Aubertin, M.; Gill, D.E.; Ladanyi, B. An internal variable model for the creep of rocksalt. *Rock Mech. Rock Eng.* **1991**, *24*, 81–97. [CrossRef]
45. Malan, D.F. Time-dependent behaviour of deep level tabular excavations in hard rock. *Rock Mech. Rock Eng.* **1999**, *32*, 123–155. [CrossRef]
46. Hagros, A.; Johansson, E.; Hudson, J.A. *Time Dependency in the Mechanical Properties of Crystalline Rocks—A Literature Survey*; Posiva Oy: Eurajoki, Finland, 2008.
47. Cristescu, N.D. Time Effects in Rock Mechanics. In *SEM Annual Conference*; Society for Experimental Mechanics Inc.: Albuquerque, NM, USA, 2009.
48. Brantut, N.; Heap, M.J.; Meredith, P.G.; Baud, P. Time-dependent cracking and brittle creep in crustal rocks: A review. *J. Struct. Geol.* **2013**, *52*, 17–43. [CrossRef]
49. Paraskevopoulou, C.; Diederichs, M. Analysis of time-dependent deformation in tunnels using the Convergence-Confinement Method. *Tunn. Undergr. Sp. Technol.* **2018**, *71*, 62–80. [CrossRef]
50. Weaver, S.H. Creep curve of steel. *Trans. Am. Soc. Mech. Eng.* **1936**, *LVIII*, 745–751.
51. Hooke, R. *De Potentia Restitutiva, or of Spring Explaining the Power of Springing Bodies*; John Martyn Printer to the Royal Society: London, UK, 1678.
52. Langer, M. The rheological behaviour of rock salt. In Proceedings of the 1st Conference on Mechanical Behaviour of of Salt, Pennsylvania State University, University Park, PA, USA, 9–11 November 1984; Trans Tech Publications Ltd.: Clausthal, Germany, 1984; pp. 201–230.
53. Patchet, S.J. Rock Mechanics Studies Associated with the Development of a Deep Potash Mine. Ph.D. Thesis, University of Newcastle, Newcastle Upon Tyne, UK, 1970.
54. Paraskevopoulou, C.; Perras, M.; Diederichs, M.S.; Amaan, F.; Loew, S.; Lam, T.; Jensen, M. The three stages of stress-relaxation—Observations for the long-term behaviour of rocks based on laboratory testing. *J. Eng. Geol.* **2017**, *216*, 56–75. [CrossRef]
55. Barla, G. Tunnelling under Squeezing Rock Conditions. Tunnelling Mechanics—Advances in Geotechnical Engineering and Tunnelling, Eurosummer-School in Tunnel Mechanics. 2000. Available online: http://scholar.google.com/scholar?hl=en&btnG=Search&q=intitle:Tunnelling+under+squeezing+rock+conditions#0 (accessed on 15 January 2021).
56. Barla, M. Tunnels in Swelling Ground. Ph.D. Thesis, University of Genoa, Genoa, Italy, 1999.

57. Terzaghi, K. Rock Tunneling with Steel Supports. In *Rock Defects and Loads on Tunnel Support*; Proctor, R.V., White, T.L., Eds.; The Commercial Shearing and Stamping Company: Youngstown, OH, USA, 1946; pp. 15–99.
58. Schowalter, W.R. *Mechanics of Non-Newtonian Fluids, 1st ed*; Pergamon: Oxford, UK, 1978.
59. Cristescu, N. Mechanics of Elastic and Inelastic Solids. In *Rock Rheology*; Springer: Dordrecht, The Netherlands, 1989; Volume 7, Available online: http://link.springer.com/10.1007/978-94-009-2554-0 (accessed on 20 January 2021).
60. Batchelor, G.K. *An Introduction to Fluid Dynamics*; Cambridge University Press: Cambridge, UK, 2000. Available online: https://www.cambridge.org/core/product/identifier/9780511800955/type/book (accessed on 20 January 2021).
61. Bingham, E.C. An Investigation of the Laws of Plastic Flow. *Bull. Bur. Stand.* **1916**, *13*, 309–353. [CrossRef]
62. Tanner, R.I. *Engineering Rheology*; Oxford University Press: Oxford, UK, 1988.
63. Roylance, D. Engineering Viscoelasticity. 2001, 1–37. Available online: https://web.mit.edu/course/3/3.11/www/modules/visco.pdf (accessed on 20 January 2021).
64. Debernardi, D. *Viscoplastic Behaviour and Design of Tunnels*; Politecnico di Torino: Torino, Italy, 2008. Available online: http://m.danieledebernardi.it/sites/danieledebernardi.it/files/Daniele_Debernardi_PhD_Thesis.pdf (accessed on 20 January 2021).
65. Itasca Consulting Group I. *FLAC*, version 7; Constitutive Models: Theory and Implementation; Itasca Consulting Group, Inc.: Itasca, IL, USA; Minneapolis, MN, USA, 2011.
66. Gioda, G.; Civdini, A. Numerical methods for the analysis of tunnel performance in squeezing rocks. *Rock Mech. Rock Eng.* **1996**, *29*, 171–193. [CrossRef]
67. Goodman, R. *Introduction to Rock Mechanics*, 2nd ed.; John Wiley & Sons: New York, NY, USA, 1989.
68. Paraskevopoulou, C.; Diederichs, M.S. A comparison of viscous models under constant strain and constant stress: Implications for tunnel analysis. In Proceedings of the World Tunnel Congress on Underground—The Way to the Future, Geneva, Switzerland, 31 May–7 June 2013.
69. Paraskevopoulou, C.; Diederichs, M.S. Long-term behaviour in tunnelling: Limitations in using creep parameters. In Proceedings of the 66th Canadian Geotechnical Conference, GeoMontreal on Geoscience for Sustainability, Montreal, QC, Canada, 3–6 October 2013.
70. Innocente, J.C.; Diederichs, M.S.; Paraskevopoulou, C.; Aubertin, J.D. Numerical investigation of the applicability of time-dependent models. In Proceedings of the International Society for Rock Mechanics Symposium—Eurock 2020—Hard Rock Engineering, Trondheim, Norway, 14–19 June 2020.
71. Perzyna, P. Fundamental problems in viscoplasticity. *Adv. Appl. Mech.* **1966**, *9*, 244–368.
72. ISRM. Suggested methods for determining the strength of rock materials in triaxial compression: Revised version. *Int. J. Rock Mech. Min. Sci. Geomech. Abstr.* **1983**, *20*, 285–290. [CrossRef]
73. *ASTM D7012-14e1*; Standard Test Methods for Compressive Strength and Elastic Moduli of Intact Rock Core Specimens under Varying States of Stress and Temperatures. ASTM: West Conshohocken, PA, USA, 2014. Available online: www.astm.org (accessed on 20 January 2021).
74. Aydan, O.; Takashi, I.; Ugur, O.; Kwasniewski, M.; Shariar, K.; Okuno, T.; Özgenoğlu, A.; Malan, D.F.; Okada, T. ISRM Suggested Methods for Determining the Creep Characteristics of Rock. *Rock Mech. Rock Eng.* **2013**, *47*.
75. Schmidtke, R.H. The Strength and Static Fatigue of a Granite, an Anorthosite, and a Limestone. Master's Thesis, University of Manitoba, Winnipeg, MB, Canada, 1986.
76. Perras, M.A.; Diederichs, M.S. A Review of the Tensile Strength of Rock: Concepts and Testing. *Geotech. Geol. Eng.* **2014**, *32*, 525–546. [CrossRef]
77. Lau, J.S.O.; Gorski, B.; Conlon, B.; Anderson, T. *Long-Term Loading Tests on Saturated Granite and Granodiorite*; Ontario Power Generation Report 06819-REP-01300-10016 R00; Nuclear Waste Management Division: Ottawa, ON, Canada, 2000.
78. Potyondy, D.O. Simulating stress corrosion with a bonded-particle model for rock. *Int. J. Rock Mech. Min. Sci.* **2007**, *44*, 677–691. [CrossRef]
79. Wang, M.; Cai, M. A grain-based time-to-failure creep model for brittle rocks. *Comput. Geotech.* **2020**, *119*, 103344. [CrossRef]
80. Mogi, K. Pressure dependence of rock strength and transition from brittle fracture to ductile flow. *Bull. Earthq. Res. Inst.* **1966**, *44*, 215–232.
81. Kaiser, P.K.; Diederichs, M.S.; Martin, C.D.; Sharp, J.; Steiner, W. Underground works in hard rock tunneling and mining. In Proceedings of the International Conference on Geotechnical and Geological Engineering; Technomic Publishing Co.: Melbourne, Australia, 2000; Volume 1, pp. 841–926.
82. Bieniawski, Z.T. Mechanism of brittle fracture of rock: Part I—Theory of the fracture process. *Int. J. Rock Mech. Min. Sci. Geomech. Abstr.* **1967**, *4*, 395–406. Available online: papers3://publication/uuid/D09BB22D-D474-4C6F-A738-F7ED091B08E0 (accessed on 20 January 2021). [CrossRef]
83. Wawersik, W.R.; Fairhurst, C. A study of brittle rock fracture in laboratory compression experiments. *Int. J. Rock Mech. Min. Sci.* **1970**, *7*, 561–575. [CrossRef]
84. Heap, M.J.; Baud, P.; Meredith, P.G.; Bell, A.F.; Main, I.G. Time-dependent brittle creep in darley dale sandstone. *J. Geophys. Res. Solid Earth* **2009**, *114*, 1–22. [CrossRef]
85. Heap, M.J.; Baud, P.; Meredith, P.G.; Vinciguerra, S.; Bell, A.F.; Main, I.G. Brittle creep in basalt and its application to time-dependent volcano deformation. *Earth Planet. Sci. Lett.* **2011**, *307*, 71–82. [CrossRef]

86. Paraskevopoulou, C.; Diederichs, M.; Perras, M.A.; Amann, F.; Loew, S. Long-term static load laboratory testing behaviour of different rock types. In Proceedings of the 68th Canadian Geotechnical Conference, Quebec City, QC, Cananda, 20–23 September 2015; p. 7. Available online: https://www.researchgate.net/publication/282247168_Long-term_static_load_laboratory_testing_behaviour_of_different_rock_types (accessed on 20 January 2021).
87. Paraskevopoulou, C.; Perras, M.; Diederichs, M.S.; Amaan, F.; Loew, S. Observations for the long-term behaviour of carboniferous limestone rocks based on laboratory testing. In Proceedings of the EUROCK Future Development of Rock Mechanics—64th Geomechanics Colloqium, Salzburg, Austria, 7–10 October 2015.
88. Paraskevopoulou, C.; Perras, M.; Diederichs, M.; Loew, S.; Lam, T.; Jensen, M. Time-Dependent Behaviour of Brittle Rocks Based on Static Load Laboratory Tests. *Geotech. Geol. Eng.* **2018**, *36*, 337–376. [CrossRef]
89. Zhang, H.; Wang, Z.; Zheng, Y.; Duan, P.; Ding, S. Study on tri-axial creep experiment and constitutive relation of different rock salt. *Saf. Sci.* **2012**, *50*, 801–805. [CrossRef]
90. Yang, S.Q.; Xu, P.; Ranjith, P.G.; Chen, G.F.; Jing, H.W. Evaluation of creep mechanical behavior of deep-buried marble under triaxial cyclic loading. *Arab J. Geosci.* **2015**, *8*, 6567–6582. [CrossRef]
91. Ngwenya, B.T.; Main, I.G.; Elphick, S.C.; Crawford, B.R.; Smart, B.G.D. A constitutive law for low-temperature creep of water-saturated sandstones. *J. Geophys. Res.* **2001**, *106*, 21811–21826. [CrossRef]
92. Wu, F.; Gao, R.; Zou, Q.; Chen, J.; Liu, W.; Peng, K. Long-term strength determination and nonlinear creep damage constitutive model of salt rock based on multistage creep test: Implications for underground natural gas storage in salt cavern. *Energy Sci. Eng.* **2020**, *8*, 1592–1603. [CrossRef]
93. Venturini, G.; Bianchi, N.; Diederichs, M. How to Quantify the Reliability of a Geological and Geotechnical Reference Model in Underground Projects. In Proceedings of the RETC 2019, Chicago, IL, USA, 16–19 June 2019; pp. 525–537.
94. Skolidis, A.; Paraskevopoulou, C.; Marinos, V.; Parsons, S. Accounting for uncertainty in geotechnical design using a probabilistic approach. In Proceedings of the WTC 2020, Word Tunnelling Congress on Innovation and Sustainable Underground Serving Global Connectivity, Kuala Lumpur, Malaysia, 11–17 September 2020.
95. Paraskevopoulou, C.; Skolidis, A.; Parson, S.; Marinos, V. Integrating uncertainty into geotechnical design of underground openings in tectonically undisturbed but lithologically varied sedimentary environments. *Tunn. Undergr. Space Technol. J.* **2021**, *113*, 103979. [CrossRef]
96. Yau, K.; Paraskevopoulou, C.; Konstantis, S. A probabilistic approach to assess the risk of liner instability when tunnelling through karst geology using geotechnical baseline reports. In Proceedings of the WTC 2019, Word Tunnelling Congress on Tunnels and Underground Cities: Engineering and Innovation Meet Archaeology, Architecture and Art, Naples, Italy, 3–9 May 2019.
97. Yau, K.; Paraskevopoulou, C.; Konstantis, S. Spatial variability of karst and effect on tunnel lining and water inflow. A probabilistic approach. *Tunn. Undergr. Space Technol. J.* **2020**, *97*, 103248. [CrossRef]
98. Van der Pouw Kraan, M.; Diederichs, M.S. Behavioural uncertainty for rock tunnels: Implications for rock mass definitions or predictions in geotechnical baseline reports. In Proceedings of the World Tunnel Congress, Iguazu, Brazil, 9–14 May 2014.
99. Paraskevopoulou, C.; Dallavalle, M.; Konstantis, S.; Spyridis, P.; Benardos, A. 2022. Assessing the Failure Potential of Tunnels and the Impacts on Cost Overruns and Project Delays. *Tunn. Undergr. Space Technol. J.* **2022**, *123*, 104443. [CrossRef]
100. Paraskevopoulou, C.; Benardos, A. Construction cost estimation for Greek road tunnels in relation to the geotechnical conditions. In Proceedings of the International Symposium Practices and Trends for Financing and Contracting Tunnels and Underground Works (Tunnel Contracts 2012), Athens, Greece, 22–23 March 2012.
101. Paraskevopoulou, C.; Benardos, A. Assessing the construction cost of tunnel projects. *Tunn. Undergr. Space Technol. J.* **2013**, *38*, 497–505. [CrossRef]
102. Paraskevopoulou, C.; Boutsis, G. Cost Overruns in Tunnelling Projects: Investigating the Impact of Geological and Geotechnical Uncertainty Using Case Studies. Special Issue Underground Infrastructure Engineering of Infrastructure Journal (MDPI). *Infrastructures* **2020**, *5*, 73. [CrossRef]
103. Benardos, A.; Paraskevopoulou, C.; Diederichs, M. Assessing and benchmarking the construction cost of tunnels. In Proceedings of the 66th Canadian Geotechnical Conference, GeoMontreal on Geoscience for Sustainability, Montreal, QC, Canada, 29 September–3 October 2013.

mining

MDPI

Article

An Investigation of Thermal Effects on Micro-Properties of Sudbury Norite by CT Scanning and Image Processing Method

Sheng-Lin Wang [1,*], Brad Simser [2], Shunde Yin [1] and Ju Huyan [3]

1 Department of Civil and Environmental Engineering, Faculty of Engineering, University of Waterloo, Waterloo, ON N2L 3G1, Canada
2 Sudbury Integrated Nickel Operations, Glencore, Sudbury, ON P0M 1S0, Canada
3 School of Transportation, Southeast University, Nanjing 210000, China
* Correspondence: s545wang@uwaterloo.ca

Abstract: Rock is constantly subjected to stress and thermal conditions. Thermal-induced micro-cracks will be generated as a result of different thermal expansion gradations between different minerals. This characteristic was investigated in this paper by studying the micro-properties of Sudbury norite via CT scanning and the image processing method. A novel filtering method, maximum–minimum shadow filtering (MMSF), was developed in this study to highlight the thermal-induced micro-cracks in Sudbury Igneous Complex (SIC) norite after different temperature treatments. Based on quantitative analysis, the areal percentages of biotite, felspar, quartz, and small amounts of metal minerals were determined. It was also found that small-scale micro-cracks were first observed in the middle of biotite grains at a temperature of 400 °C. The cracks further propagated and extended with the temperature increase. In addition, the orientations of cracks either remained at the same distribution or became more evenly distributed with the rising temperature. A linear relationship was found between the average porosity of SIC norite and the temperature. Moreover, the anisotropic properties between vertical and horizontal directions of norite were also noticeable. Overall, the paper presented a quantitative study on the effects of thermal treatment and the anisotropic properties of SIC norite. Methodology and findings from this paper will be a significant reference for future studies regarding the thermal impacts on norite and similar rocks.

Keywords: Sudbury norite; thermal effect; image processing; quantitative analysis; micro-crack

1. Introduction

Rock, as a collection of minerals, is constantly subjected to stress and thermal conditions [1–4]. Different minerals will have different thermal expansion rates under temperature rising, thus causing thermal stress under high temperatures. During the process of crustal movement and mankind's activities, such as drilling, mining, and earthquakes, rocks may expose to higher-than-normal temperatures [5]. Such thermal and exhumation effects could induce the appearance of micro-cracks which were reported in granite and sandstone especially when the condition temperature exceeds 400 °C [1,6,7]. Based on recent studies, three major thermal damage stages were indicated [8,9]: (1) <400 °C, a release of free water, crystal and structural water causes micro-cracks and voids to emerge; (2) 400–600 °C, the mineral composition starts to change and the transition of quartz from α phase to the β phase occurs, increasing micro-cracks, (3) 600–1000 °C, severe damage of rock mass occurs due to mineral expansion, decomposition, metallic bond breaking, and melting. Recently, with developed data visualization methods, more quantitative analysis, and Computed Tomography (CT) scanning were conducted. Table 1 summarizes the recent methodology and key findings for evaluating the thermal effects on several types of rocks.

Table 1. Recent literature on the thermal effects of rocks.

Recourse	Rock Type	Heating Method	Key Findings
[6]	Sandstone	25–600 °C Rate: 5 °C/min Condition: 2 h	Between 400 and 600 °C, the pore volume, porosity and peak strain increased rapidly, while the strength decreases rapidly.
[1]	Granite	25–800 °C Rate: 2.5 °C/min Condition: 6 h	The heterogeneity coefficient reached its peak at around 500 °C. The anisotropy coefficient remained steady from 20 °C to 500 °C. It then decreased sharply after 600 °C.
[10]	Granite	100–1000 °C Rate: 5 °C/min Condition: 24 h	The pore network models (PNMs) of thermally treated rock were developed and enabled the quantifications of the size and distribution of the pores in granite.
[7]	Shale rock	25–500 °C Rate: 1 °C/min Condition: 24 h	P-wave velocity and brittleness index of shale decreased due to dehydration and organic matter burning. In addition, the temperature rising made the shale more homogeneous and less anisotropic.
[8]	Red sandstone	10–1300 °C Rate: 10 °C/min Condition: 1 h	The porosity of sandstone increased at temperatures starting from 500 °C and peaking at 1000 °C.

Norites are found in a variety of environments on earth, ranging from oceanic crust [11] to layered intrusions [12], and found to impact melt sheets, such as the Sudbury Igneous Complex (SIC) [13,14]. In particular, SIC norite contains high absolute rare-earth element (REE) contents and is usually observed in Nickel (Ni) and Copper (Cu) ores. Therefore, research into the SIC norite holds both scientific and economic importance. Past literature mainly focused on its origins, chemical evolutions, and REE characterizations, and is dated. In contrast, studies on microstructure, thermal resistance property and crack propagation characteristics of norite are rarely seen.

In this regard, this paper investigated the effects of elevated temperature on the heterogeneity and anisotropy of SIC norite based on CT scanning and image processing methods. The thermal-induced cracks were reconstructed and analyzed. In addition, the heterogeneity and anisotropy properties of norite were quantified based on a 3-dimensional reconstruction method. The experimental method and study provided a significant reference for the characterization of microstructure and micro-cracks in SIC norite under high-temperature conditions.

2. Specimen and Laboratory Test

2.1. Specimen

Norite tested in this study was sampled from a Ni ore in Sudbury, Ontario, Canada. The 1.85 Ga old Sudbury structure is an elliptical structure formed by explosive volcanism combined with mafic intrusion [15]. In addition, an impact in such an area was suggested during orogenesis, and such impact could be the origin of REEs in SIC norite [14]. Norite cores from this study were taken 1985 m below the surface of the mining hole collar, with an average diameter of 47.6 mm. SIC norite (hereinafter will be called norite) consists mainly of quartz, feldspar, and biotite which together form an interlock structure. To conduct the CT scanning, cubic specimens were carefully cut from the cores, with approximately 10 mm on each side. The integrity and homogeneity of the rock samples were carefully examined before the thermal treatment.

2.2. Thermal Treatment

The prepared norite cubes were heated at a heating rate of 2.5 °C/min in a high-temperature furnace, with an accuracy of ±1 °C. Thermal treatments were classified into four different groups (200 °C, 400 °C, 600 °C, and 800 °C). The slow heat rate was adapted from past literature [1] to avoid thermal shock and generate a homogeneous thermal field in norite. The heated norite specimens were then kept at the condition temperature for 16 h. Afterward, they were cooled down in the furnace before being moved to room temperature.

2.3. CT Scan

The CT scanning was conducted by a sub-micron CT scanner located at the multi-scale additive manufacturing (MSAM) lab, University of Waterloo. The scanner includes an X-ray launcher, a specimen platform, and a flat panel detector that receives the X-rays passing through the specimen, see Figure 1. After scanning, three cross-sections in the directions of x, y and z were collected and analyzed.

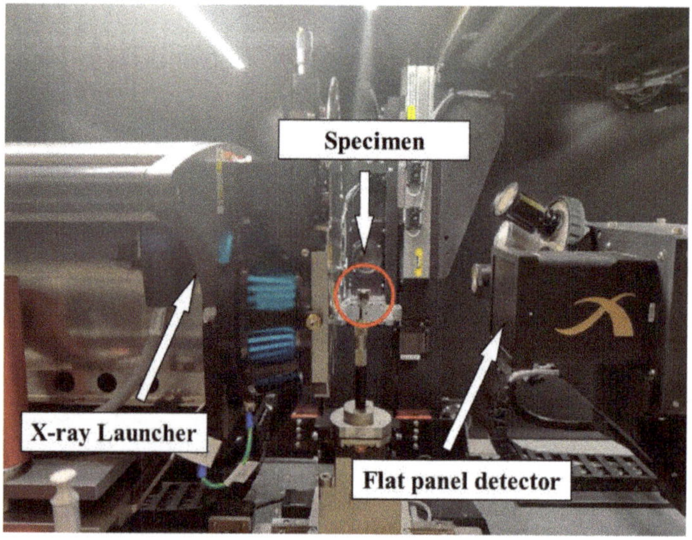

Figure 1. CT Scanning set-up for norite after thermal treatment.

2.4. *Image Processing Method*

After acquisitions of CT images, they were preprocessed and analyzed. The major procedure includes image acquisition, image enhancement, shadow elimination, and crack characterization. MATLAB together with ImageJ were used for the image processing. Figure 2 illustrates the general procedure for each specimen. Important steps are introduced respectively in the following sections.

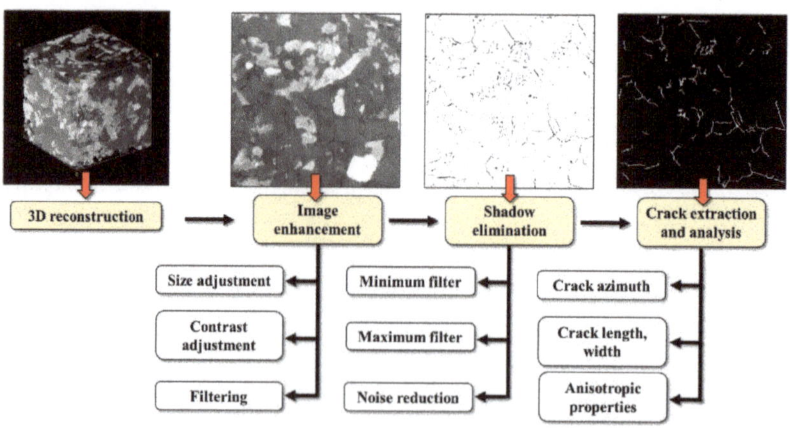

Figure 2. Scheme for the quantitative image processing for CT images.

2.4.1. Three-Dimensional Reconstruction

The scanned images were cropped to have the same size (650 $\times$ 650 pixels); afterward, a 3-dimensional cube was reconstructed by image stacks. Concerning the scanning setup, one pixel represented 10.02 μm or approximately 10 μm. Therefore, the analyzing cube had a size of 6.5 mm on each side. Figure 3 presents an example of a 3D reconstructed norite cube and its orthogonal sides. Each cube has three observation planes: *XY* plane (original scanning plane, parallel to ground level and perpendicular to the *z*-axis), *XZ* plane (perpendicular to the *y*-axis), and *YZ* plane (perpendicular to the *x*-axis). It should be noted that the coordinate system and the *x*, *y*, and *z* directions are consistent for all the specimens.

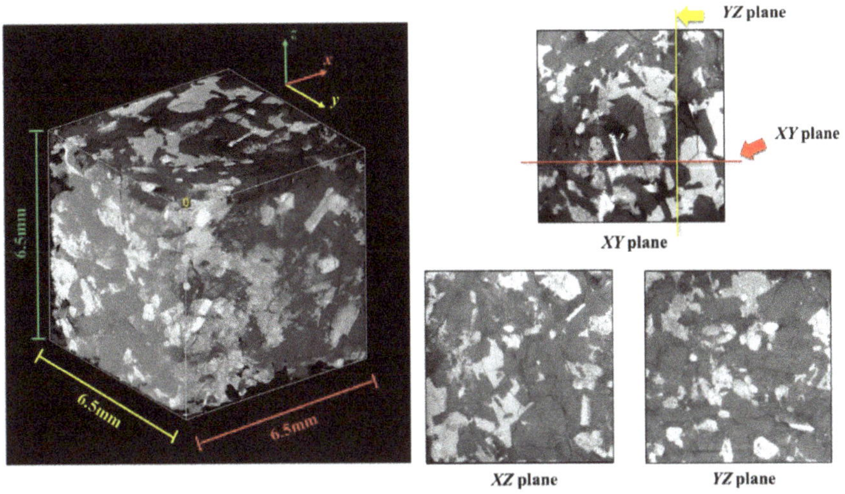

Figure 3. Three-dimensional reconstructed norite and its orthogonal slices.

2.4.2. Maximum–Minimum Shadow Filtering (MMSF)

The most used method for image enhancement includes contrast, brightness, and sharpness adjustment. Such adjustment enables further segmentation and crack extraction [16]. However, since the grayscale levels for cracks and some dark minerals are similar, such conventional adjustments only highlighted the cracks but also included the dark minerals and their boundaries with quartz, see Figure 4b).

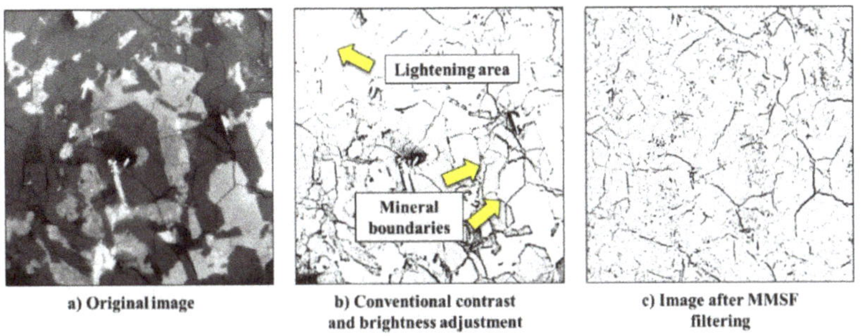

Figure 4. Comparisons between (**a**) original image, (**b**) conventional image enhancement, and (**c**) shadow elimination (MMSF filtering).

Therefore, a new filtering method was developed in this study called maximum–minimum shadow filtering (MMSF). The original image was called "O". First, for each pixel (*x*, *y*), the algorithm searched its surrounding area (N × N pixels) and found the maximum greyscale level, P. The pixel (*x*, *y*) was rewritten as P. The returned matrix, A, was called the maximum filter image. Based on A, a similar method was conducted, whereas this time, the algorithm picked the minimum greyscale level, Q, from each pixel surrounding area. Therefore, A was transferred to B, which is called the maximum–minimum filter image. To have the final filtered image, I, the original image was subtracted by the filtered image, B. See the equation below for details.

$$I = O - B \quad (1)$$

Figure 4c presents the image after MMSF, it could be seen that the effects of mineral boundaries and the dark connected domain were eliminated. Instead, the image only presents linear-shaped cracks and some scattered noises. Such scattered noises will be removed in the next step. However, it should be noted that the filtering also eliminated some minor cracks and broke some thin cracks into segments. Those cracks will be reconnected and restored in the next step.

2.4.3. Crack Extraction and Analysis

During the previous steps, the contrast and sharpness of the image were adjusted, thus generating scattered noise. In addition, cracks with less width might be broken into disconnected segments. In this regard, some image restoration process was conducted, this included, see Figure 5:

(1) Image binarization and invert. Use adaptive thresholding to binarize the image and highlight the thermal cracks in white;
(2) Image close. It performed the morphological closing on the binary image using the structuring element, SE. SE is a linear element used to connect broken segments with a length of three pixels;
(3) Bridge of unconnected segments. Automatically connect the broken segments with infinite calculations until the image became stable;
(4) Small objects delete and minor branch prune. Clear the noise and cut the minor branches generated during the previous image restore steps. Figure 5 illustrates the details of the image restoration process.

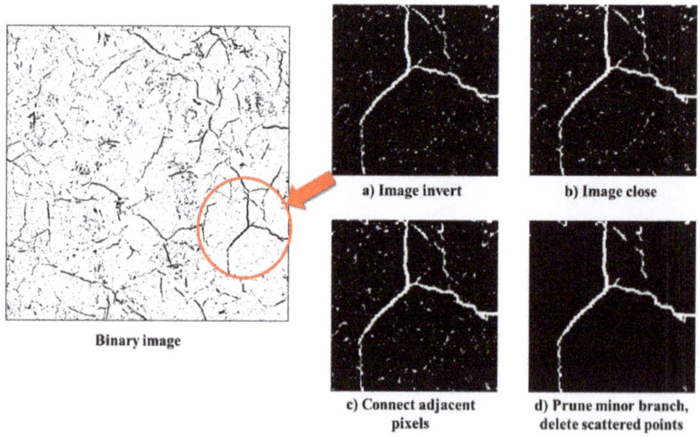

Figure 5. Adjustment of MMSF image, prior to crack extraction. Steps include (**a**) image convert; (**b**) image close; (**c**) scattered pixels connection; and (**d**) prune cut.

Finally, the characteristics of each crack were evaluated using the "imageRegionAnalyzer" toolbox in MATLAB. It should be noted that the final output of length and width are in the unit of pixels. Then, they were transferred into μm (1 pixel = 10 μm from the image). The length was defined as the maximum spine length of each crack. On the other hand, orientation was defined as the angle between the line connecting the start and end point of each crack, and the horizontal axis in each image. Orientation ranges between 0° and 360°.

3. Results and Discussion

3.1. Structure of Norite

Figure 6 presents the structure of the norite after a 200 °C and 850 °C treatment, respectively. Based on different greyscales, the norite was separated into several sub-regions: biotite grains (lighter region), combined quartz and feldspar (dark grey region), and micro-cracks (linear-shaped dark grey and black region). Similar mineral types and identifications were addressed in past studies for Sudbury norite and granite [10,13]. In addition, metal minerals such as Ni were frequently observed in CT scan images. This intermediate-sized mineral has a pure white color and a clear boundary with other minerals. The introduction of Ni and Cu in SIC norites was also documented and presented in past literature [14].

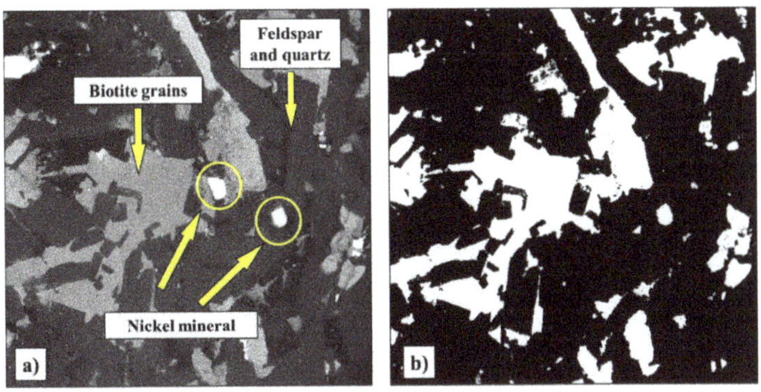

Figure 6. Norite after (**a**) 200 °C treatment and (**b**) binary image, at *XY* plane.

To quantify the percentage of minerals, CT images were binarized using the threshold obtained from the Otsu algorithm. The "white" area indicates the biotite grain and the Ni, whereas the "black" area denotes the feldspar and the quartz. In particular, Figure 7 summarizes the percentage of the three mineral categories, based on pixels counting.

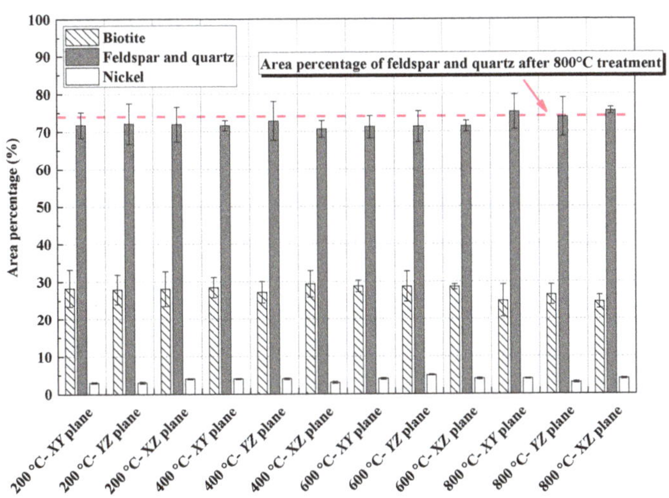

Figure 7. Area percentage of different minerals.

It can be seen that the area percentage of biotite ranged between 22% and 27%. For felspar and quartz, they made up between 70% and 75%. In addition, metal minerals (Ni) occupied approximately 1% to 3% of the total area; however, this might take up more than 5% in some images. Such ratios for each mineral in SIC norite corresponded with the results by physical characterizations from past studies [14,17].

Overall, the percentage for each mineral category in different directions remained consistent. In addition, feldspar and quartz pixel counts in 600 °C- and 800 °C-treated norite were slightly larger than those in lower temperature-treated norite. The reason may be due to the increase in thermal cracks which were showing a similar color compared with dark minerals.

3.2. *Effects of Thermal Treatment*

Figure 8 presents the CT scanning images for SIC norite under 10.0 μm resolution. According to the images, norite subjected to a temperature equal to and under 200 °C showed no significant micro-cracking. With the elevation of temperature (>400 °C), small-scale micro-cracks started to emerge. Such thermal cracks were first observed in the middle of biotite grains (see Figure 9a) with sparse micro-cracks generated due to differences in thermal expansion rates [10]. Further increase in temperature would lead to more micro-cracks in not only biotite but also in quartz and feldspar. The current micro-cracks would also propagate into the adjacent minerals. In addition, thermal-induced tension also generated cracks between different minerals.

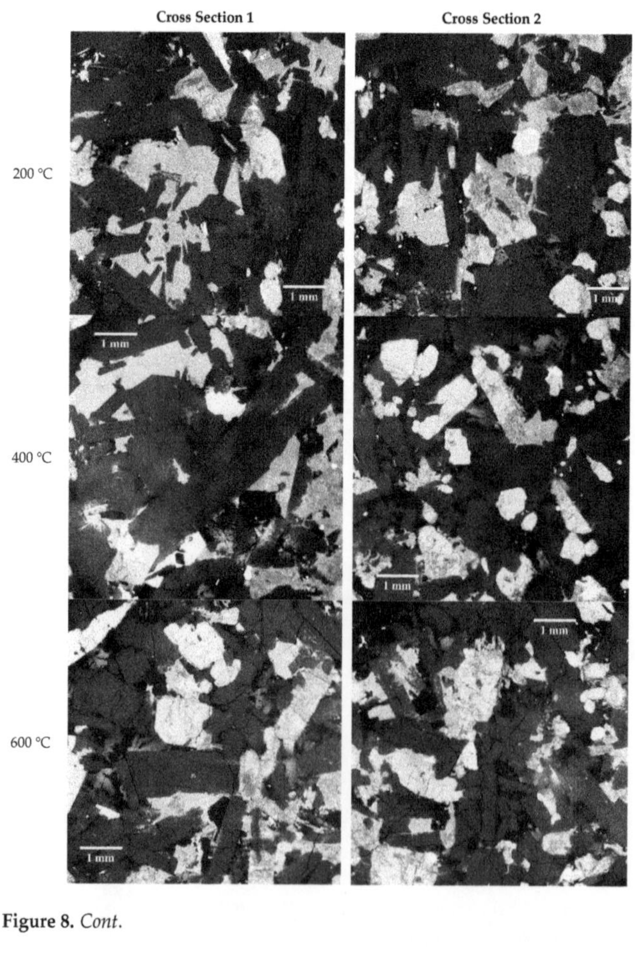

Figure 8. *Cont.*

850 °C 

Figure 8. CT images of high temperature treated norite, *XY* plane (parallel to ground surface).

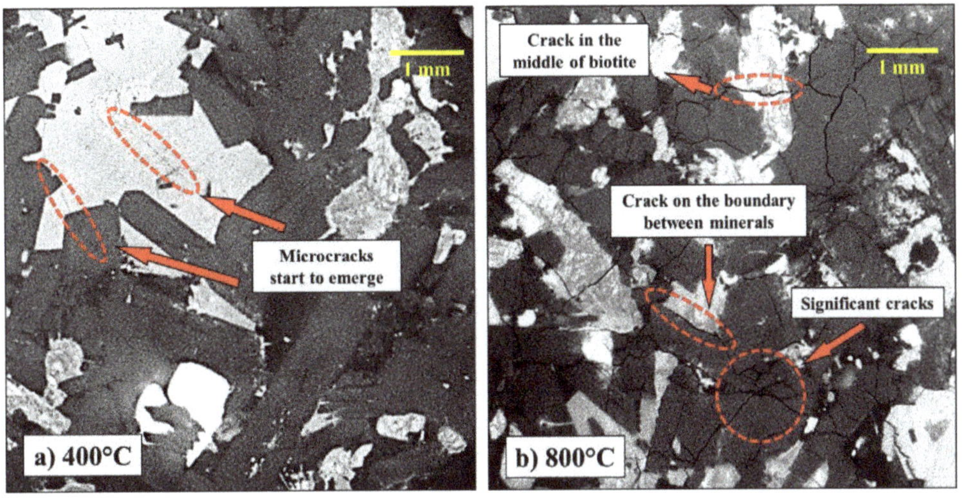

Figure 9. Details of thermal cracks on norite after (**a**) 400 °C and (**b**) 800° C treatment.

Visual observation indicated that the length and width of thermal cracks had increased compared to those of the specimens preheated by 200 °C and 400 °C. With the preheating temperature further developed to 800 °C, different micro-cracks started to connect, thus dividing the integral mineral into smaller segments. Figure 9b presents this morphology. Such facts in SIC norites generally correlate with the phenomenon introduced previously in granite, sandstone, and shale rocks [1,6,8]. Nevertheless, the decomposition of carbonate and the clay minerals had not been seen in CT images.

3.3. Quantitative Analysis of Micro-Cracks

The following figures (Figures 10–12) summarize the distribution of length, width, and orientations for micro-cracks. Each specimen has three planes (*XY*, *YZ*, and *XZ*). For each plane, five different images were selected from different locations and were analyzed. It should be noted that the quantification for norite after 200 °C was not successful since the micro-cracks were not visible. Objects with a maximum length of fewer than 15 pixels or 150 µm were not taken into account to eliminate the effects of scattered small objects.

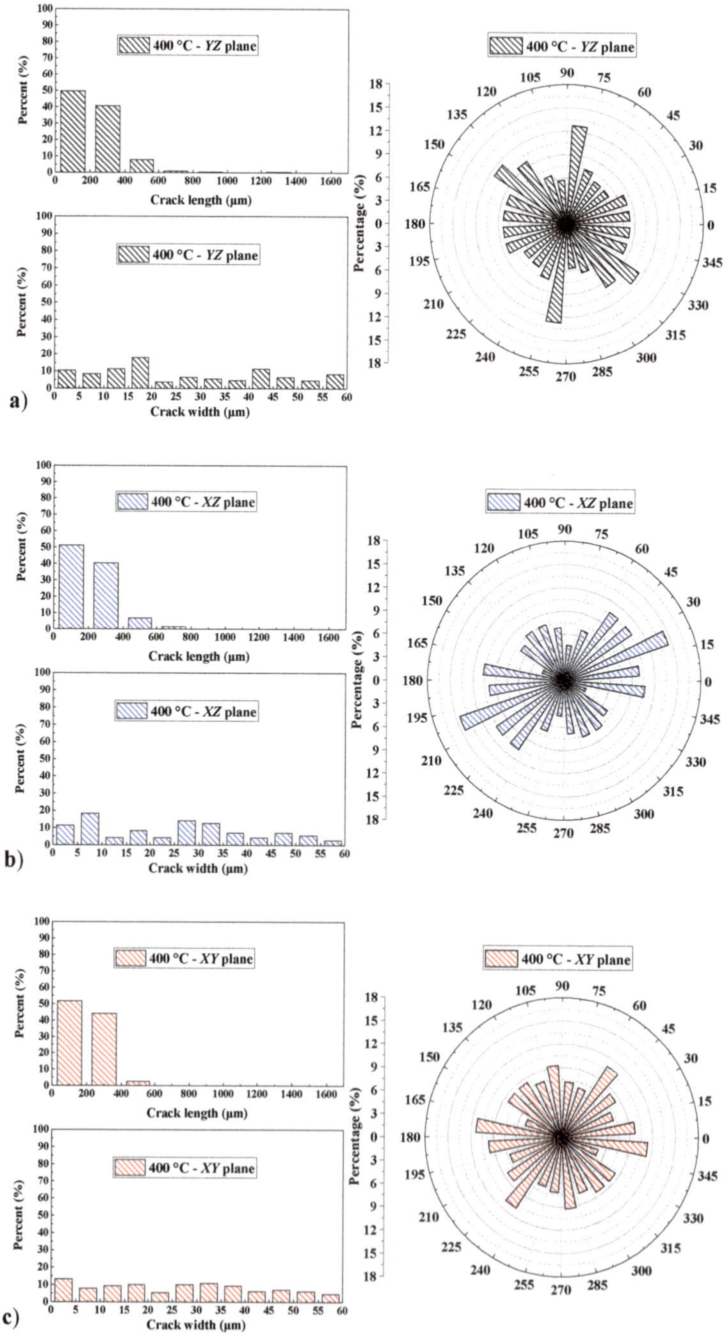

Figure 10. Characteristics of cracks for norite after 400 °C treatment, in (**a**) *YZ* plane; (**b**) *XZ* plane; and (**c**) *XY* plane.

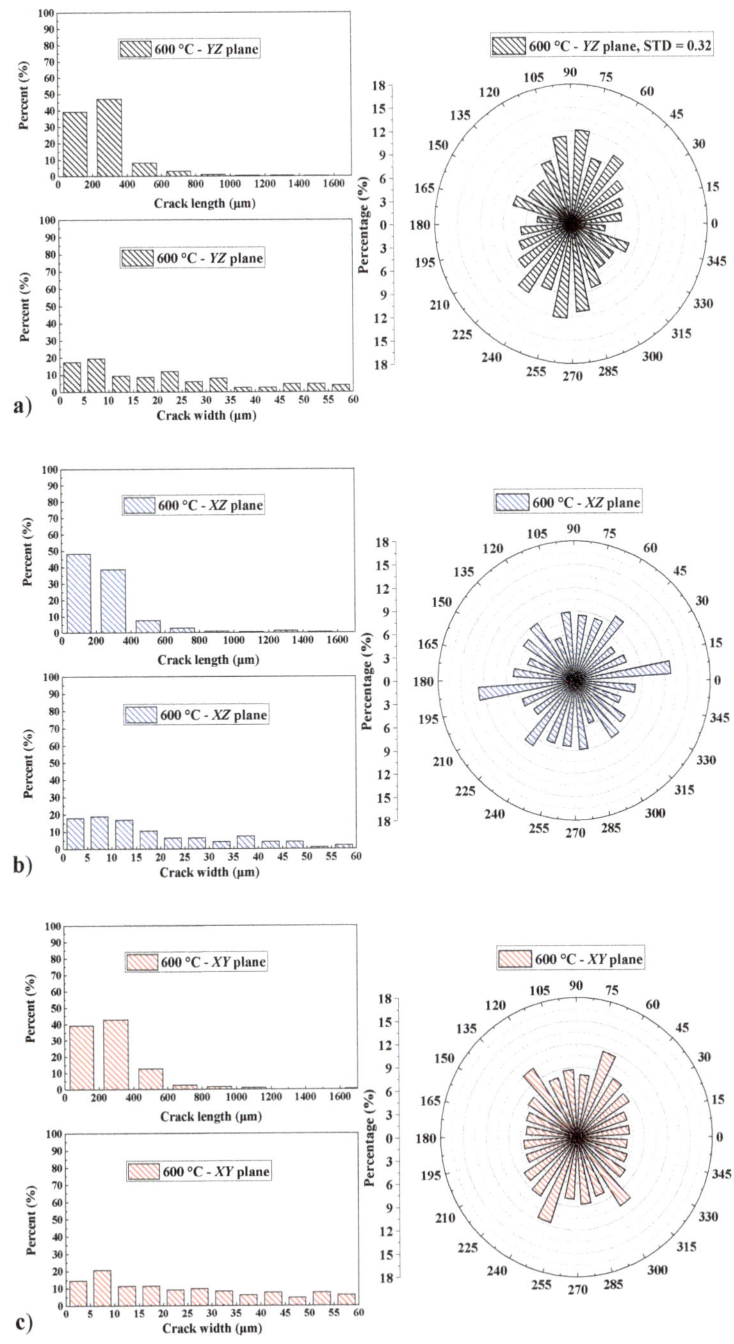

Figure 11. Characteristics of cracks for norite after 600 °C treatment, in (**a**) *YZ* plane; (**b**) *XZ* plane; and (**c**) *XY* plane.

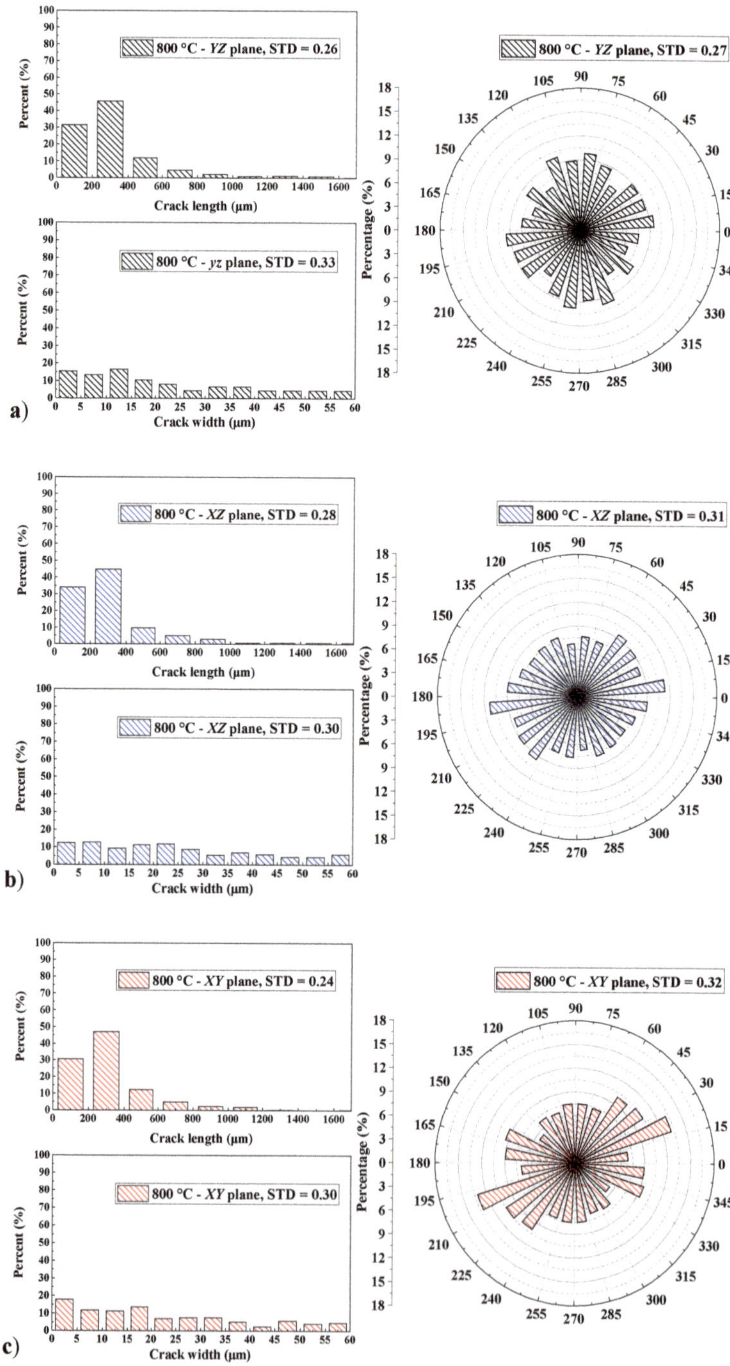

Figure 12. Characteristics of cracks for norite after 800 °C treatment, in (**a**) *YZ* plane; (**b**) *XZ* plane; and (**c**) *XY* plane.

The figures indicate that, in 400 °C-treated norite, more than 90% of micro-crack length falls into the two categories (0~200 μm and 200~400 μm). In addition, 5% to 10% of the micro-cracks have a length making up between 400 and 600 μm. On the other hand, the cracks are more likely to have their width ranging between 0 and 15 μm; such micro-cracks are very thin in width and could not be seen by the naked eye. One could also observe the resemblance of crack length and width in three different planes. Thus, it indicates that there were no significant anisotropic properties in 400 °C-treated norite.

With the heating temperature further rising to 600 °C and 800 °C, one of the most distinct facts is the extension of thermal-induced cracks. Such effect was due to the significant volume expansion in quartz minerals associated with the α-to-β transition at the temperature of 576 °C [9,10]. In 600 °C-treated norite, 200 μm to 400 μm long cracks became the most proportioned thermal-induced cracks. In addition, the increased percentages of longer cracks (crack length ranges between 400 μm and 1200 μm) were also notable. Such a trend was due to the tension increase based on further elevated heating.

On the right side of each figure, the overall orientation of each micro-crack was summarized. The consistency of the crack orientation could be seen from each plane along with different heating temperatures. For example, the most proportioned crack orientation in the *XZ* plane was between 15° and 30° after 400 °C. This range changed slightly to 15° and 30° for both 600 °C- and 800 °C-treated norites. The trend, the growth and propagation of cracks going in the same direction, could be due to the effects of the deformation twin [18]. Another notable phenomenon is the even distribution trend of crack orientations with higher temperature-treated norites, which is particularly noticeable from the *YZ* planes. The reason for this could be that, with a higher temperature treatment, more cracks propagate in the norite. In a generally homogeneous structure, the minor micro-cracks would occur in any direction, therefore, adjusting the overall distribution of crack orientations. Similar findings were reported by the research on the heterogeneity properties of heated granite [1]. It should be noted that the development of crack length and orientations could be more evident by performing continuous CT scanning on specimens under temperature growth. Moreover, the pretreatment of CT images and the image processing method, especially MMSF filtering, might slightly change the overall results on crack distributions. The effects of such methodology and further validations could be investigated by introducing more types of rocks in the future.

3.4. Anisotropic Analysis of Treated SIC Norite

The areal porosity, $D_{i,n}$, of high temperature-treated SIC norite was defined in the following Equation (2). It should be noted that once the first surface $D_{z,1}$ was calculated, then $D_{x,1}$ and $D_{y,1}$ were acquired from their corresponding orthogonal slices, as is presented in Figure 3. Quantification of $D_{i,n}$ was conducted by using the "bwarea" function in MATLAB. The results of $D_{i,n}$ are summarized in Figure 13.

$$D_{i,n} = \frac{S_{i,n}}{S_{total}} \times 100\% \quad (2)$$

where

$S_{i,n}$ = area of cracks and voids;
$S_{i,n}$ = total area of each image;
i = direction of each plane = x (*YZ* plane), y (*XZ* plane), z (*XY* plane);
n = number of each image at the same direction = 1, 2, 3, ...

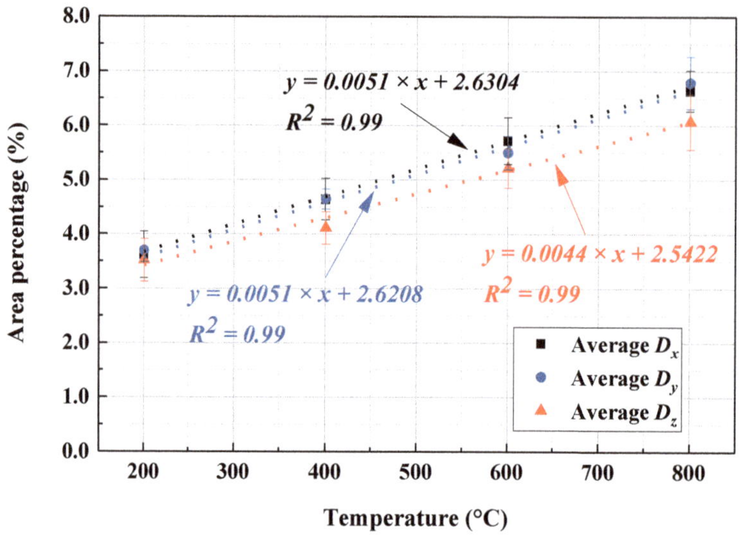

Figure 13. Average porosity of SIC norite in each direction.

Not surprisingly, an increase in porosity was observed with the increase in treatment temperature, indicating the growth of cracks and voids. From this study, however, a linear relationship between average areal porosity $D_{i,n}$ and the treatment temperature was shown, with a high coefficient of determination values. This linear relationship was also reported in past literature, especially when the temperature rose from 400 °C to 1000 °C [8,10]. In addition, the rate and porosity of D_x and D_y were very similar to each other, indicating a homogeneous structure in the horizontal plane. On the other hand, D_z was a little far from D_x and D_y, denoting a slight difference in voids and cracks between vertical and horizontal directions.

Furthermore, the anisotropy coefficient, A, is introduced to evaluate the anisotropy of SIC norite after thermal treatment. The definition of A for each direction was based on Equation (3) [1]. Results were plotted in Figure 14.

$$A_{x/z} = \frac{|D_{x,n} - D_{z,n}|}{D_{z,n}};\ A_{y/z} = \frac{|D_{y,n} - D_{z,n}|}{D_{z,n}};\ A_{x/y} = \frac{|D_{x,n} - D_{y,n}|}{D_{y,n}} \tag{3}$$

where $D_{x,n}$, $D_{y,n}$, $D_{z,n}$ = areal porosity at the slice of $D_{z,n}$, and its corresponding $D_{x,n}$ and $D_{y,n}$; $A_{x/z}$ = anisotropy coefficient for x-direction compared with z-direction.

From 200 °C to 800 °C, the anisotropy coefficient for all the slices ranged between 0% and 20%. In general, the anisotropy coefficients $D_{x,n}$ and $D_{y,n}$ peaked at around 16% at the temperature of 400 °C, then they decreased slightly to between 10% and 15%, at the temperature of 600 °C and 800 °C. Such a trend was also reported recently [1]. The reason may be due to the intensive generation of cracks between 400 °C and 600 °C. However, with further temperature increases, the $D_{x,n}$ and $D_{y,n}$ remained constant. It should also be noted that, at high temperatures (>600 °C), different slices have more scattered anisotropy coefficient values. On the other hand, $D_{x,y}$ remained constant throughout the temperature change. Overall, the anisotropic properties of SIC norite were noticeable between the vertical and horizontal directions. However, within the horizontal plane, the structure was more homogeneous (between XZ plane and YZ plane).

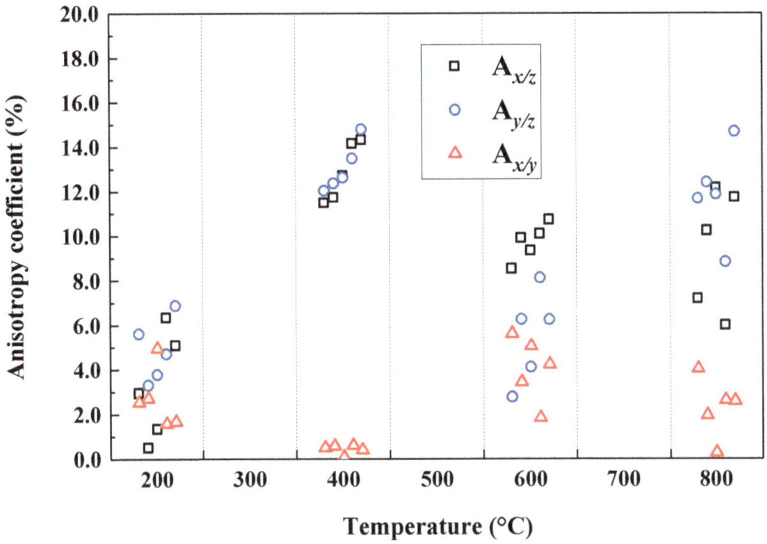

Figure 14. Anisotropic coefficients of SIC norite after treatment.

It was indicated from the literature that the orientation of micro-cracks in rocks would significantly influence the anisotropic properties of strength and modulus [19–21]. The longer the crack, the more important effect it may have on the formation and development of the failure pattern of rocks. During these studies, cracks that have lengths of around 500 μm (0.5 mm) were found to have the highest frequency. Therefore, the following Figure 15 presents the orientations of "major micro-cracks" (crack length longer than 500 μm) after 600 °C and 800 °C. It should be noted that the number of major micro-cracks generated at lower temperatures was too low, so these "short" length cracks were not analyzed.

As Figure 15 indicates, the orientations of major micro-cracks in the *XY* plane and *XZ* plane were generally consistent between 600 °C and 800 °C. Their most proportioned crack orientations accounted for between 0° and 60° and 120° and 195°, respectively. Such orientation ranges were slightly different from those illustrated in Figures 11 and 12. On the other hand, the distribution of orientations in the *XZ* plane became greater at 800 °C, and such a trend aligned with the situations in Figures 11 and 12.

In general, SIC norite did not show significant anisotropic properties with the increase in treatment temperature. However, several orientations have higher frequencies for thermal-induced micro-cracks, especially in *XY* and the *XZ* planes thus could further influence the damage pattern of norite.

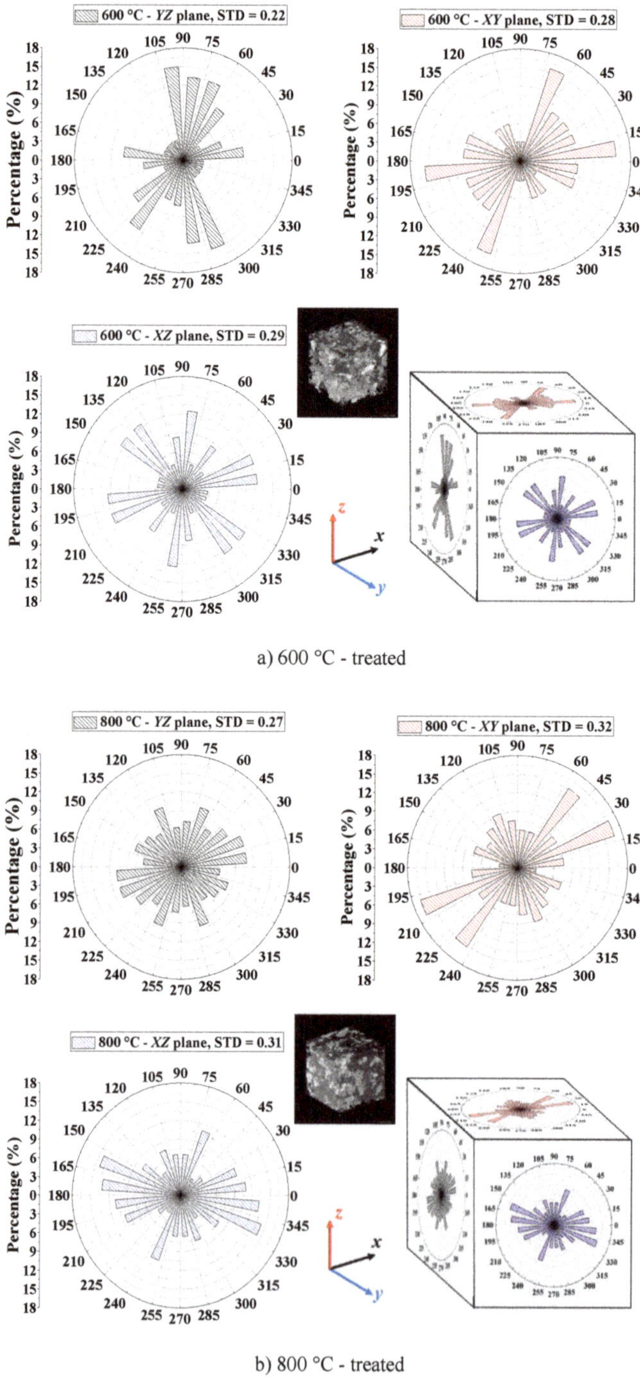

Figure 15. Orientations of major micro-cracks in (**a**) 600 °C and (**b**) 800 °C-treated SIC norite.

4. Conclusions

Based on the laboratory tests and analysis of the data, several specific conclusions and discussions can be addressed, as follows:

(1) The areal percentage of biotite ranged between 22% and 27%. The rest was occupied with felspar, quartz, and a small amount of metal minerals. Small-scale micro-cracking was first observed in the middle of biotite grains at the temperature of 400 °C. A further increase in temperature would generate more micro-cracks not only in biotite but also in quartz and feldspar. With the preheating temperature further developed to 800 °C, different micro-cracks started to connect, thus dividing the integral mineral into smaller segments;
(2) The image processing method together with the MMSF introduced in this study clearly identified the thermal-induced cracks. Quantitative analysis of pre-treated images indicated that most micro-crack lengths fell into the two categories (0~200 μm and 200~400 μm). On the other hand, the overall length of micro-cracks developed with the increase in preheating temperature, especially when it reached and exceeded 600 °C;
(3) The orientation of cracks exhibited two major trends with the increase in temperature: they followed a similar distribution pattern (in the XZ and XY planes) or became more evenly distributed (in the YZ plane);
(4) A linear relationship between the average porosity of SIC norite and the treatment temperature was found in each direction. Such relationships for the YZ and XZ planes were very similar, with a slight difference from plane XY. This indicated the anisotropic properties between vertical and horizontal directions of norite. This fact was further confirmed by the anisotropic coefficients which soared sharply from 200° C to 400 °C and then declined at 600 °C and 800 °C.

Overall, the paper presented a comprehensive study on the effects of thermal treatment and the anisotropic properties of SIC norite. The image processing tools were developed and used for the quantitative study of thermal-induced micro-cracks generated at different temperatures. It should be noted that the MMSF filtering treatment and image processing methods should be further validated by performing an investigation on other types of rocks. In addition, the crack propagation characteristics would be more evident under scanning observations during the heating process. Such limitations could be improved by future microscopic investigations on different rock samples, and by introducing advanced CT scanning equipment. Nevertheless, the methodology and findings from this paper will be an important reference for future studies regarding the thermal effects on norite and similar rocks.

Author Contributions: Conceptualization, S.-L.W. and S.Y.; methodology, S.-L.W. and S.Y.; software, S.-L.W. and J.H.; validation, S.Y. and B.S.; formal analysis, S.-L.W.; investigation, S.-L.W.; resources, B.S.; data curation, S.-L.W.; writing—original draft preparation, S.-L.W.; writing—review and editing, S.Y. and J.H.; visualization, S.-L.W.; supervision, S.Y.; project administration, S.Y. and B.S. All authors have read and agreed to the published version of the manuscript.

Funding: This research received no external funding.

Data Availability Statement: Data are contained within the article. MATLAB and ImageJ codes are available on reasonable request from the corresponding author.

Acknowledgments: This research received no specific grant from any funding agency from the public, commercial, or not-for-profit sectors.

Conflicts of Interest: The authors declare no conflict of interest.

References

1. Fan, L.; Gao, J.; Wu, Z.; Yang, S.; Ma, G. An investigation of thermal effects on micro-properties of granite by X-ray CT technique. *Appl. Therm. Eng.* **2018**, *140*, 505–519. [CrossRef]
2. Ashraf, U.; Zhang, H.; Anees, A.; Mangi, H.N.; Ali, M.; Zhang, X.; Imraz, M.; Abbasi, S.S.; Abbas, A.; Ullah, Z.; et al. A Core Logging, Machine Learning and Geostatistical Modeling Interactive Approach for Subsurface Imaging of Lenticular Geobodies in a Clastic Depositional System, SE Pakistan. *Nonrenew. Resour.* **2021**, *30*, 2807–2830. [CrossRef]
3. Ullah, J.; Luo, M.; Ashraf, U.; Pan, H.; Anees, A.; Li, D.; Ali, M.; Ali, J. Evaluation of the geothermal parameters to decipher the thermal structure of the upper crust of the Longmenshan fault zone derived from borehole data. *Geothermics* **2022**, *98*, 102268. [CrossRef]
4. Zhang, X.; Zhang, H.; Chang, F.; Xie, P.; Li, H.; Wu, H.; Ouyang, C.; Liu, F.; Peng, W.; Zhang, Y.; et al. Long-range transport of aeolian deposits during the last 32 kyr inferred from rare earth elements and grain-size analysis of sediments from Lake Lugu, Southwestern China. *Palaeogeogr. Palaeoclim. Palaeoecol.* **2021**, *567*, 110248. [CrossRef]
5. Riaz, M.S.; Bin, S.; Naeem, S.; Kai, W.; Xie, Z.; Gilani, S.M.M.; Ashraf, U. Over 100 years of faults interaction, stress accumulation, and creeping implications, on Chaman Fault System, Pakistan. *Int. J. Earth Sci.* **2019**, *108*, 1351–1359. [CrossRef]
6. Zhang, Y.; Sun, Q.; He, H.; Cao, L.; Zhang, W.; Wang, B. Pore characteristics and mechanical properties of sandstone under the influence of temperature. *Appl. Therm. Eng.* **2017**, *113*, 537–543. [CrossRef]
7. Suo, Y.; Chen, Z.; Rahman, S.S. Changes in Shale Rock Properties and Wave Velocity Anisotropy Induced by Increasing Temperature. *Nonrenew. Resour.* **2020**, *29*, 4073–4083. [CrossRef]
8. Zhang, H.; Sun, Q.; Jia, H.; Dong, Z.; Luo, T. Effects of high-temperature thermal treatment on the porosity of red sandstone: An NMR analysis. *Acta Geophys.* **2021**, *69*, 113–124. [CrossRef]
9. Zhou, C.; Xie, H.; Zhu, J.; Zhou, T. Failure criterion considering high temperature treatment for rocks from a micromechanical perspective. *Theor. Appl. Fract. Mech.* **2022**, *118*, 103226. [CrossRef]
10. Isaka, B.A.; Ranjith, P.; Rathnaweera, T.; Perera, M.; De Silva, V. Quantification of thermally-induced microcracks in granite using X-ray CT imaging and analysis. *Geothermics* **2019**, *81*, 152–167. [CrossRef]
11. Nguyen, D.K.; Morishita, T.; Soda, Y.; Tamura, A.; Ghosh, B.; Harigane, Y.; France, L.; Liu, C.; Natland, J.H.; Sanfilippo, A.; et al. Occurrence of Felsic Rocks in Oceanic Gabbros from IODP Hole U1473A: Implications for Evolved Melt Migration in the Lower Oceanic Crust. *Minerals* **2018**, *8*, 583. [CrossRef]
12. Waterton, P.; Hyde, W.R.; Tusch, J.; Hollis, J.A.; Kirkland, C.L.; Kinney, C.; Yakymchuk, C.; Gardiner, N.J.; Zakharov, D.; Olierook, H.K.H.; et al. Geodynamic Implications of Synchronous Norite and TTG Formation in the 3 Ga Maniitsoq Norite Belt, West Greenland. *Front. Earth Sci.* **2020**, *8*, 562062. [CrossRef]
13. Riller, U.; Schwerdtner, W.M. Mid-crustal deformation at the southern flank of the Sudbury Basin, central Ontario, Canada. *GSA Bull.* **1997**, *109*, 841–854. [CrossRef]
14. Lightfoot, P.C.; Doherty, W. Chemical Evolution and Origin of Nickel Sulfide Mineralization in the Sudbury Igneous Complex, Ontario, Canada. *Econ. Geol.* **2001**, *96*, 1855–1875. [CrossRef]
15. Deutsch, A.; Grieve, R.A.F.; Avermann, M.; Bischoff, L.; Brockmeyer, P.; Buhl, D.; Lakomy, R.; Müller-Mohr, V.; Ostermann, M.; Stöffler, D. The Sudbury Structure (Ontario, Canada): A tectonically deformed multi-ring impact basin. *Geol. Rundsch.* **1995**, *84*, 697–709. [CrossRef]
16. Wang, S.-L.; Lv, Q.-F.; Baaj, H.; Li, X.-Y.; Zhao, Y.-X. Volume change behaviour and microstructure of stabilized loess under cyclic freeze–thaw conditions. *Can. J. Civ. Eng.* **2016**, *43*, 865–874. [CrossRef]
17. Chai, G.; Eckstrand, R. Rare-earth element characteristics and origin of the Sudbury Igneous Complex, Ontario, Canada. *Chem. Geol.* **1994**, *113*, 221–244. [CrossRef]
18. Rybacki, E.; Evans, B.; Janssen, C.; Wirth, R.; Dresen, G. Influence of stress, temperature, and strain on calcite twins constrained by deformation experiments. *Tectonophysics* **2013**, *601*, 20–36. [CrossRef]
19. Heap, M.J.; Coats, R.; Chen, C.-F.; Varley, N.; Lavallée, Y.; Kendrick, J.; Xu, T.; Reuschlé, T. Thermal resilience of microcracked andesitic dome rocks. *J. Volcanol. Geotherm. Res.* **2018**, *367*, 20–30. [CrossRef]
20. Nadan, B.J.; Engelder, T. Microcracks in New England granitoids: A record of thermoelastic relaxation during exhumation of intracontinental crust. *GSA Bull.* **2009**, *121*, 80–99. [CrossRef]
21. Takemura, T.; Golshani, A.; Oda, M.; Suzuki, K. Preferred orientations of open microcracks in granite and their relation with anisotropic elasticity. *Int. J. Rock Mech. Min. Sci.* **2003**, *40*, 443–454. [CrossRef]

mining

MDPI

Article

Stability Assessment of the Slopes of an Oceanside Coral Limestone Quarry under Drawdown Condition of Semidiurnal Ocean Tides

Joan Atieno Onyango [1,2,*], Takashi Sasaoka [1], Hideki Shimada [1], Akihiro Hamanaka [1] and Dyson Moses [1,3]

1 Department of Earth Resources Engineering, Faculty of Engineering, Kyushu University, Fukuoka 819-0395, Japan
2 Department of Mining, Materials and Petroleum Engineering, Jomo Kenyatta University of Agriculture and Technology, Nairobi P.O. Box 62000-00200, Kenya
3 Department of Geography, Earth Science, and Environment, School of Pure and Applied Science, University of Malawi, Zomba P.O. Box 280, Malawi
* Correspondence: jaonyango@jkuat.ac.ke

Abstract: Vipingo quarry in Kilifi county, Kenya, is one of the quarries supplying coral limestone for limestone manufacturing in the coastal region. Due to its close proximity to the Indian ocean, the semidiurnal ocean tides tend to have an influence on the stability of the quarry slopes adjacent to the shoreline. Finite element numerical analysis using the generalized Hoek–Brown criterion is conducted to assess the stability condition of the slopes followed by slope-angle optimization to determine the safest overall slope angle as well as analyzing the stability of the slopes due to action of varying ocean tides. The optimum overall slope angles for various excavation depths are found to be 52° for 20 m, 46° for 30 m, 42° for 40 m, and 39° for 50 m, which are the same even with varying distance of the slope face from the shoreline. A parametric analysis shows that there is no significant effect of the tides on slope stability for excavations above the water table, but as the quarry gets deeper, the slope stability is affected. A sensitivity factor (ζ) is introduced, being a measure of how much the slope safety factor is reduced as a result of the semidiurnal tidal action.

Keywords: slope stability; finite element analysis; generalized Hoek–Brown criterion; semidiurnal ocean tides; sensitivity factor

1. Introduction

Slope engineering practices has over time experienced instability in natural and artificial slopes, such as landslides after heavy or prolonged rain. Failure of reservoir embankments as a result of sudden decline in water levels (also known as rapid drawdown) and due to prolonged immersion are also common phenomena. Instability caused by action of water seepage forces is mostly associated with soil slopes and weak rock slopes. Seepage is directly responsible for instability in slopes, because in wet conditions, the shear strength of the geomaterials is greatly reduced. This is corroborated by laboratory experiments by various researchers that show a reduction in rock mechanical strength properties in a wet/saturated state [1–3]. In the design of slopes in weak rock masses, groundwater seepage is one of the most important parameters that have to be considered [4]. This is because in weak rocks, the weakly cemented rock fabric is already at a higher potential for collapse and is aggravated by water seepage. In weak carbonate rock masses, for instance, the high porosity and high permeability of the rock permits easier seepage through the rock, resulting in significant impact on the rock mass strength. Preexisting joint apertures in the rock mass may open or close under the action of water, thus inducing a rapid change in the hydraulic transmissivity [5]. With the alteration of joint apertures also comes a change in joint hydraulic pressure, thus alteration of rock structure [6].

Analysis methods for evaluating seepage effects on slope stability have evolved over the years, from the use of flow nets developed by Philipp Forchheimer in the early 1900s [7], to limit-equilibrium analysis methods [8]. The computer age has made it easier to use more robust numerical analysis methods that can accommodate even more complex seepage boundary conditions. Numerous researchers have employed a wide range of numerical analysis methods to evaluate seepage effects on slope stability of open pits, considering such water sources as rainfall, dams, and lakes. Drawdown analysis is always used for the case of ponded water that considers only the case of decrease in water levels i.e., lowering the water table. The term "rapid drawdown" has been traditionally used with earth dam reservoir levels, meaning the duration of decrease in water level is short relative to the life span of the dam. Fan et al. [9], for instance, using FLAC^{3D}, evaluated the impact of an adjacent lake on the stability of an open pit mine by studying the effect of decreasing water levels after rainy seasons. Moosavi et al. [10] used the discrete element numerical modeling software UDEC to evaluate the drawdown effects of groundwater aquifers on the stability of open pit slopes as the excavation advanced to greater depths. There are two classic procedures for analyzing drawdown effects and the resulting stability conditions, grouped as the stress-based undrained approach recommended for impervious materials and the pure flow (drained) approach recommended for rigid pervious materials, typically granular soils [11]. Wong, Szeto, and Zhang [12], in studying the failure modes of porous rocks, reiterated that in some ways the phenomenology of porous rock deformation parallels that in soil mechanics. A slope generally fails in saturated conditions as a result of development of positive pore-water pressure occasioned by high porosity of the slope material [13].

Flow analysis considers modeling three types of drainage conditions: drained, undrained, and fully drained [14]. The drained condition is where fluid can move into and out of the model and the total volume can change. It is a transient pore-pressure state where drainage is a function of time and excess pore pressure can exist. In the undrained condition, fluid cannot move into or out of the model, there is no volume change, and excess pore pressure is a function of applied loads. The fully drained condition is a steady state characterized by zero excess pore pressure and drainage analysis is at infinite time.

Analysis methods for water seepage effects can be either coupled or uncoupled. Uncoupled analysis is the case of total stress analysis where deformation is a result of changes in total stress, i.e., changes in pore pressure do not affect deformation and changes in loading or deformation do not affect pore pressure. Coupled analysis, which is based on Biot theory, is used for cases where changes in pore pressure can affect deformation and changes in loading or deformation can affect pore pressure. Two types of coupling, namely, direct and indirect coupling, were discussed by Rutqvist and Stephansson [6]. Direct coupling occurs through deformation and pore-fluid interactions where changes in the pore volume result in a change to the mechanical properties. Direct coupling is in two forms: solid to fluid coupling, where a change in the stress causes a change in the pore-fluid pressure, and fluid to solid coupling where a change in the pore-fluid pressure causes a change in the rock mass volume.

Indirect coupling, on the other hand, is where changes in the mechanical or hydraulic processes affect each other through changes in mechanical and hydraulic properties. Two forms of indirect coupling are solid-to-fluid coupling, where a change in the stress causes a change in the hydraulic properties of the rock mass, and fluid-to-solid coupling, where a change in the pore-fluid pressure causes a change in the mechanical properties of the rock mass. According to C. Huimei, Z. Zhiye and S. Jianping, [5], flow can be considered to be through the porous media if there are significant variably oriented and connected fractures in the rock mass. This is the assumption that is considered by continuum methods of coupled numerical hydromechanical modeling, such as the finite difference Method (FDM) and finite element method (FEM) [15,16].

In the analysis of seepage effects on slope stability, engineers are always confronted with the tough choice of when to consider a steady-state seepage condition or transient seepage. The difference between the two conditions is that in a steady state, seepage parameters, such as hydraulic head, flow rate, or rock/soil hydraulic property, are constant, whereas in a transient state, the properties change with time [17]. Transient seepage has been applied mostly to soil slopes, because in soil, the pore pressure response to changes in seepage parameters is controlled by the hydraulic conductivity and compressibility of soil. Pauls et al. [18] conducted a transient analysis to analyze the impact of drawdown after flooding on highly plastic clay. They found that transient analysis gave a more realistic estimate of safety factors compared with filed observations of piezometric data, while the steady-state analysis result was more exaggerated. Seepage effects on rock slopes are often analyzed for slopes in heavily jointed rock and those that possess a small finite number of well-developed jointing pattern with relatively good spacing [19]. Steady-state seepage analysis is more common in rock slopes due to low permeability of rocks as a result of fines. Martin and Stacey recommend drained transient analysis for rock formations whose permeability is greater than $\sim 10^{-8}$ m/s. According to the US Army Corps of Engineers report by Tracy et al. [17], transient analysis is best used as a verification of piezometer readings and when it is necessary to determine the time it takes to achieve a steady state. Transient analysis depends on the computer program's capacity to run a transient analysis.

Analyses of seepage effects on slope stability have not addressed conditions unique to oceanic tidal variations. In this paper, a situation where the water level rises and falls with two high levels and two low levels every lunar day is presented. An oceanside coral limestone quarry is assessed for stability under the influence of semidiurnal ocean tides as the excavation progresses to greater depths. The changes in tidal levels subject the adjacent quarry slopes to a phenomenon that can be described as a cycle of daily lifts and drawdowns where the slopes experience the impacts of rising and falling external water levels. Alonso and Pinyol [11] and Cen et al. [20] describe two consequences of reduction in water level that is an unloading effect of removing water. When the water level is high, hydrostatic pressure helps in stabilizing the slope, and so a reduction in the water level results in the reduction of the stabilizing external hydrostatic pressure, as well as causing a modification of the internal pore-water pressure due to phreatic fluctuations that produce effective stress changes within the ground. The stability of slopes can therefore be affected by the drawdown phenomenon [21] created by the rising and falling of ocean semidiurnal tides.

The objective of this work is to assess how the daily rising and falling of water table levels occasioned by tidal variations impact on the stability of slopes of a quarry adjacent to the ocean. Finite element numerical analysis using Phase 2 Version 7 software developed by RocScience (Toronto, ON, Canada) is used in the study. The analysis steps include first determining the optimum slope angles for various excavation depths followed by assessing the impact of tidal changes on the slope stability. The groundwater analysis and stress analysis modules of the software make it a suitable tool to enable the goals of this study. Parametric analysis on interaction of slope at different excavation depths and the adjacent ocean waters is done to evaluate and give a reasonable guide for determining the suitable distance of the slope face from the shoreline.

2. Engineering Geological Background of Area under Study

2.1. Physical Location

Vipingo coral limestone quarry is situated in Kilifi county along Kenya's coastal plain (Figure 1a), which lies along the Indian Ocean seashore and reaches 3–8 km inland, with a relatively flat topography and elevation mostly less than 33 m above sea level. The plain comprises mainly Pleistocene deposits, with Pleistocene coral reef being the most important for Vipingo quarry. Previous geological exploration estimated the Pleistocene coral reef deposit to be approximately 1–4 km wide and up to 100 m thick, extending along the whole length of the Kenyan coast. The quarry is adjacent to the Indian Ocean at about 50 m from the shoreline, extending in a strip of almost 4 km long and about 2 km wide, as shown in Figure 1b.

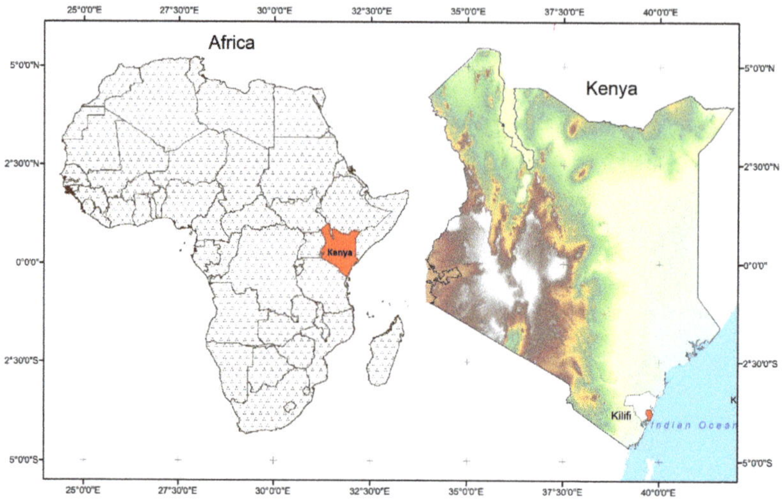

(**a**)

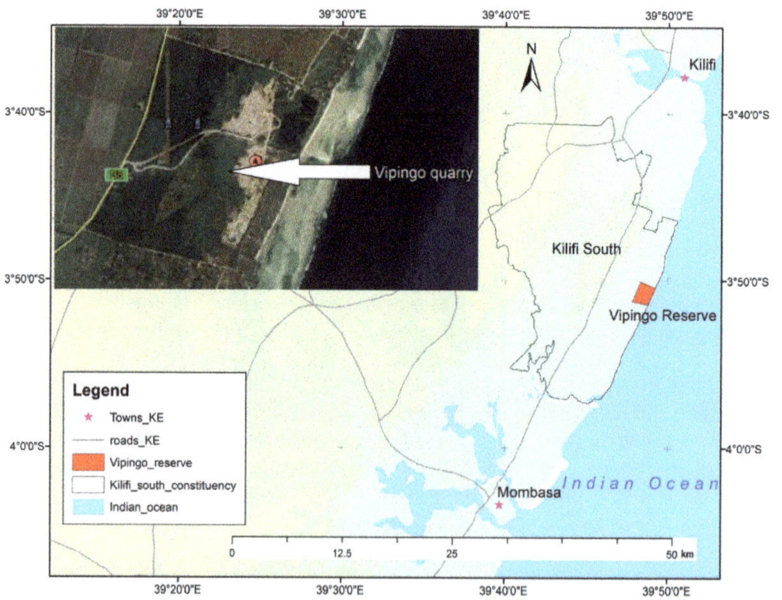

(**b**)

Figure 1. Location of study area. (**a**) Location of Kilifi county. (**b**) Location of Vipingo quarry (maps created using QGIS version 3.16.5 LTR available on https://www.qgis.org, accessed on 29 March 2021).

2.2. Geology and Hydrogeology

Most of the area lies on coral limestone of the Pleistocene age, a fossil coral reef, acquired as limestone reserve, and is covered with shallow but fertile soil. The Pleistocene period was characterized by fluctuating sea levels resulting in various types of sediments. The reef would grow upwards during rising sea levels and expand laterally during stagnation, causing a heterogeneous mix of weathered limestone, clays, and sand. Low water levels during glacial periods would expose the reef surface to erosion, weathering, and karstification (formation of sinkholes and caves through dissolution of limestone by carbonic acid present in fresh rainwater with high CO_2 and O_2 content). The sediments underlying the coastal plains are mainly sands and fossil coral limestone, both highly porous and permeable, which allows for higher aquifer transmissivity and favorable recharge conditions. The groundwater table gently slopes towards the coast, and groundwater flows from the foot plateau through the sandy back-reef deposits into the limestone reef, before ultimately discharging into the sea. All rainwater is drained naturally through the highly porous coral limestone to sea level.

The Indian Ocean experiences semidiurnal tides, that is, two high and two low tides of approximately equal size every lunar day. The tidal level is measured with the coastline as the datum. The state of Oregon [22] defines the coastline as the line of mean low water along that portion of the coast that is in direct contact with the open sea and the line marking the seaward limit of inland waters. Based on this definition, the tidal parameters for Vipingo area as documented by the Kilifi county government are as follows: mean high (MH) water = 2.96 m, mean low water = 0.71 m, mean higher high water = 3.15 m, mean lower high water = 2.78 m, mean higher low water = 0.8 m, mean lower low water = 0.65 m [23]. This places the low tide at 0.71 m and high tide at 2.96 m above sea level.

2.3. Engineering Background

Vipingo quarry is one of the major quarries supplying coral limestone for cement production in the coastal region of Kenya. The elevation of the quarry pit is currently maintained above sea level, the maximum pit depth so far being 20 m. The pit, however, gets deeper as excavation progresses inland. The rock mass has no distinct beddings or foliations. The top soil is thin, and together with weathered rock in some sections makes a band of weak top layer averaging 3 m, forming low-grade limestone. Some sections have coral sand of average thickness of 4 m. The eastern edge of the pit is situated just 50 m from the shoreline of the Indian Ocean.

Despite excavation being done above the water table, during high tide, water level rises and can be seen on the pit floor, as in Figure 2. Instabilities on the slopes have previously been reported, especially during rainy season, but no study has been done to relate slope stability to ocean-water seepage. As the excavation advances, it is important to evaluate the impact of the ocean waters on the stability of the pit slopes and make foolproof slope design judgments. This will also be crucial at the end of the quarrying life, because there is a rehabilitation plan in place and so it is important to consider all factors affecting slope stability in the final pit design.

Figure 2. Ocean water visible in the quarry floor during high tide.

3. Methodology

3.1. The Workflow Process

This work is part of an ongoing slope design analysis. The field investigation and laboratory tests have previously been done. The scope of this work involves numerical modeling and analysis of stability condition of the quarry slopes under the effects of seepage of ocean waters. The workflow process is as shown in Figure 3.

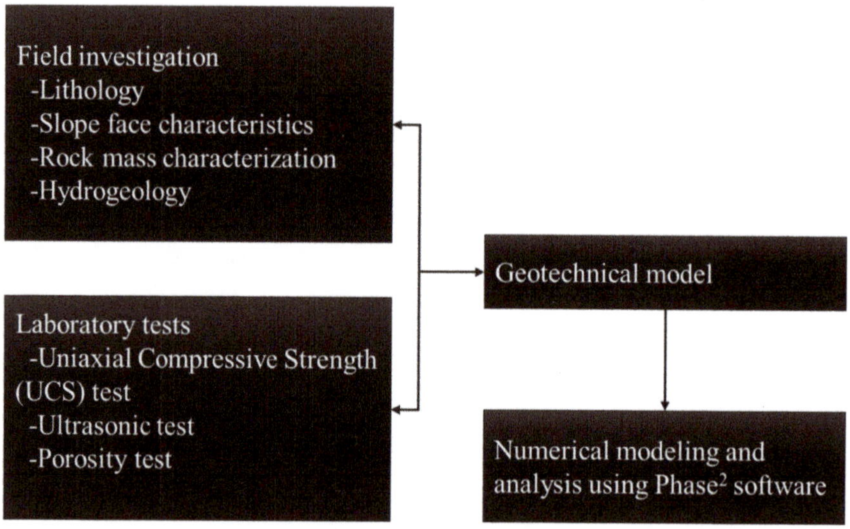

Figure 3. The workflow process.

3.2. Slope Model Construction

Figure 4 shows the environment around the quarry, showing the position of the quarry relative to the Indian Ocean waters.

Figure 4. The quarry environment.

In this work, a model of the high-grade limestone region suffices for simulation based on the objective of the study. The regions with weathered rock and coral sand are not considered in this study, since they happen to form the topmost layers of the rock mass and are not in contact with ocean water. The model boundary in Figure 5 guides the development of the geometry of the geomechanical prototype for this analysis. The model geometries are quite simple, so they were generated by entering coordinates directly into the user interface. The analysis considers slopes at excavation depths of 20 m, 30 m, 40 m, and 50 m, with varying distances of 50 m, 75 m, and 100 m from the shoreline.

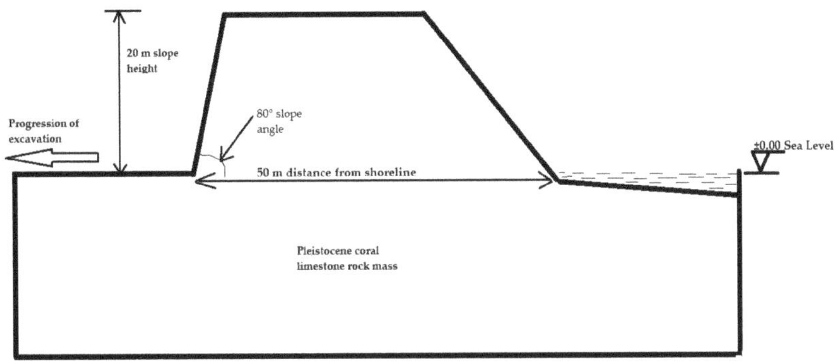

Figure 5. Model and boundary of slope section A-A under consideration.

3.3. Numerical Modeling

Due to the homogeneous nature of the rock mass under study, finite element analysis by shear strength reduction (SSR) using RocScience's Phase 2 Version 7 software was found to be sufficient for carrying out the numerical modeling and analysis for the area under study. The SSR concept entails systematically using finite element solutions to reduce the shear strength envelope of material by a strength reduction factor (SRF) or factor of safety until solutions do not converge. That is, the deformations become unacceptably large, thus bringing the slope to the verge of failure (within a specified tolerance). The generalized Hoek–Brown criterion is the strength model applied in this analysis. Based on the shear-normal stress envelopes developed by Balmer [24] for the Mohr-Coulomb criterion, Hammah et al. [25] formulated the generalized Hoek–Brown criterion by relating σ_n and τ to σ_1 and σ_2 in the following equations:

$$\tau = (\sigma_1 - \sigma_3)\frac{\sqrt{1 + am_b\left(m_b\frac{\sigma_3}{\sigma_{ci}} + s\right)^{a-1}}}{2 + am_b\left(m_b\frac{\sigma_3}{\sigma_{ci}} + s\right)^{a-1}} \tag{1}$$

$$\sigma_n = \frac{1}{2}(\sigma_1 + \sigma_3) - \frac{1}{2}(\sigma_1 - \sigma_3)\frac{am_b\left(m_b\frac{\sigma_3}{\sigma_{ci}} + s\right)^{a-1}}{2 + am_b\left(m_b\frac{\sigma_3}{\sigma_{ci}} + s\right)^{a-1}} \tag{2}$$

where σ_1 and σ_3 are the major and minor effective principal stresses at failure, σ_{ci} is the uniaxial compressive strength of the intact rock material, and a, m_b and s are material constants.

In conducting analysis by shear strength reduction, the Equation (1) is simply divided by the reduction factor F. In Phase 2 software, the strength reduction occurs automatically until the critical SRF is obtained, which gives the safety factor of the slope.

Analyzing the slope stability under the influence of seepage forces in Phase 2 makes use of the finite element solid–fluid coupling which refers to the interaction between fluid pore pressure and solid stresses and deformation [26]. The two analysis options are coupled and uncoupled. In uncoupled analysis, an undrained simulation is first conducted to compute the excess pore pressure, after which a solid analysis is carried out to calculate the change in the stress caused by the changes in pore pressure. Coupled analysis is based on Biot's theory and is applicable to time-dependent analysis problems where changes in pore pressure can affect deformation and changes in loading or deformation can affect pore pressure. Phase 2 makes use of Biot's equation in coupled analysis and is formulated as [14]:

$$\frac{K'}{\gamma_w}\left[k_x\frac{\partial^2 u_w}{\partial x^2}+k_y\frac{\partial^2 u_w}{\partial y^2}+k_z\frac{\partial^2 u_w}{\partial z^2}\right]=\frac{\partial u_w}{\partial t}-\frac{\partial p}{\partial t} \quad (3)$$

where K' is the bulk modulus of the material, γ_w is the unit weight of water, k_x, k_y, and k_z are the permeabilities in different directions, u_w is the pore-water pressure, p is the mean total stress, and t denotes time.

When 2D equilibrium is considered, the gradient of effective stress is augmented by the gradients of fluid pressure u_w, leading to the equilibrium and continuity equations below:

$$KMr + Cu_w = f \quad (4)$$

$$C^T\frac{dr}{dt} - KPu_w = 0 \quad (5)$$

where KM and KP are the elastic solid and fluid matrices, f is the external loading vector, and C is a rectangular coupling matrix.

3.4. Boundary Conditions

In numerical modeling, the displacement boundary conditions are set to ensure zero velocity in the model. The bottom boundary of the model is pinned/fixed in the X–Y direction. The left and right sides of the model have a rolling boundary i.e., fixed in X direction and free to move in Y direction. The slope bottom face (quarry floor), the slope face, the slope top, and the area facing the ocean have a free boundary. Discretization and meshing of the model are achieved by employing graded three-node triangular elements with 200 nodes on external boundary. The slope-face mesh is refined by single refinement i.e., one round of mesh densification.

Seepage analysis is done in a fully drained, uncoupled, steady-state condition. The rate of change of tidal levels is not quite clearly defined, but the high and low tidal levels are definite, so analysis in the steady state suffices to fulfill the objective of the study. The semidiurnal tides mean the pit slopes experience two high and two low head levels daily. At high tide, ocean water is usually visible in the pit floor. The slope therefore experiences alternating rapid drawdowns and "drawups", and subsequent rapid hydrostatic and pore pressure changes. The analysis is implemented by running a finite element groundwater analysis with changing boundary conditions, in this case being the changes in the total head. The analysis has three stages, the first stage being the low-tide condition, the second the rise to high-tide condition, and the third the falling back to low tide again.

The infiltration faces of the model are the ocean floor up to the tidal level, and the left and right boundaries. The tidal level is defined by the total head as the hydraulic boundary condition and the weight of the ocean water is defined by adding a ponded water distributed load to the model. The outflow faces are the slope face and the slope bottom face. The bottom boundary is impermeable. Unknown boundary condition is set for the excavation face, the bottom of the model, the top part, and the rest of the sloping section of the model facing the ocean.

3.5. Input Parameters

The rock mass properties required for analysis by the generalized Hoek-Brown criterion are given in Table 1, being adapted from data previously published by Joan et al. [27]. The hydraulic conductivity value was obtained from literature on Pleistocene coral limestone on the tropical island aquifers of Barbados by Jones and Jay [28].

Table 1. Input parameters.

Parameter	Symbol	Unit	Value
UCS	σ_{ci}	MPa	17.20
Elastic Modulus	E_i	GPa	32.72
Porosity	n	%	8.95
Hoek-Brown Parameters	GSI		20
	m_i		9
	D		0.70
	a		0.54
	s		9.22×10^{-6}
	m_b		0.11
Hydraulic conductivity	K	m/s	1.62×10^{-4} m/s

GSI = geological strength index; m_i = intact rock constant; D = disturbance factor; a, s, m_b are Hoek-Brown constants for the rock mass.

4. Numerical Modeling Results

4.1. Slope Angle Optimization

The quarrying activity is done with near-vertical benches, which results in precarious slopes, as seen in Figure 6. The slope stability assessment therefore commenced with determining the safest overall slope angle. Overall slope angle optimization was done by reducing the slope angle systematically from 90° to 30° and noting the associated factor of safety. Figure 7 shows the resulting probable deformation profile (represented by the gray boundary line) for a steep slope of 80°. It can be seen that the upper part of the slope is at a high risk of falling over the slope toe. Shear strain is also highest at the slope toe. It is therefore necessary to unload the slope by removing the material to leave a gentler and stable slope, which is the basis for slope angle optimization. Slope angle optimization was done based on the design concept of an allowable safety factor. An allowable design safety factor value of 1.2 was considered reasonable for the case under study. It is worthy to note that at every excavation depth, the distance from the shoreline to the slope face was maintained at 50 m, 75 m, and 100 m, and only the slope face angle was altered, leading to the differences in model shapes.

Figure 6. Near-vertical slopes from excavation.

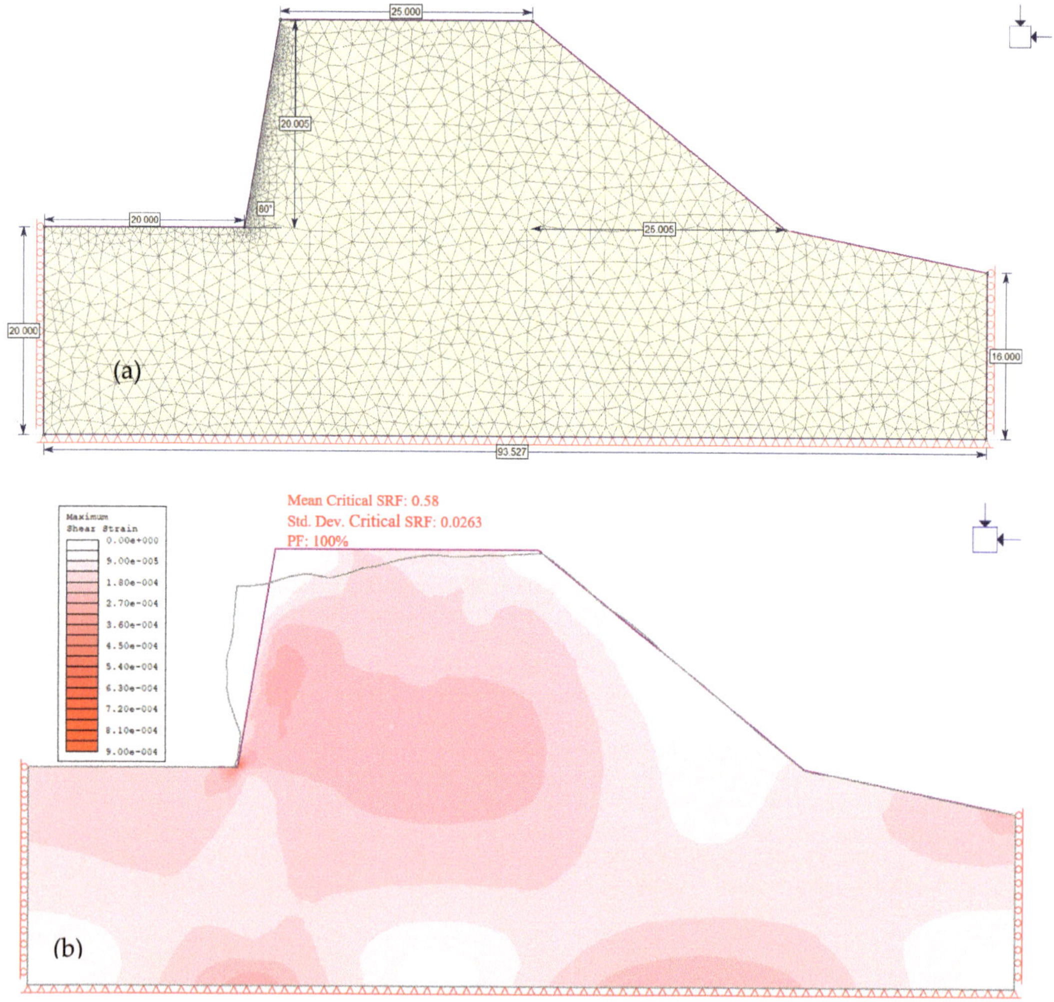

Figure 7. (**a**) Dimension and mesh of slope at 80° angle; (**b**) probable resultant deformation of steep quarry slope.

The optimum slope angle of 52° corresponding to a safety factor of 1.2 was obtained for an excavation depth of 20 m at 50 m from the shoreline. Following the optimization of the slope angle at 20 m excavation depth, the other slopes at deeper excavations were determined from a smaller range of trials. Optimum slope angle for each excavation depth was determined by starting from the slope angle of the prior excavation depth, then reducing the slope angle till the angle corresponding to a safety factor of 1.2 was attained. The overall slope angles for various excavation depths were as follows: 46° for 30 m depth, 42° for 40 m depth, and 39° for 50 m depth, all at 50 m distance from the shoreline, as shown in Figure 8.

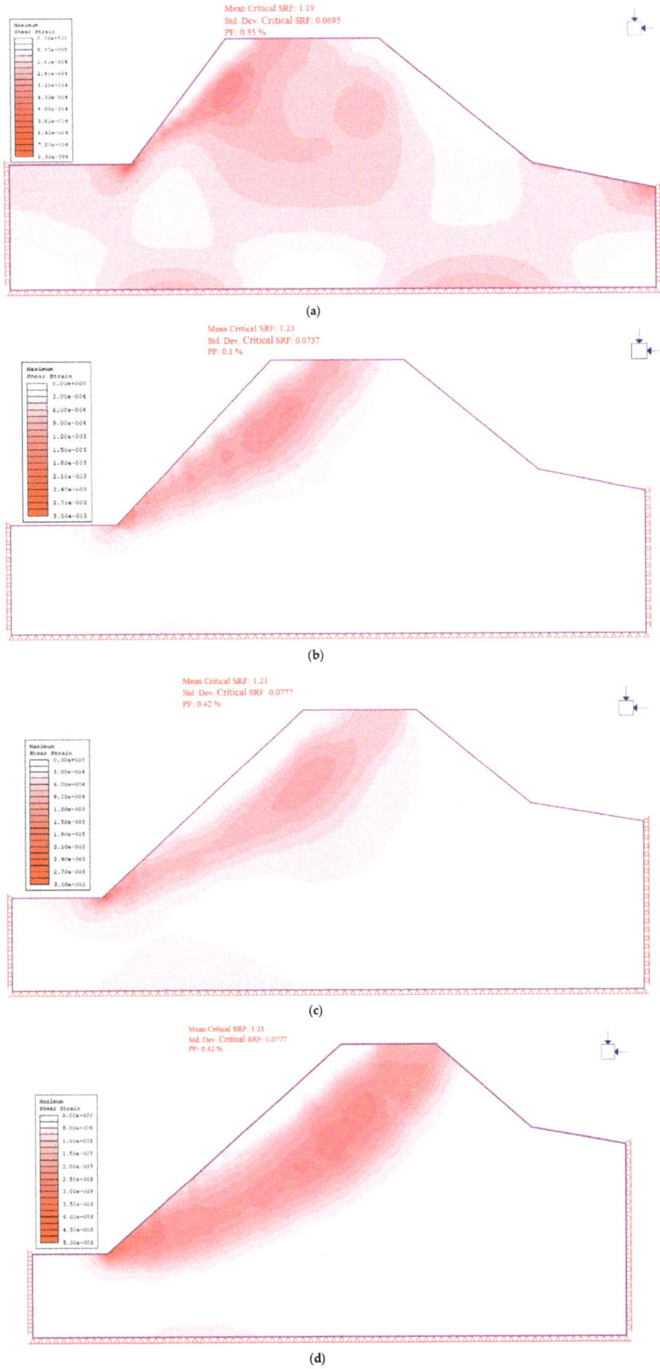

Figure 8. Optimized slope angles for various excavation depths at 50 m distance from the shoreline. (**a**) 20 m depth, 52° slope angle; (**b**) 30 m depth, 46° slope angle; (**c**) 40 m depth, 42° slope angle; (**d**) 50 m depth, 39° slope angle.

The stability analysis also incorporated the statistical analysis function of Phase 2 to evaluate the slope probability of failure (PF). The PF is generated by the analysis software when the statistical option is selected in project settings. A standard deviation is set for material properties to allow the statistical calculation. From Figure 8, the probability of failure is high for steep slope angles between 90° and 60°, after which it significantly reduces when the safety factor is ≥1. The graph gives a good visual impression of the stability of the slope, in that even when the safety factor is high, the steep slope angles present high probabilities of failure. In the dry condition, there is no significant difference in slope angle relative to distance from the shoreline. The data in Table 2 were used to plot the graph of variation of safety factor with slope angles, as shown in Figure 9.

Table 2. Slope angle optimization.

Slope Angle (Degrees)	CSRF	PF (%)
0	2.07	1.67×10^{-14}
35	1.79	5.65×10^{-11}
40	1.62	1.55×10^{-8}
45	1.43	3.12×10^{-5}
50	1.26	0.02
55	1.11	2.84
60	0.97	68.08
65	0.91	96.28
70	0.78	100
75	0.65	100
80	0.58	100
85	0.53	100
90	0.45	100

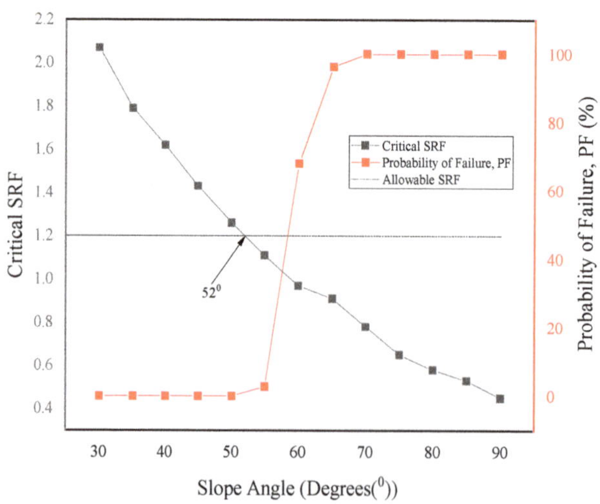

Figure 9. Graph of slope angle vs. strength reduction factor (FSRF) and probability of failure (PF).

4.2. Pore Pressure Computations

In Phase 2 finite element analysis, groundwater seepage analysis is integrated with stress analysis. This means that the groundwater seepage analysis is first computed and then the pore pressure results are automatically used in stress analysis. In this section, the ocean water seepage is first separately computed so as to have an idea of how the pore pressures in the slope vary with variation in tidal levels. Analysis of seepage of ocean waters through the slope was done as a drawdown analysis in three stages. The first stage was low tide, the second high tide, and the third low tide to simulate the daily rising and falling of ocean tides. The total head levels used were 0.71 m for low tide and 2.96 m for high tide, giving a head difference of 2.25 m. Plots of variation of pore pressures in the slope due to changing tides are shown in Figure 10 for the slope at 20 m excavation depth. The flow lines mark the phreatic surface (line of zero pore pressure) in the slope due to ocean-water seepage. Above the line, the pore pressure is negative and below the line, the pore pressure is positive. At low tide, the discharge velocity to the seepage face is low, while at high tide, the discharge velocity increases with positive pore pressure developing in the lower sections of the slope. At high tide, the water table in the slope also rises, and this explains why seawater is seen on the quarry floor during high tide.

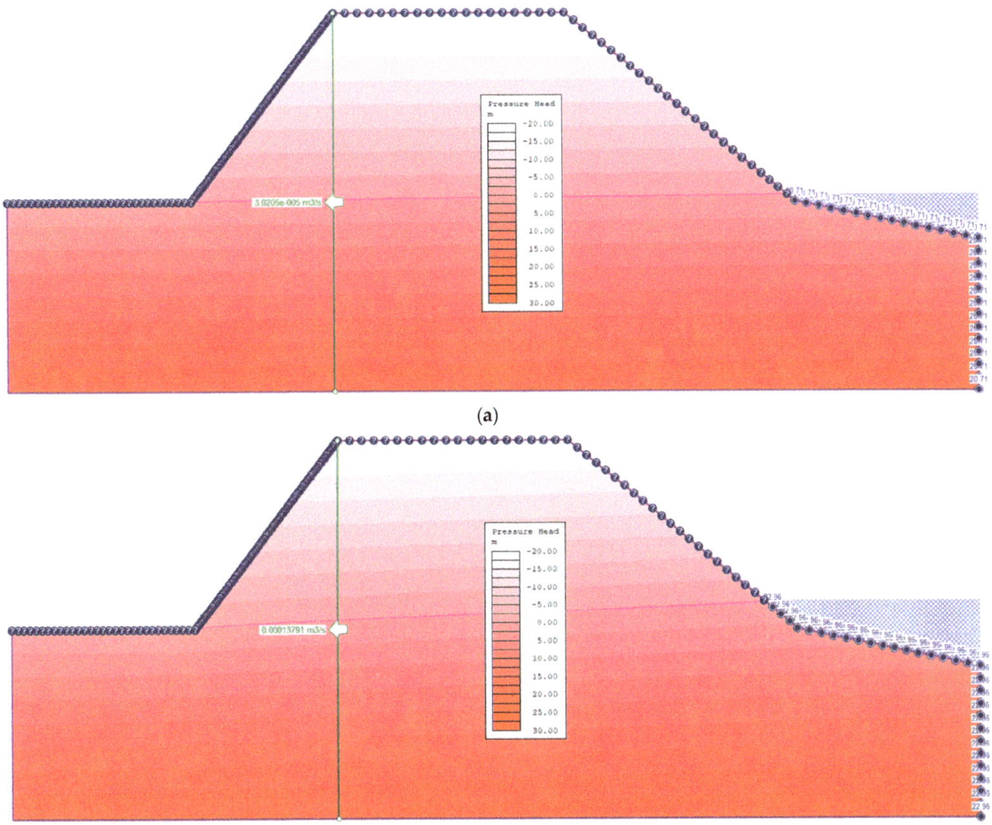

(**a**)

(**b**)

Figure 10. Pore pressure within the slopes at low and high tides (20 m excavation depth 50 m from shoreline). (**a**) Low tide level 0.71 m; (**b**) high tide level 2.96 m.

A parametric analysis was conducted on the variation in ocean-water seepage with excavation depth and distance from the shoreline. The nephograms in Figure 11 show the various pore pressure plots. At the same distance from the shoreline, the pore pressure in the slope and discharge velocity increase with increasing excavation depth. On the other hand, when the excavation depth is held constant, the discharge velocity decreases with increasing distance from shoreline and the pore pressure intensity also reduces. An observation of the trend of the phreatic surface shows that the farther the slope face is from the shoreline, the smaller the slope section exposed to ocean-water intrusion.

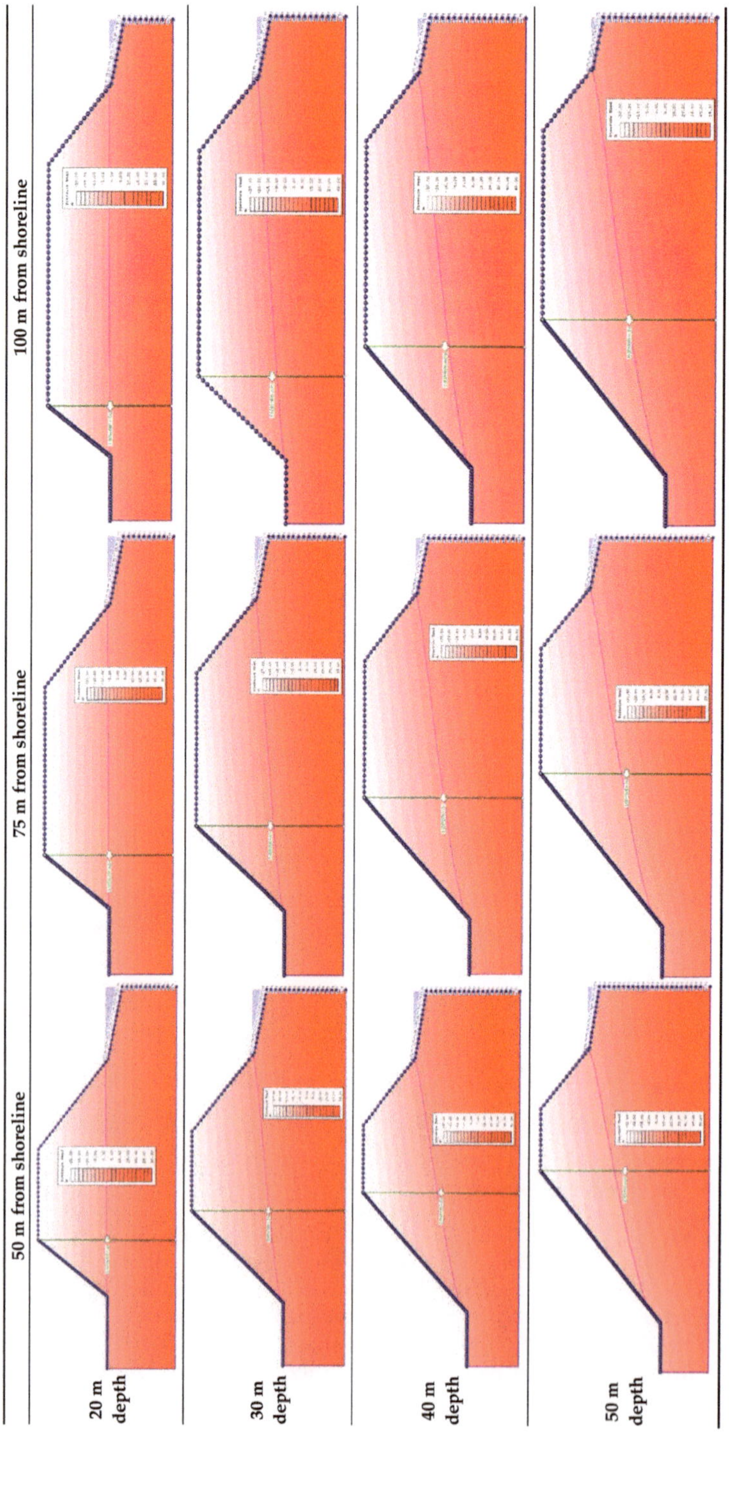

Figure 11. Variation of pressure heads with excavation depth and distance from shoreline.

4.3. Slope Stability Due to Seepage of Ocean Water

Laboratory experiments on this rock mass previously done by Joan et al. [3] indicated a general reduction in strength when the rock is saturated. Water seepage leads to strength reduction in weak rocks because an increase in pore pressure leads to subsequent decrease in the shear strength. On this basis, it is imperative to analyze the behavior of the rock slopes under the influence of ocean-water seepage occasioned by the rising and falling tidal levels, which subject the slopes to changes in pore pressures. In Phase 2 software, the groundwater analysis module computes pore pressures, then the results are automatically used for stress analysis. Due to the high porosity and high permeability of the rock mass, a fully drained uncoupled analysis was done. A parametric analysis was conducted to analyze the stability for varied excavation depths and distances from the shoreline. The results are summarized in Table 3, showing the difference in safety factor between a dry slope and the case where the slope is subjected to total head variations due to changes in tidal levels. It is clear that the presence of ocean water in close proximity to the quarry affects the stability of the slopes as a result of the tendency for water seepage to reduce the strength of rocks, hence subsequent reduction in the safety factor. The images in Figures 12–14 show the nephograms of variations in slope safety factor and shear strain due to changes in pore pressure occasioned by changes in ocean tides. The data in Table 3 were used to plot the graphs in Figures 15 and 16 in order to analyze the variation in slope safety factor with distance from shoreline and excavation depths, respectively.

Table 3. Variation of slope safety factor due to seepage.

Distance from Shoreline	Excavation Depth	Safety Factor		Percentage Change
		Dry	Ponded	
50 m from shoreline	20 m	1.19	1.17	1.68
	30 m	1.23	1.19	3.25
	40 m	1.21	1.13	6.61
	50 m	1.21	1.05	13.22
75 m from shoreline	20 m	1.20	1.19	0.83
	30 m	1.22	1.2	1.64
	40 m	1.19	1.15	3.36
	50 m	1.19	1.09	8.40
100 m from shoreline	20 m	1.2	1.18	1.67
	30 m	1.21	1.18	2.48
	40 m	1.22	1.14	6.56
	50 m	1.21	1.11	8.26

The graph in Figure 15 shows the trend of variation of the safety factor with excavation depth. Under the influence of the adjacent ocean water, with the distance from the shoreline held constant, there is a trend of the safety factor increasing with increasing excavation depth from 20 m to 30 m, probably due to the stabilizing effects of the hydrostatic pressure. Beyond 30 m, however, the safety factor decreases sharply as the excavation depth increases. This can be attributed to the reduction in effective stress due to a larger portion of the slope being under positive pore pressures. In Figure 16, the trend is clearly that of the safety factor increasing with increasing distance of the excavation face from the shoreline, due to a decrease in the intensity of the pore pressure away from the shoreline. The simulated seepage also shows the trend of seepage velocity increasing with increasing excavation depth and decreasing with increasing distance of slope face from the shoreline, further explaining the declining safety factor with excavation depth.

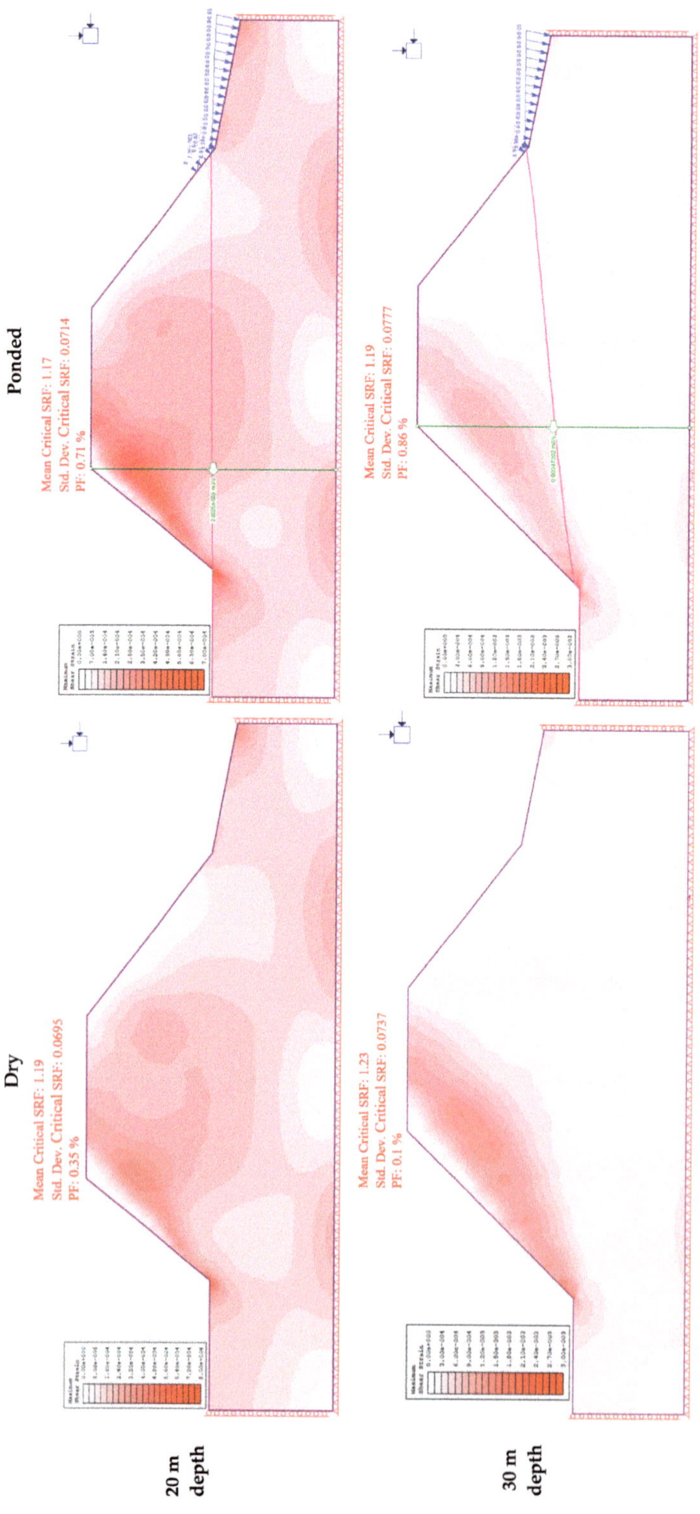

Figure 12. *Cont.*

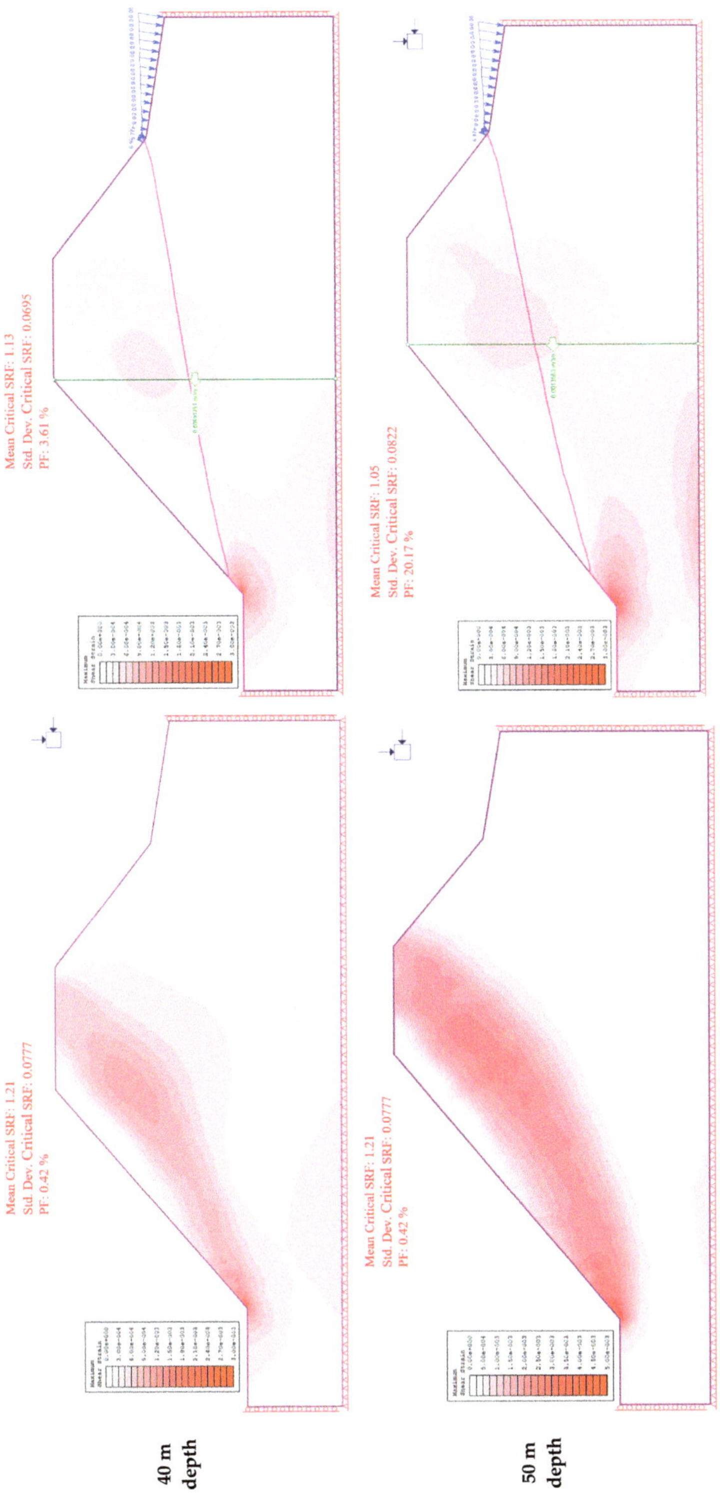

Figure 12. Changes in slope safety factor at 50 m from the shoreline.

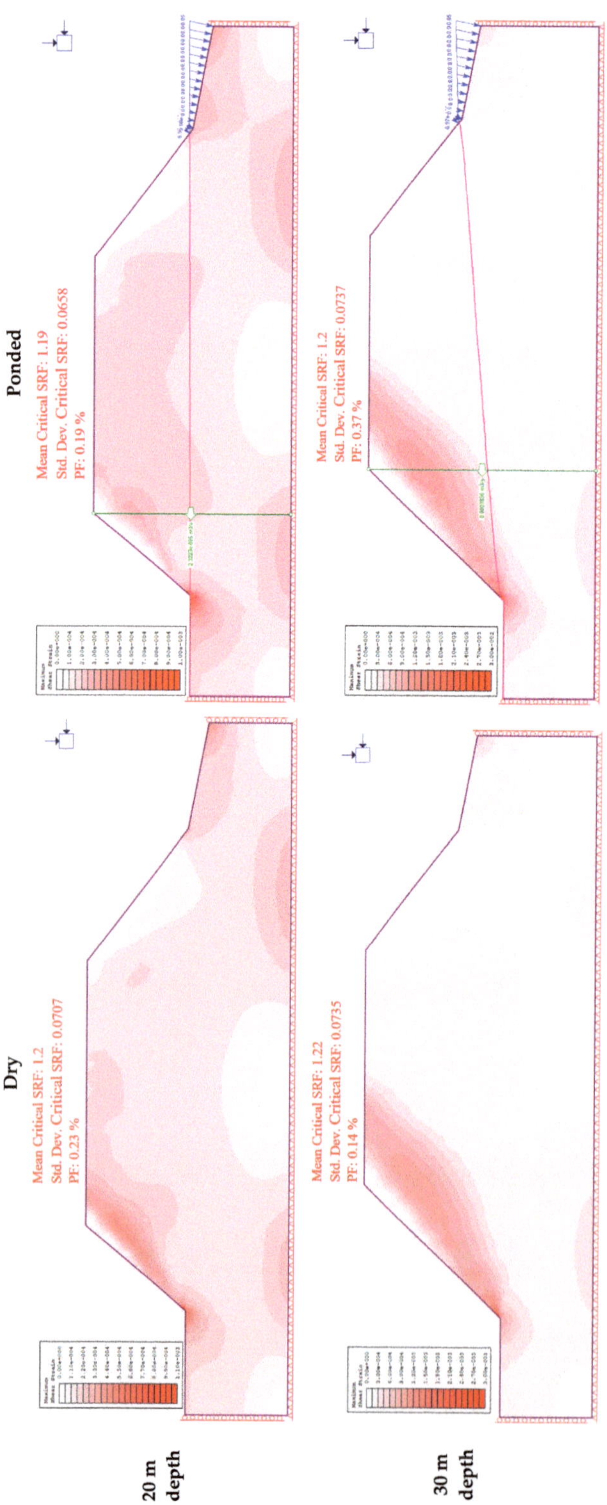

Figure 13. *Cont.*

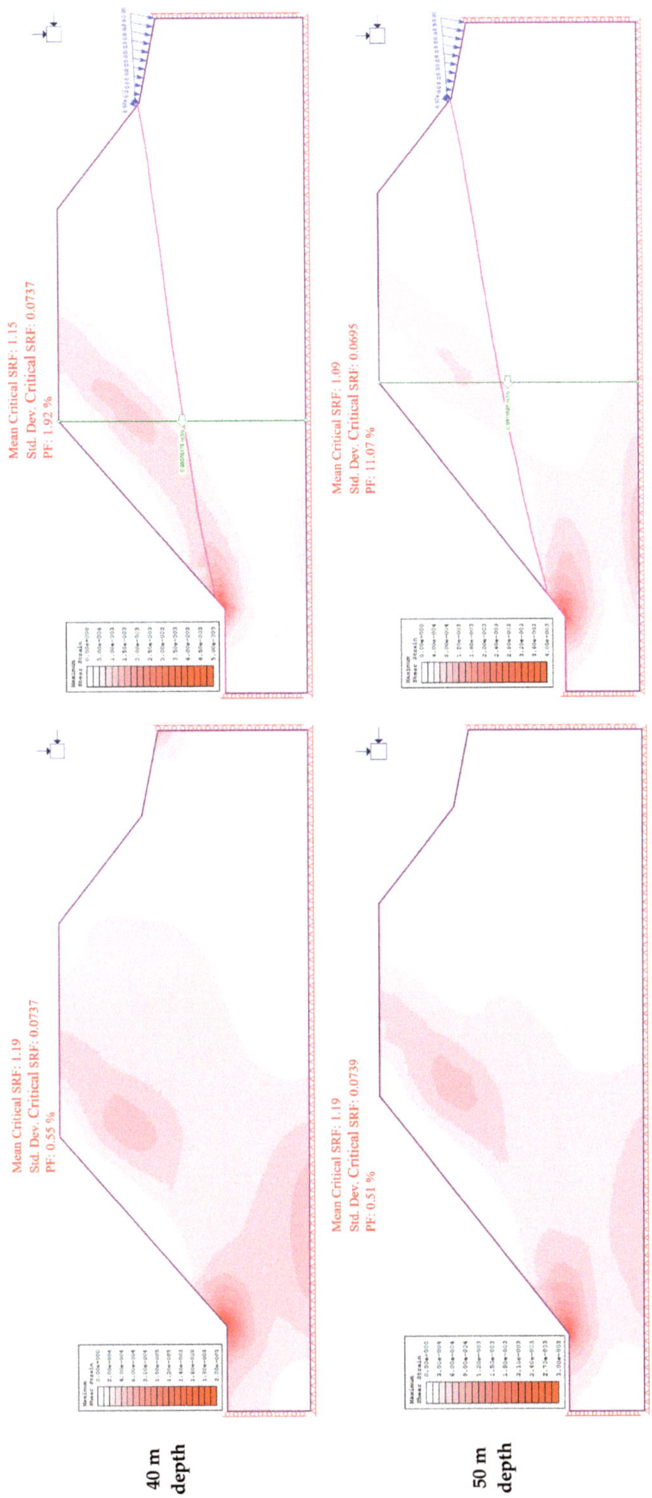

Figure 13. Changes in slope safety factor at 75 m from the shoreline.

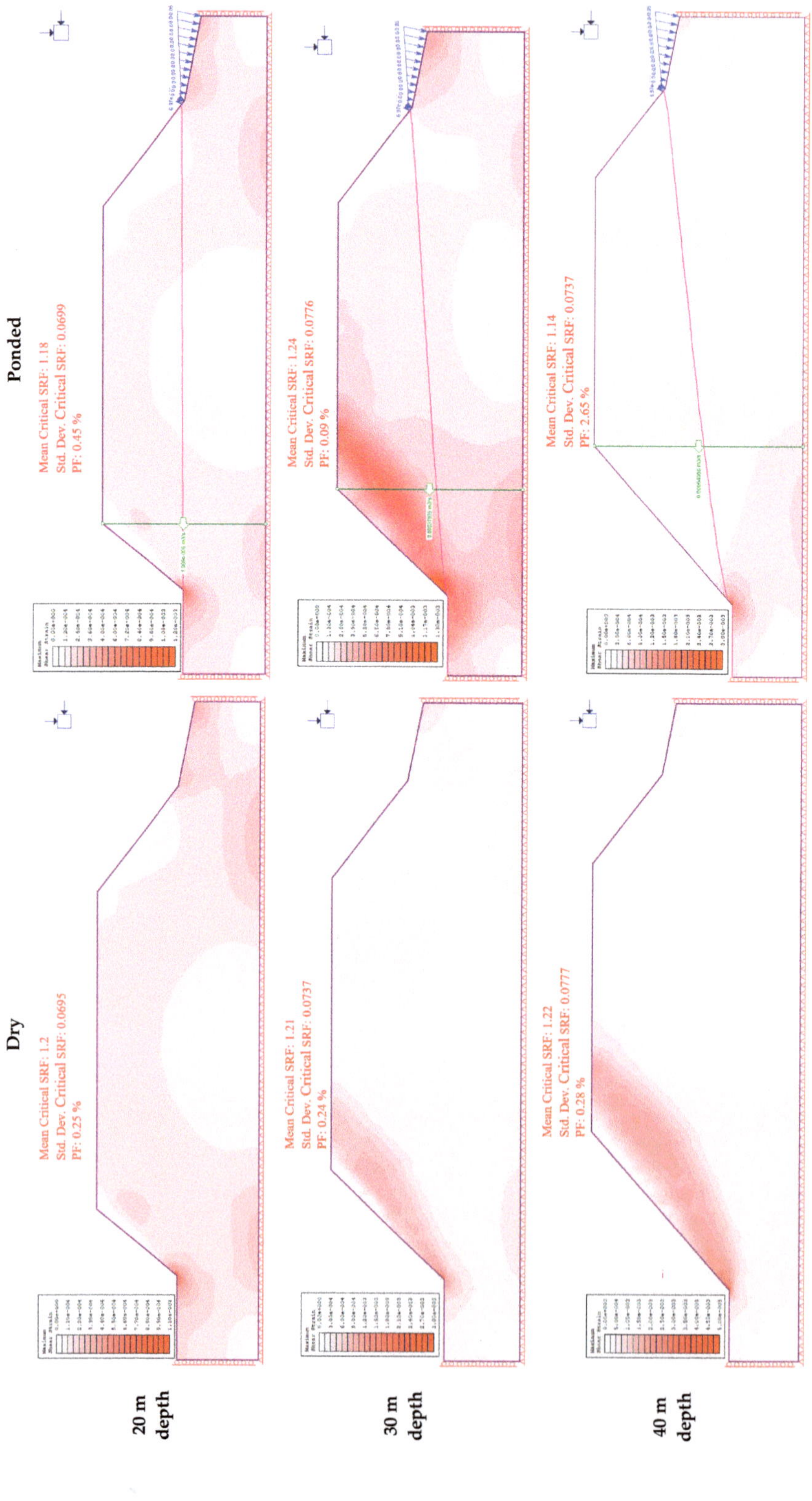

Figure 14. *Cont.*

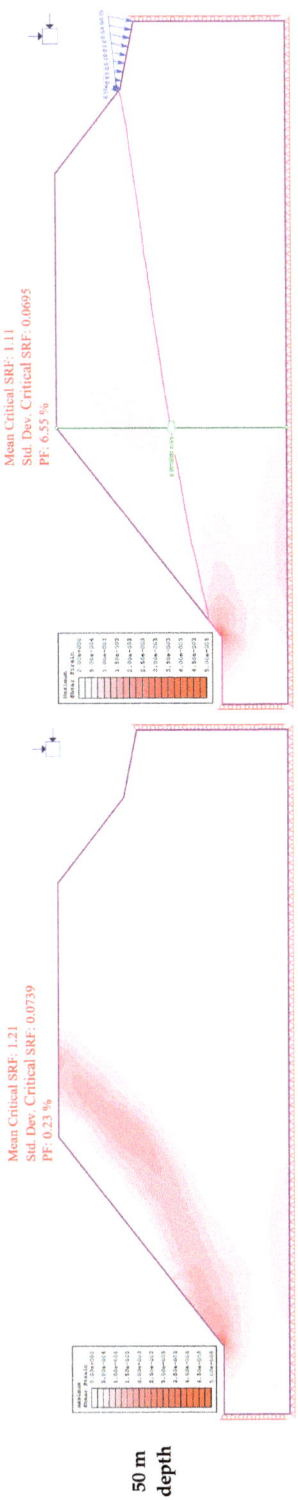

Figure 14. Changes in slope safety factor at 100 m from the shoreline.

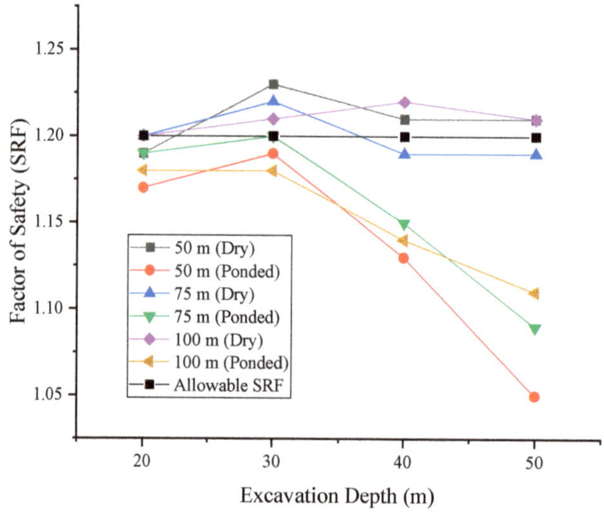

Figure 15. Variation in safety factor with excavation depth.

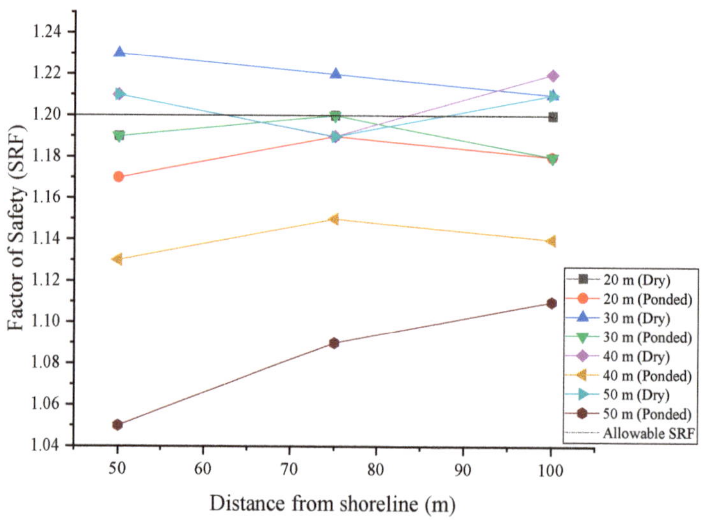

Figure 16. Variation in safety factor with distance from shoreline.

5. Discussion

The rock mass under study is classified as weak based on ISRM standards. The stability of slopes in this rock mass is therefore largely dependent on the rock mass strength. Slopes engineered on weak rock need to have the overall safe slope angle determined first, followed by the other slope design parameters [4]. In Vipingo quarry, mining is done with steep near-vertical slopes, where instabilities have been recorded in the past, even with shallow excavation. It is therefore imperative to determine the optimum overall slope angle to ensure safe operations, even as the excavation advances to greater depth. Based on rock strength of 20 on GSI scale, it was determined that the safe overall slope angle (allowable safety factor of 1.2) at 50 m from the shoreline is 52° for 20 m excavation depth, 46° for 30 m depth, 42° for 40 m depth, and 39° for 50 m depth for dry slopes. The slope angles were the same even for slope face at 75 m and 100 m from the shoreline.

The analysis also incorporated the probability of failure analysis. The graph of safety factor and probability of failure against slope angle in Figure 8 shows that the probability of failure is high for steep slope angles above 60°, after which it significantly reduces when the safety factor is ≥1. In the discussion on slope stability-acceptance criteria for open cast mine design, Adams [29] and Tapia et al. [30] outline the various acceptable failure probabilities as dependent on the slope scale (either bench, inter-ramp, or overall slope) and level of severity of the consequences of failure (low, medium, or high). The allowable probability of failure (PF) in this study is ≤10%. All optimized slope angles meet the criteria for allowable probability of failure, being less than 10%. With the optimum slope angles achieved for various depths, the nephograms of shear strain show a significant reduction in strains at the slope toe. The stable slopes have shear planes lying deep within the slope along non-circular failure surfaces through the slope toe. As the excavation depth increases, the strains on the slope toe generally increase, and this can be attributed to the overlying weight of the slope. A combination of the two parameters, i.e., safety factor and probability of failure, thus, gives a robust design judgment.

The daily rising and falling of semidiurnal tides of the Indian Ocean with a head difference of 2.25 m creates a drawdown effect on the adjacent quarry slopes. At 20 m and 30 m excavation depths, there is no significant variation in pore pressure with distance from shoreline. As the excavation depth increases beyond 30 m, there is substantial decrease in pore pressure with increasing distance from the shoreline. At any given distance from the shoreline, the discharge velocity towards the slope face increases with increasing depth of excavation and decreases with increasing distance from the shoreline. In consequence, the pore pressures in the slope increase with increasing excavation depth and decrease with the excavation face getting further from the shoreline. Positive pore pressures develop in the slope at high tide for shallow excavations and as the excavation gets deeper, positive pore pressures continue to build up in the slope as the phreatic surface rises in the slope. Positive pore pressures contribute to compromising the stability condition of slopes due to the consequent decrease in effective stress, which leads to changes in rock mass strength [31,32].

Generally, the slope safety factor is reduced as a result of the shear strength reduction effects of increasing pore pressure in the slope. A computation of the percentage change in slope safety factor due to ocean water seepage is given in Table 3. The percentage decrease in safety factor due to seepage increases with increasing excavation depth. The distribution of shear strains in the slope due to seepage is also noteworthy in that a potential circular slip surface is observed in all the analyses. A notable difference, however, exists in the depth of the slip surface from the slope face. In dry conditions, the slip surfaces are shallower and close to the slope face. However, with the pore pressure added to the analysis, the shear planes are deeper in the slope with shear strains concentrated on the toe of the slope. This isan indication of significant reduction in shear strength of the rock mass due to seepage, and the increased shear strain at the slope toe is because of the slope toe being a seepage face experiencing the highest pore pressures. This is observed irrespective of the distance of the excavation face from the shoreline. Similar observations of reduction in slope safety factor due to pore pressure changes have been observed in the analysis of Chengmenshan copper mine by Fan et al. [9]. Other researchers investigating pore pressure variations in rock slopes, such as Saada et al. [33] and Preisig et al. [34], also made similar observations. The variations in shear strain patterns depicting the critical slip surfaces are also corroborated by results reported in the analysis by Wu et al. [35].

The reduction in slope safety factor due to pore pressure in the slopes leads to the introduction of a measure of slope sensitivity to ocean-water seepage, the sensitivity factor (ζ), which represents the percentage change in slope safety factor due to seepage. The slope sensitivity is expressed as:

$$\zeta = \frac{SRF_{dry} - SRF_{ponded}}{SRF_{dry}} \tag{6}$$

where SRF_{dry} is the safety factor of the dry slope and SRF_{ponded} is the reduced safety factor of the slope due to ponding effect of ocean water. ζ is the slope sensitivity factor.

Graphical representations of sensitivity of the slope to seepage forces are given in Figures 17 and 18 based on the data on Table 3. The sensitivity of the slope safety factor to ocean water seepage increases with excavation depth. On the other hand, the slope sensitivity to seepage decreases with increasing distance of excavation face from the shoreline. Clearly, the sensitivity of the slope to seepage of the adjacent ocean waters depends on both the excavation depth and distance of the slope face from the shoreline. From the analysis of the effects of ocean tides, it is evident that slope design in this condition has to take into consideration the slope sensitivity to seepage effects. The design safety factor should therefore be above the allowable safety factor of 1.2 by an amount equal to the sensitivity factor. Equation (6) can therefore be rewritten to give the design safety factor for dry slope that will give a factor of 1.2 under seepage condition:

$$SRF_{dry} = \frac{1.2}{1-\zeta} \quad (7)$$

where SRF_{dry} is the design safety factor and ζ is slope sensitivity factor.

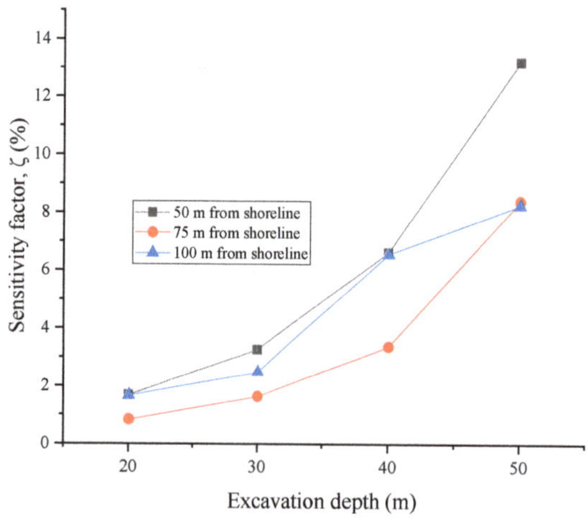

Figure 17. Sensitivity factor vs. excavation depth.

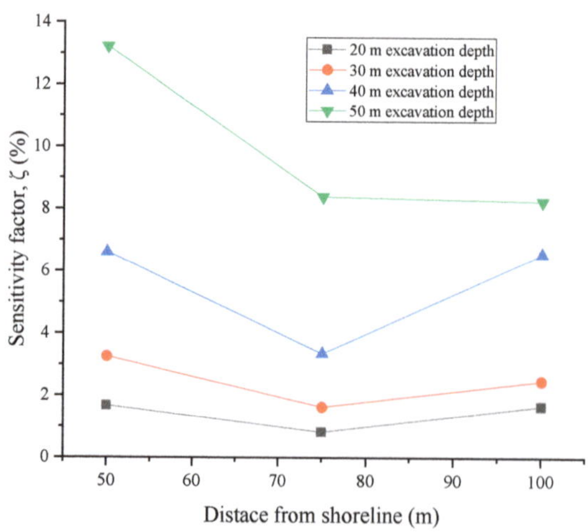

Figure 18. Sensitivity factor vs. distance from shoreline.

The analysis in this study considered the rock slope strength condition at a GSI value of 20. The slope sensitivity factors from Figures 17 and 18 can therefore be used in Equation (7) to determine the appropriate overall slope design parameters for the rock mass.

Progression of the excavation below sea level makes the quarry a sink for ocean-water intrusion, as seen by the orientation of the phreatic surface in pore pressure computations, which agrees with field observations of water seeping into the quarry floor during high tides. The carbonate rock mass is also prone to karstification, and so water seepage is an enemy to the stability of slopes in this rock mass. It is therefore crucial to protect the slope toe from erosion. The hydrogeology of the area is such that the natural drainage of surface waters is into the ocean. The slope toe can be protected by constructing drainage ditches to collect the water at the slope toe and enhancing drainage into the ocean by creating water systems to minimize erosion of the slope toe. The mined-out slopes and slope toe should then be protected from subsequent erosion by water in the drainage ditches. Drainage ditches constructed in such a porous rock mass with high hydraulic conductivity need further controls in place to handle the flow of water. Vegetation such as vetiver grass also has the ability to hold water by its thick foliage, thereby reducing the flow velocity and enhancing downward seepage flow velocity and minimizing surface runoff that erodes slopes. In consequence, drainage of water from the slope towards the ocean is improved, leaving the slope face free of eroding effects of water. The thick plant roots would also improve cohesion on the surface layers, thus minimizing further erosion.

6. Conclusions

Stability of the quarry slopes has been assessed in both dry conditions and under the action of ocean tides by finite element analysis. The slope sensitivity to the changing tidal levels has also been evaluated for both varying excavation depths and distance from the shoreline. Shallow slopes up to 20 m depth do not seem to be much affected by the tides when the excavation is above the water table. A slope sensitivity factor (ζ) is introduced to represent the percentage change in slope safety factor due to seepage. The slope sensitivity to seepage increases with increasing excavation depth and closer proximity to the shoreline. The slope design in this condition therefore considers both the excavation depth and distance of slope face from the shoreline as crucial parameters. The appropriate distance of the slope face from the shoreline depends on the target excavation depth: shallower pits can be closer to the shoreline. Hence, the deeper the final pit, the further the slope face should be from the ocean. An equation is given that uses the parameter of sensitivity factor to guide the decision on the design safety factor that will give the allowable safety factor of 1.2 under the strength degradation of seepage condition. The probabilistic analysis combined with the safety factor also give a robust design judgment. The graphs of the slope sensitivity factor are a good guide for determining the slope sensitivity factor to use in a design for the rock mass condition under study.

Author Contributions: Conceptualization and methodology were done by J.A.O., D.M. and J.A.O. wrote the manuscript text. T.S. and H.S. were responsible for supervision and funding for the research work. A.H. was in charge of study materials and computing resources. All authors have read and agreed to the published version of the manuscript.

Funding: The research work is part of a doctoral program supported by JICA.

Institutional Review Board Statement: Not applicable.

Informed Consent Statement: Not applicable.

Data Availability Statement: All data generated or analyzed during this study are included in this published article.

Acknowledgments: Sincere gratitude to JICA for scholarship funding that has facilitated this research work. The contribution of Vipingo quarry in availing the information and data that enabled the conceptualization of this research work is sincerely appreciated.

Conflicts of Interest: The authors declare that there are no conflict of interests regarding the publication of this paper.

References

1. Hashiba, K.; Fukui, K.; Kataoka, M. Effects of water saturation on the strength and loading-rate dependence of andesite. *Int. J. Rock Mech. Min. Sci.* **2019**, *117*, 142–149. [CrossRef]
2. Hamzaban, M.T.; Büyüksağiş, I.S.; Manafi, M.; Touranchehzadeh, A. The effect of saturation on the physical and mechanical behavior of some rock samples. *Civ. Eng. Beyond Limits* **2021**, *3*, 23–31. [CrossRef]
3. Onyango, J.A.; Mwanzia, D.K.; Moses, D.; Oo, C.T.; Batsaikhan, U.; Phaisopha, S.; Krop, I.T. Laboratory Investigation of the Physico-Mechanical Properties of Coral Limestone of Vipingo Area in Kenya's Coastal Region. *Int. Res. J. Innov. Eng. Technol.* **2022**, *6*, 42–53. [CrossRef]
4. Martin, D.; Stacey, P. *Guidelines for Open Pit Slope Design in Weak Rocks*; CRC Press Taylor & Francis Group: New York, NY, USA, 2018.
5. Chen, H.; Zhiye, Z.; Jianping, S. Coupled hydro-mechanical model for fractured rock masses using the discontinuous deformation analysis. *Tunn. Undergr. Space Technol.* **2013**, *38*, 506–516. [CrossRef]
6. Rutqvist, J.; Ove, S. The role of hydromechanical coupling in fractured rock engineering. *Hydrogeol. J.* **2003**, *11*, 7–40. [CrossRef]
7. Reddi, L.N. *Seepage in Soils: Principles and Applications*; John Wiley & Sons: Hoboken, NJ, USA, 2003.
8. Iverson, R.M.; Reid, M.E. Gravity-driven groundwater flow and slope failure potential: 1. Elastic Effective-Stress Model. *Water Resour. Res.* **1992**, *28*, 925–938. [CrossRef]
9. Fan, W.; Cao, P.; Zhang, K.; Li, K.; Chen, C. Stability Assessment and Optimization Design of Lakeside Open-Pit Slope considering Fluid-Solid Coupling Effect. *Math. Probl. Eng.* **2015**, *2015*, 691826. [CrossRef]
10. Moosavi, E.; Shirinabadi, R.; Gholinejad, M. Prediction of seepage water pressure for slope stability at the Gol-E-Gohar open pit mine. *J. Min. Sci.* **2016**, *52*, 1069–1079. [CrossRef]
11. de Agreda, E.A.P.; Puigmartí, N.M.P. Slope Stability under Rapid Drawdown Conditions. In Proceedings of the First Italian Workshop on Landslides, Naples, Italy, 8 June 2009; pp. 11–27.
12. Wong, T.F.; Szeto, H.; Zhang, J. Effect of Loading Path and Porosity on the Failure Mode of Porous Rocks. *Appl. Mech. Rev.* **1992**, *45*, 281–293. [CrossRef]
13. Acharya, K.P.; Netra, P.B.; Ranjan, K.D.; Ryuichi, Y. Seepage and slope stability modelling of rainfall-induced slope failures in topographic hollows. *Geomat. Nat. Hazards Risk* **2016**, *7*, 721–746. [CrossRef]
14. Rocscience. Groundwater Seepage. 2021. Available online: https://www.rocscience.com/help/rs2/tutorials/groundwater-seepage/coupled-drawdown-analysis (accessed on 7 March 2022).
15. Narasimhan, T.; Witherspoon, P. An integrated finite difference method for analyzing fluid flow in porous media. *Water Resour. Res.* **1976**, *12*, 57–64. [CrossRef]
16. Koyama, T.; Nishiyama, S.; Yang, M.; Ohnishi, Y. Modeling the interaction between fluid flow and particle movement with discontinuous deformation analysis (DDA) method. *Int. J. Numer. Anal. Methods Geomech.* **2011**, *35*, 1–20. [CrossRef]
17. Tracy, F.T.; Brandon, T.L.; Corcoran, M.K. *Transient Seepage Analyses in Levee Engineering Practice*; U.S. Army Corps of Engineers: Washington, DC, USA, 2016.
18. Pauls, G.J.; Sauer, E.K.; Christiansen, E.A.; Widger, R.A. A transient analysis of slope stability following drawdown after flooding of a highly plastic clay. *Can. Geotech. J.* **1999**, *36*, 1151–1171. [CrossRef]
19. Freeze, A.R.; Cherry, J.A. *Groundwater*; Prentice Hall: Hoboken, NJ, USA, 1979.
20. Cen, W.; Li, D.; Wang, H. Impact of transient seepage on slope stability of earth-rock dams with geomembrane barrier defects. *Environ. Geotech.* **2019**, *7*, 581–590. [CrossRef]
21. Song, D.; Ailan, C.; Zhu, C.; Xiurun, G. Seismic stability of a rock slope with discontinuities under rapid water drawdown and earthquakes in large-scale shaking table tests. *Eng. Geol.* **2018**, *245*, 153–168. [CrossRef]
22. State of Oregon. *Oregon Territorial Sea Plan*; Department of Land Conservation and Development: Salem, OR, USA, 1994.
23. Tides4Fishing. Tides and Lunar Charts-Kilifi. 17 12 2021. Available online: https://tides4fishing.com/af/kenya/kilifi (accessed on 16 March 2022).
24. Balmer, G. A general analytical solution for. *Am. Soc. Test. Mater.* **1952**, *52*, 1260–1271.
25. Hammah, R.E.; Yacoub, T.E.; Corkum, B.C.; Curran, J.H. The Shear Strength Reduction Method for the Generalized Hoek-Brown Criterion. In Proceedings of the 40th US Symposium on Rock Mechanics (USRMS), Anchorage, AK, USA, 25–29 June 2005.
26. Rocscience. RocScience. 2021. Available online: https://www.rocscience.com/help/rs2/documentation/rs2-model/project-settings/general/general-settings (accessed on 5 March 2022).
27. Onyango, J.A.; Sasaoka, T.; Shimada, H.; Hamanaka, A.; Moses, D.; Tumelo, D. Evaluating Rock Mass Properties of Vipingo Coral Limestone Quarry Based on a Modified Geological Strength Index (GSI) and State of Karstification. *Open J. Geol.* **2022**, *12*, 57–79. [CrossRef]
28. Jones, I.C.; Jay, L.B. Hydrogeologic and climatic influences on spatial and interannual variation of recharge to a tropical karst island aquifer. *Water Resour. Res.* **2003**, *39*, 1253. [CrossRef]
29. Adams, B.M. Slope Stability Acceptance Criteria for Opencast Mine Design. In Proceedings of the 12th Australia New Zealand Conference on Geomechanics, Wellington, New Zealand, 22–25 February 2015.

30. Tapia, A.; Contreras, L.F.; Jefferies, M.; Steffen, O. Risk Evaluation of Slope Failure at the Chuquicamata Mine. In Proceedings of the 2007 International Symposium on Rock Slope Stability, Perth, Australia, 12–14 September 2007.
31. Rinaldi, M.; Casagli, N.; Dapporto, S.; Gargini, A. Monitoring and modelling of pore water pressure changes and riverbank stability during flow events. *Earth Surf. Processes Landf.* **2004**, *29*, 237–254. [CrossRef]
32. Abramson, L.W.; Thomas, L.S.; Sunil, S.; Boyce, M.G. *Slope Stability and Stabilization Methods*; John Wiley & Sons: Hoboken, NJ, USA, 2001.
33. Saada, Z.; Maghous, S.; Garnier, D. Stability analysis of rock slopes subjected to seepage forces using the modified Hoek–Brown criterion. *Int. J. Rock Mech. Min. Sci.* **2012**, *55*, 45–54. [CrossRef]
34. Preisig, G.; Eberhardt, E.; Smithyman, M. Hydromechanical Rock Mass Fatigue in Deep-Seated Landslides Accompanying Seasonal Variations in Pore Pressures. *Rock Mech. Rock Eng.* **2016**, *49*, 2333–2351. [CrossRef]
35. Wu, D.; Wang, Y.; Zhang, F.; Qiu, Y. Influences of Pore-Water Pressure on Slope Stability considering Strength Nonlinearity. *Adv. Civ. Eng.* **2021**, *2021*, 8823899. [CrossRef]

mining

MDPI

Article

Numerical Modelling of Blasting Fragmentation Optimization in a Copper Mine

Vidal Félix Navarro Torres [1], Cristian Castro [2], María Elena Valencia [2], Janine Rodrigues Figueiredo [1,*] and Leandro Geraldo Canaan Silveira [3,4]

1 Instituto Tecnológico Vale, Ouro Preto 35400-000, MG, Brazil
2 Itasca Chile, Santiago de Chile 7500524, Chile
3 Vale S.A., Nova Lima 34000-000, MG, Brazil
4 Department of Mining Engineering, Campus Morro do Cruzeiro, Universidade Federal de Ouro Preto, Ouro Preto 35400-000, MG, Brazil
* Correspondence: janine.figueiredo@itv.org

Abstract: The blasting operation considerably influences the overall productivity of opencast mines, especially when blasting results in oversized fragments that impact the operations ranging from excavation to milling. In this work, a numerical analysis of blasting performance was implemented to optimize the blasting parameters and improve the fragmentation of the hard rock in a copper open pit mine site in Brazil. In this paper, the methodology comprised data collection, 3D numerical model construction for blasting optimization using Blo-Up software, calibration with historical data, and predictive analysis, including testing two different blast designs. With the objective of achieving a desired P_{80} size of the blasting fragmentation, the results indicate an optimized calibrated model with an overall error equal to 4.0% using a Swebrec distribution fitted to the model data. The optimal P_{80} size of the resulting muckpile was equivalent to ~0.53 m for the hard rock copper fragments, which was close to the desired P_{80} size.

Keywords: blasting optimization; numerical modelling; fragmentation; hard rock mine; optimal size

1. Introduction

In opencast mine operations, the comminution process comprises the majority of the total energy consumption. This process also impacts mine productivity, considering the size of the muckpile fragments resulting from the blast. For this reason, the subject of rock fragmentation, namely particle size distribution (PSD), during blasting operations has been considered to improve the comminution efficiency [1], especially at hard rock mines [2–4].

Rock fragmentation is controlled by several factors related to the rock mass properties and drilling and blasting design parameters [5]. A rock mass is described by the physical and mechanical properties of both the intact rock and discontinuities, which can be used to characterize the fragmentation strength of the rock mass during blasting operations. The drilling and blasting design parameters include burden distance, drillhole spacing, stemming height, and borehole properties. The explosive properties also impact the rock fragmentation quality, such as the explosive type, heat and specific charge and other controllable parameters that affect the energy efficiency in rock blasting [6].

The optimization of drilling and blasting parameters in terms of rock fragmentation can contribute to improving the productivity of mine-to-mill operations (loading, transport and milling) and minimizing the total operation costs [1,7]. For instance, the data originating during the drilling of drillholes are an important data source for estimating intact rock properties and impact of the comminution process [3]. Blasting operations have been optimized through several methods. Leng et al. [8] applied different diameters of blastholes to minimize oversized boulders and toes using a nonideal detonation model and a statistical damage constitutive model. Amoako et al. [9] studied the prediction of blasting fragmentation distribution based on the blast design parameters using machine learning techniques. A multilayered artificial neural network and support vector regression models were used to predict mean rock fragment size, which could be used to improve loading and excavation productivity.

The rock fragmentation is represented by modelling or simulation (i) by the muckpile shape as the position and the shape of the fragment; (ii) by blasting design parameters; and (iii) rock mass properties [5,10]. The particle size distribution is defined by sieving to obtain mass weights of the remaining material on each sieve deck, and it is then plotted in a histogram through weights vs. mesh size. Further, a cumulative distribution function or a relative mass passing function (P) might be plotted, in which the particle size (X) is against the cumulative undersize, and P represents the probability that a fragment is smaller than X [10,11]. Alternatively, the PSD can be obtained through the processing of a digital image or using lasers and sensors based on artificial intelligence algorithms [9,12–14].

1.1. Particle Size Distribution Models

Some empirical methods for predicting PSD of rock fragmentation by explosives are described in the literature, mostly using two-parameters: the size coefficient and the distribution coefficient [11]. The Rosin–Rammler model [15] has been widely accepted by the scientific community and industry. In this model, the PSD of rock fragmentation from blasting is expressed in terms of a factor, the median fragment size (X_{50}), when the breakage model is known. Although, different model distributions have been used to determine the PSD curve using the 80% passing size (X_{80}) [10,16].

The Rosin–Rammler model is written as:

$$\begin{aligned} P_{RR(x)} &= 1 - \exp-(X/X_c)^n = 1 - \exp(-\ln 2 * (X/X_{50})^n \\ &= 1 - 2\exp-(X/X_{50})^n = \left(1 - 5\exp-(X/X_{80})^n\right) \end{aligned} \tag{1}$$

with X_c as the characteristic size, and X_{80} the 80% passing size. According to Ouchterlony and Sanchidrián [10], through a series expansion of Equation (1), when $X \ll X_{50}$, then $P_{RR} \approx \ln 2 * (X/X_{50})^n$, and that in a log-log diagram, this reduces to a straight line of slope n. In this way, X_c might be written as:

$$X_c = (\ln 2)^{1/n} X_{50} \tag{2}$$

The PSD might be represented by three parameters, as well as the Swebrec function [10,11] given by:

$$P_{Swebrec(x)} = 1/1 + \left(\ln\left(\frac{X_{max}}{X}\right) / \ln\left(\frac{x_{max}}{x_{50}}\right)\right)^b \tag{3}$$

where P(x), is the cumulative undersize (%); X_{max} is the maximum size of particles (mm); X is the particle size (mm); X_{50} is the sieve size that retains 50% of the material (mm); and b is the shape parameter, called the undulation exponent or sometimes the natural breakage characteristics (NBC) exponent [17]. This parameter might be between 1 and 2, where the inflection point tends towards $X = X_{max}$ and $X = X_{50}$, respectively [11].

The aim of this study was to optimize the blasting parameters through a numerical analysis of a copper open pit mine in hard rock. Oversized fragments formed and impacted the overall mine production, mostly crushing operations. The design blasting parameters were evaluated and calibrated to achieve the best rock fragmentation size for the mining system. For this purpose, the specific objectives were (i) the data collection and 3D model construction in Blo-Up software; (ii) the calibration of historical data; and (iii) predictive analysis.

1.2. Blo-Up Software

Blo-Up is a numerical model that combines 3D continuum and discontinuum numerical methods (Figure 1) to reproduce general trends in nonideal detonation, rock fracturing and muckpile formation. The first model component, "Programmed Burn", is used to model the detonation process and axial blasthole flow. For the second component, a continuum method is used to represent the near-blasthole volume and the detonation process. The third component adopts a discrete element method (DEM) to represent the rock mass and model the wave propagation and initial fragmentation through muckpile formation [9].

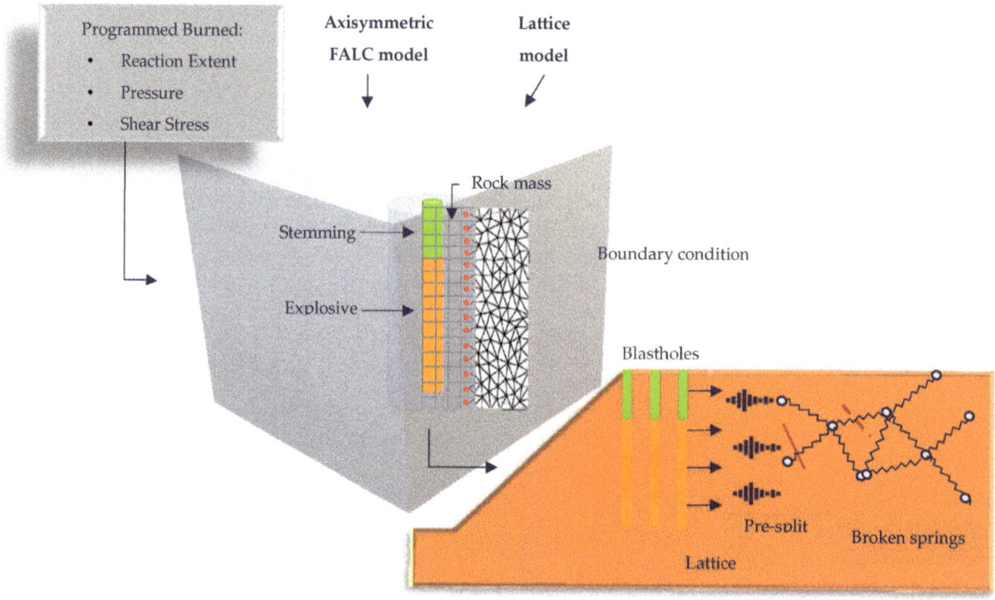

Figure 1. Representation of numerical components in Blo-Up. Adapted from [18].

1.2.1. Explosive Model

According to Castro and Valencia [18], the numerical modelling process involves modelling using the data blasting parameters (density, energy, reaction extent, equations of state of the products and velocity of detonation-VOD). The model used is a nonideal detonation model [19]. These data are represented in the central zone of a continuum FLAC model with axial symmetry, developed by Itasca [20]. Then, the detonation process, expansion and axial flow of the detonation products are simulated through a programmed detonation algorithm [21]. The energy is released to the zones representing the explosive based on a predetermined VOD.

1.2.2. Main Rock Body Representation

The FLAC zones are united in a simulation based on the "lattice" DEM. The rock mass is represented as a set of nodes connected by elastic springs with tensile strength and a viscous damping mechanism [18] (Figure 2).

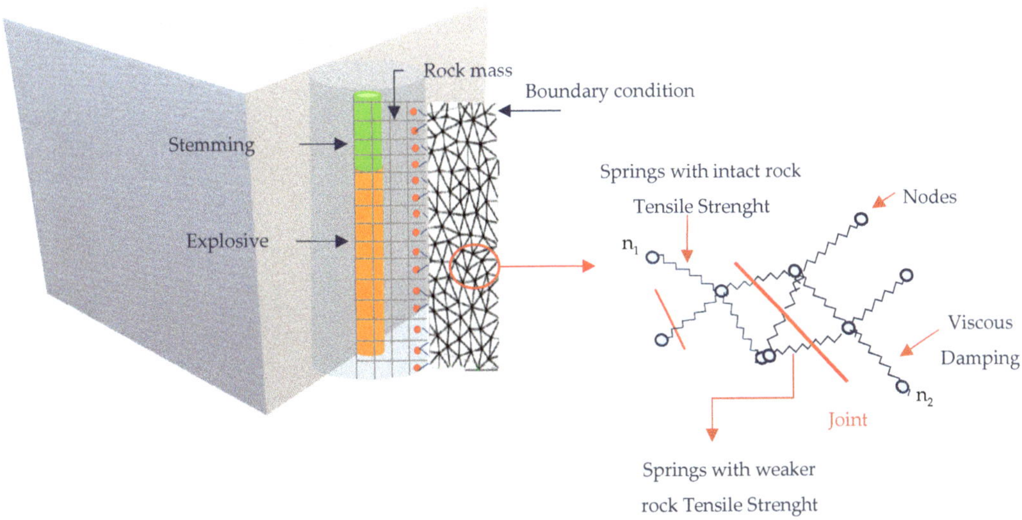

Figure 2. Main rock body representation in Blo-Up. Adapted from [18].

The energy dissipation that occurs in real rock masses is allowable by this viscous damping mechanism and is dependent on a factor "α", with $0 \leq \alpha \leq 1$. Although Blo-Up explicitly incorporates the presence of node sets in the rock mass, these are planes in which the tensile strength of the springs that cut through the plane is decreased by a factor "F", with $0 \leq F \leq 1$ [18].

1.2.3. Fracture Mechanics in Blo-Up

The fracture mechanics and wave propagation in the rock mass are represented in the Blo-Up model. During the energy transmission through the rock mass, motion equations are generated for each point in the mass as a function of the sum of unbalanced forces [18], as shown in Equation (1).

$$\sum \vec{f} = \vec{f_c} + \vec{f_d} + \vec{f_g} = m\,(\ddot{X} - \vec{g}) \tag{4}$$

where f_c is the force from lattice springs, f_d is the viscous damping force and f_g is the gravitational force.

1.2.4. Fragment Size Calculation

As stated by Castro and Valencia [18], in the size distribution calculation, the lattice is divided into fragments to define the fragment sizes; when fracturing occurs, the springs connecting the nodes are broken. Blo-Up defines these fragments and their total volume to maintain different sizes. The size of the fragments is the cube root of the volume of a fragment. The volume of a fragment is the number of nodes that make up the fragment times the nodal volume. The nodal volume is the lattice resolution cubed. The minimum fragment size found always corresponds to the actual lattice resolution. For fragment sizes below the lattice size, the PSD can be reflected by the adjusted probability distribution equations of Rosin–Rammler (Equation (1)) or Swebrec (Equation (3)) [13].

2. Background Review

The study area consists predominantly of the biotite-garnet schist (BDX) in an open pit mine located in southwestern Pará state in Brazil (Figure 3).

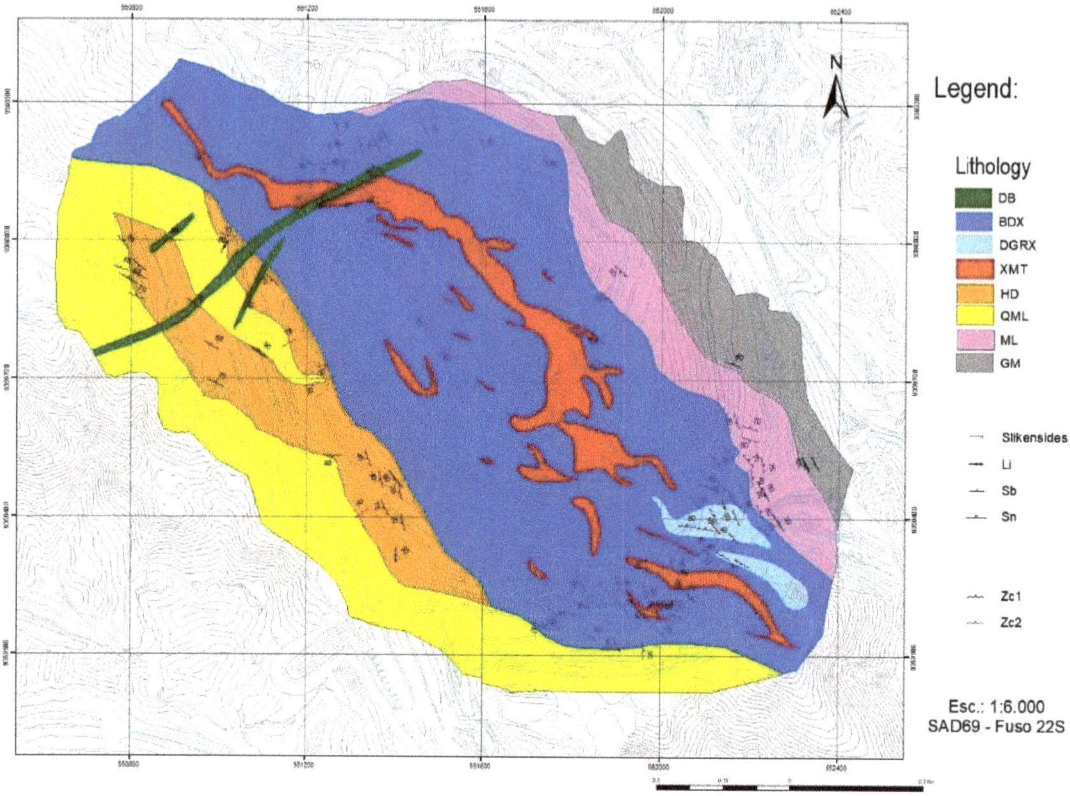

Figure 3. Lithology of the study area. Reproduced with permission from Vale S.A. [22].

The intact rock properties for this lithology are generalized to the mean values among rock classes I, II and III (Table 1), as they represent the most common rock classes in the mine [22]. The Young's modulus of the BDX rock mass was estimated through the empirical approach proposed by Hoek and Diederichs [23]. The shear zone of the mine area and the intact rock properties for the BDX are shown in Figure 4 and Table 1.

Table 1. Intact rock properties and rock mass classification of BDX.

Lithology	RMR Classification	Tensile Strength (Mpa)	Cohesion (kPa)	Hoek and Brown Classification				Density (t/m³)
				GSI	m_i	UCS (MPa)	E_i (GPa)	
BDX	I	9.0	130	83	12	90	63.4	3.2
	II	9.0	130	66.5	12	90	63.4	3.2
	III	9.0	130	50	12	90	63.4	3.2
	IV	2.5	70	25	12	25	9.4	2.7
Average BDX	I, II and III	9.0	130	50–83	12	90	63.4	3.2

Rock mass rating (RMR); material constant (m_i); geological strength index (GSI); uniaxial compressive strength (UCS); elastic modulus (E_i).

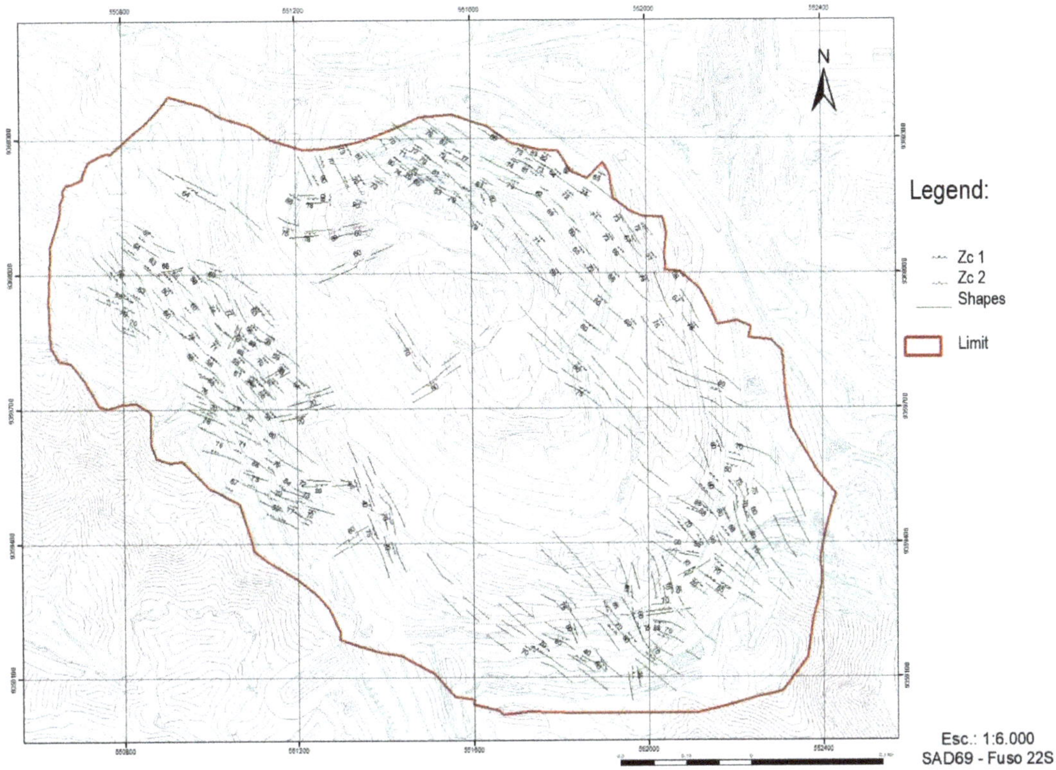

Figure 4. Shear zone of the hard rock of a copper mine. Reproduced with permission from Vale S.A. [22].

Discontinuities in the form of shear zones, fractures and foliations can be recognized all along the mine, reflecting only one structural domain in the entire mine [22]. The orientations of the discontinuities that were included at the model stage are shown in Table 2.

Table 2. Parameters of BDX discontinuities along the mine study area.

Structure	Dip (°)	Dip Direction (°)	Spacing (cm)	Persistence (m)
Foliation	68	243	70.4	20
Fractures Set 1	75	53	89.4	12.7
Fractures Set 2	80	143	89.4	12.7
Fractures Set 3	87	297	89.4	12.7
Fractures Set 4	75	4	89.4	12.7
Primary Shear Zone (ZC1)	75	219	135.1	17.8
Secondary Shear Zone (ZC1)	75	332	135.1	17.8

All the available background information about the intact rock and rock mass characterization was used as inputs for the calibration process.

3. Materials and Methods

The optimization process developed in this study is summarized in the flowsheet of Figure 5.

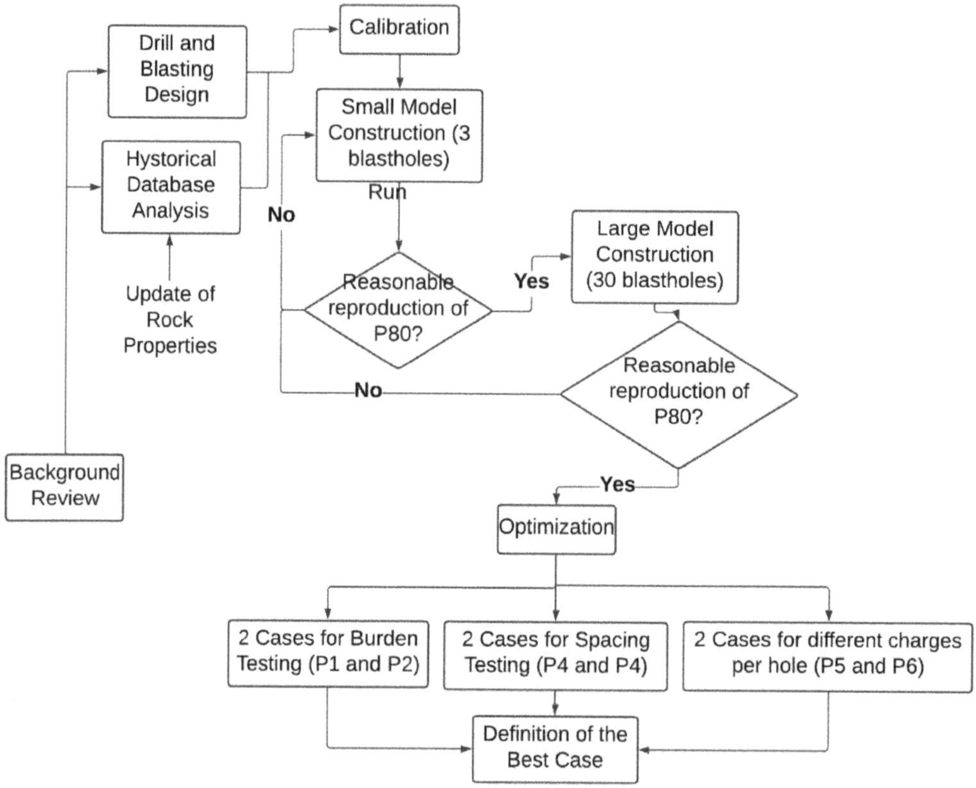

Figure 5. Optimization methodology flowchart.

3.1. *Drilling and Blasting Design*

The drilling and blasting design considered the standard parameters for the drilling pattern, explosives characterization, blasthole, charge information and detonation sequence in the BDX. These parameters were addressed in terms of mean values. A blasting polygon (L3_127_005_R00) performed in the mine [24] was taken into account in the calibration process. The drilling pattern was characterized by a burden of 5.0 m, spacing of 5.6 m, area per blasthole of 28.0 m^2, blasthole diameter of 12 1/4, drill length of 16.1 m, subdrill length of 1.0 m and dip angle of 90°.

The explosives used in the mine were provided by Orica [25]. For these drilling patterns, the blasthole charge included two types of emulsions: Fortis Advantage System 70 (FDS 70) and Fortis Advantage System 100 (FDS 100), which main performance parameters are reported in Table 3.

Table 3. Parameters of BDX discontinuities along the mine study area.

Explosive Parameter	FDS 70	FDS 100
Reported Density (g/cm^3)	1.15–1.25	1.15–1.25
Reported VOD (m/s)	3500–6000	3500–6000
RWS (%)	1.07	0.97
RBS (%)	1.6	1.45
CO_2 Output (kg/ton)	159	201
Mean Density (g/cm^3)	1.2	1.2
Last Measured VOD (m/s)	5636	5636

Velocity of detonation (VOD); relative weight strengths (RWS).

The drilling pattern considered the use of gravel as the stemming of 5.5 m, and the blasthole charge information, in terms of mean magnitudes, was as follows: charge length of 10.2 m, charge weight of 966.2 kg, and specific charge of 655.6 g/t.

For the calibration of the Blo-Up model, the stemming was considered ideal, providing full confinement. The booster was located 1.0 m above the level of the drillhole bottom. For the blasting sequence, the delay between blastholes in a single row was 5.0 ms, and the delay between rows was 85.0 ms.

3.2. Historical Database Analysis

A database with 1596 historical blasting designs and 80.0% passing size (P_{80}) was filtered to select the data of the predominant lithology as a case study. Finally, the historical database analysis included 284 blasting designs with fragmentation sizes between 2015 and 2019. The P_{80} data were considered in a descriptive statistics analysis.

A drilling and blasting design (L3_127_005), performed in the BDX, was also used in the calibration process; the data included the drilling patterns, blasthole charge information, detonation sequence and PSD.

3.3. Calibration

As shown in Figure 5, the calibration stage involved the definition of the model resolution, construction and calibration of the small model and large model, and finally the optimization of the designs studied.

The model resolution was 0.16 m, considering the Blo-Up software runtime limitations and the PSD. The PSD indicates that the passing percentage for a size of 0.16 m is 64.5% [24]. This means that Blo-Up was modelled with the upper 35.5% of the size distribution. Afterwards, two models were constructed and calibrated: a small model with three blastholes and a large model with 30 blastholes.

The small model was aimed to produce a fast response in the exploratory stage of the calibration and was run up to 50.0 ms, considering up to 40.0 ms after the last blasthole detonation. The large model was developed to validate the results of the small model for a more realistic blasting scenario and was run up to 600.0 ms. In these models, the detonation time effectively represents the fragmentation due to detonation and blasting.

In the construction of the models, the boundary conditions were established to represent a bench of 15.0 m in height. The blastholes were located half a spacing distance and half a burden distance from the edges in the model area that represents the free face. The parameters of the blasting designs are shown in Table 4, and the model geometries are shown in Figure 6.

Table 4. Blasting designs of the small and large models.

Model	Free Face Distance (m)	Burden (m)	Spacing (m)	Bench Height (m)	Hole Depth (m)	Blasthole Diameters (inch)	Stemming (m)	Charge Length (m)	Subdrill (m)
Small	2.8	5.6	-	15.0	16.0	12 1/4	5.5	10.5	1.0
Large	2.5	5.0	5.6	15.0	16.1	12 1/4	5.5	10.6	1.1

In both models, the discontinuities observed for the BDX were included. Figure 7 shows the foliation, fractures (4 sets), shear zones (2 sets) and joints (7 sets).

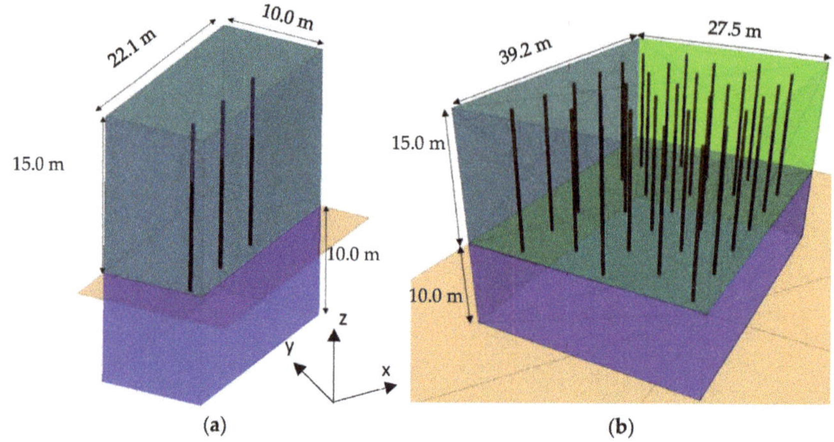

Figure 6. (**a**) Isometric view of the small model and (**b**) the large model in Blo-Up.

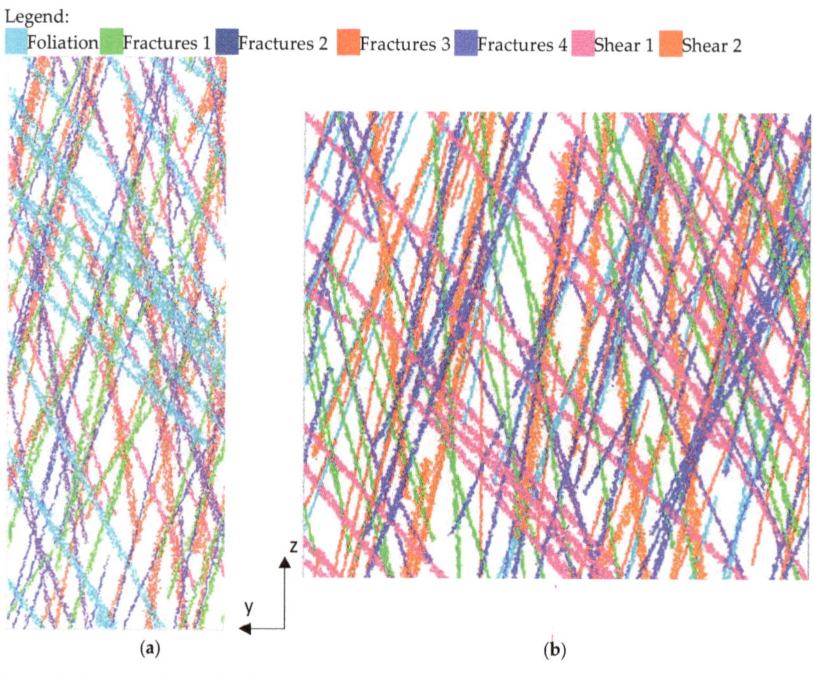

Figure 7. Cross-section of the discontinuities of the (**a**) small model and (**b**) large model.

3.4. Optimization

After the calibration, the optimal blasting design was chosen for the optimization process in terms of the desired fragmentation, which is a P_{80} of 0.55 m. To determine the optimal blasting, a total of 6 cases were tested (Table 5).

Table 5. Cases for design optimization.

Model	P1	P2	P3	P4	P5	P6
Number of Blastholes	30	30	30	30	30	30
Burden(m)	6	6	6.5	5.5	5	6
Spacing (m)	9	8	8	8	5.6	8.5
Blasthole Length (m)	16	16	16	16	16	16
Charge Length (m)	10.5	10.5	10.5	10.5	11.0	10.5
Subdrill (m)	1.0	1.0	1.0	1.0	1.0	1.0

The P1, P2, P3 and P4 cases had the same charge per blasthole and explosive, according to the L3_127_005 blasting design. For all of these cases, the blasthole diameter was 12¼″.

4. Results and Discussions

4.1. Historical Database Analysis

The analysis of the 284 blasting designs with a P_{80} size are consolidated in the histogram of P_{80} frequency in Figure 8.

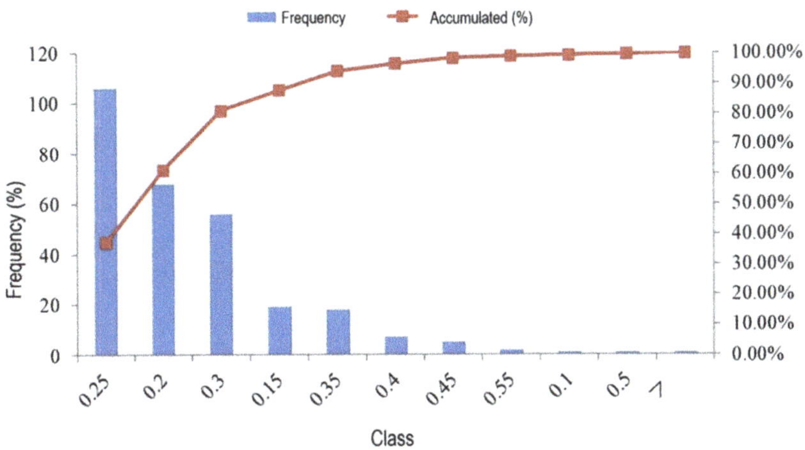

Figure 8. Histogram for P_{80} in production blasting in BDX.

Over the five years analysed, the average P_{80} was equal to 0.25 m with a standard variation of 0.07 m and a median of 0.22 m in the BDX. These rock fragmentation results are due to different magnitudes of different drilling patterns, blasthole charges, and other parameters. The analysis of Figure 8 allows us to infer that the historical database has 95% designs with a $P_{80} \leq 0.32$ m, even though there were cases with $P_{80} \geq 0.55$ m, but those were a minority in the data analysed.

4.2. Model for Fragmentation Analysis

Considering the old model for fragmentation, which had a P_{80} close to 0.80 m, and the reports from mine operations related to boulders from the run of mine (ROM) with oversize > 1.0 m were investigated. As mentioned before, this work aimed to optimize mine blast design with an optimum size of P_{80} equal to 0.55 m using Blo-Up software. Figure 9 shows the PSD of the L3_127_005 blast and the old model for fragmentation provided by the mine company.

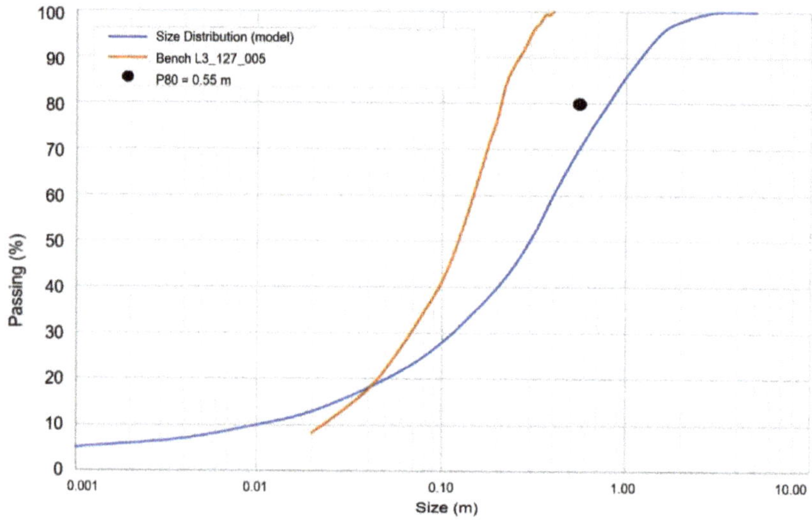

Figure 9. Particle size distribution: model for fragmentation and L3_127_005 blasting.

Through the analysis of Figure 9 and taking into account the desired P_{80} size of 0.55 m, it can be observed that the L3_127_005 blasting had a smaller P_{80} and P_{100} smaller than 0.40 m. Analyzing these two PSD, the L3_127_005 presents a finer size, considering the size results of the muckpile. The blasting pattern used in this blast was equal to 5.0 m × 5.6 m and was considered regular blasting because it has similar results in terms of P_{80} and the same pattern.

The calibration stage aimed to reproduce the L3_127_005 blasting to obtain the fragmentation results of the BDX as observed at the site. For the optimization stage, two different burdens, two spacings and two specific charges were assessed. With these parameters, the target was increasing the P_{80} from ~0.21 m of the L3_127_005 blasting to achieve the desired P_{80} of 0.55 m [9].

4.3. Calibration Results

At the calibration stage, four models were considered. These models included (1) the best case of the small model from Blo-Up and (2) the best case of the large model from Blo-Up. The minimum size of these models was limited by the model resolution. Two models based on a best-fit exercise through the Swebrec-type PSD better fit the data from the Blo-Up model. These models included (1) the best case of the corresponding small model and (2) the best case of the corresponding large model. The PSD obtained in these models are shown in Figure 10.

The PSD showed that the models had some P_{80} variability related to their scale and compared to the observed PSD, which could be explained by data unavailability from particle sizes larger than 0.2 m. The large models had coarser fragment sizes than the small models, and we noticed more conservative values in the range of sizes between 0.2 and 0.4 m, but a reasonable representation for the rest of the fine curves. The Swebrec distribution models the whole range of sizes, with an error of 4.0% compared to the observed P_{80} of L3_127_005_R00 blasting. As was found by Coello-Velázquez et al. (2019), the Swebrec distribution provides the best fit compared to the observed values.

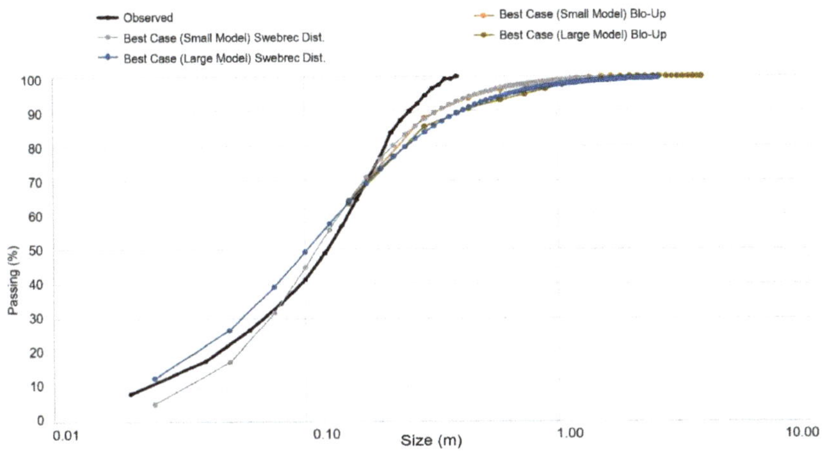

Figure 10. Particle size distribution of the calibration stage: Observed in the mine vs. predicted by Blo-Up (best case).

To achieve the best cases above, vertical blast-induced damage was introduced, and in the first run, it was exported. The result of this run was assumed to represent microcracks, which occur within the rock mass due to vibrations and blasting. In the second run, the vertical component of the damage was applied to the same model. Furthermore, it was considered that the extension of the induced fractures and induced damage from blasting occurred down to a depth of 3.0 m below the toe of the bench. This region was modelled assuming that the tensile strength of the rock mass was 1.0 MPa weaker.

To better describe these different zones of the rock mass, the tensile strength was divided into fresh rock, damaged rock and joints. The calibrated geomechanical properties for the Blo-Up model are shown in Table 6.

Table 6. Calibrated properties for the Blo-Up model.

Calibrated Properties	Value
Density (kg/m^3)	3.2
Young Modulus (Pa)	1.0×10^{10}
Poisson's Ratio	0.24
UCS (MPa)	90
Tensile Strength of Rock (MPa)	4.6
Tensile Strength of Damaged Rock (MPa)	3.5
Tensile Strength of Joints (MPa)	3.2
Friction Angle (°)	32
Damping	0.3

The distribution of fragments was represented in the contour of the blasted rock fragments over 0.5 m. Figure 11a shows the fragment contour for the BC Large Model resulting from calibration in Blo-Up. Figure 7b shows the vertical cross-section along the height of the bench. The analysis of these contours provides evidence that larger fragments were concentrated in the upper part of the bench, including the damaged area and the area below, and were located at the boundary conditions.

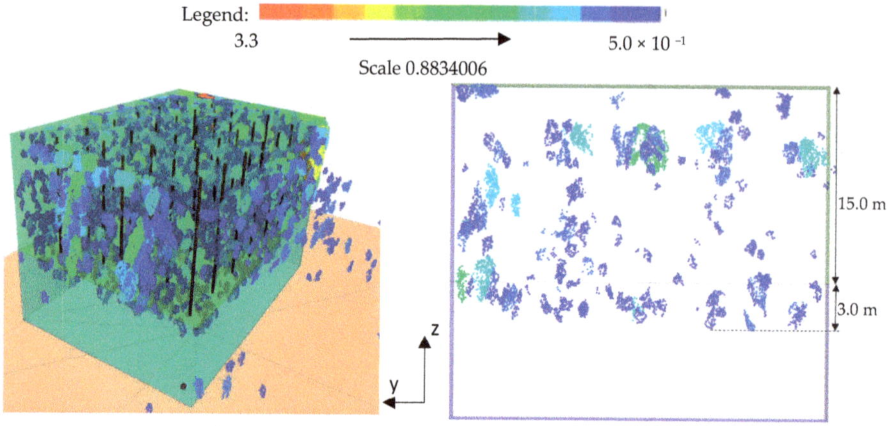

(**a**) (**b**)

Figure 11. Fragment contour over 0.5 m: (**a**) modelled by Blo-Up and (**b**) cross-section of the large model in Blo-Up.

4.4. Optimization Results

Considering the spacing optimization with cases of varying spacing, a constant burden of 6 m, and a charge length of 10.5 m for each blasthole, large models P1 and P2, with spacings of 9.0 m and 8.0 m, respectively, had a runtime of 550.0 ms (30 blastholes). Figure 12 shows the PSD results for each model.

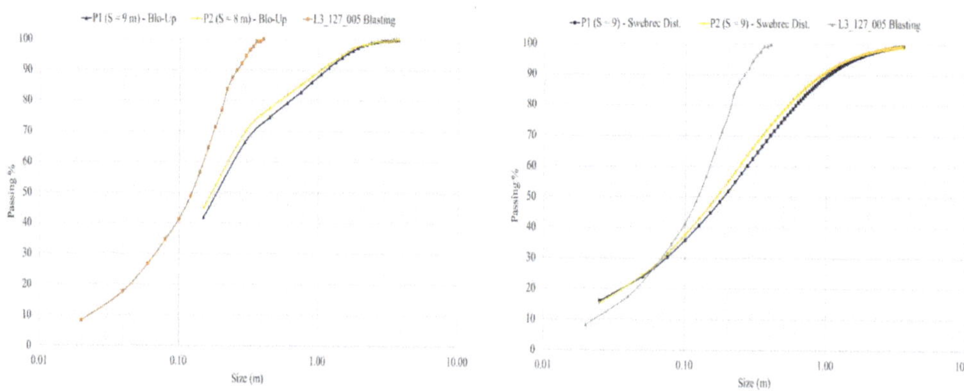

(**a**) (**b**)

Figure 12. Spacing optimization results: (**a**) Blo-Up model and (**b**) best fitted Swebrec distribution.

From the PSD, model P2 achieved the best P_{80} of ~0.53 m, which was close to the desired P_{80}. This result was obtained with a drilling pattern of 6.0 m of burden and 8.0 m of spacing and with the same charges per blasthole as blasting design L3_127_005.

For the burden optimization, cases P3 and P4 had different burdens, a spacing of 8.0 m and a charge length of 10.5 m. Models P3 and P4 had burdens of 6.5 m and 5.5 m, respectively, and the results are shown in Figure 13.

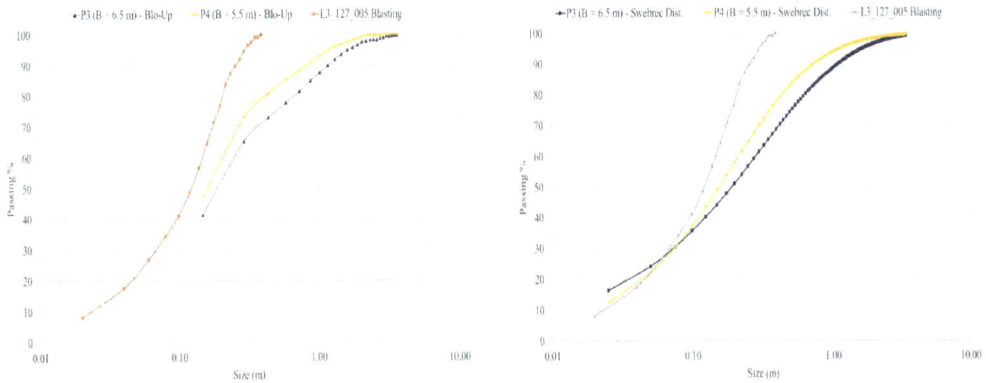

(a) (b)

Figure 13. Burden optimization results: (**a**) Blo-Up model and (**b**) best fitted Swebrec distribution.

The results of burden optimization failed to achieve P_{80}~0.55 m, and model P3 had a coarse size distribution; on the other hand, the P4 model shows finer results than expected. This finding indicates that a burden of 6 m is enough to achieve the desired P_{80} of 0.55 m.

Finally, the charge length optimization considered the P5 models with burden and spacing maintained at 5.0 m and 5.6 m and charge lengths of 10.0 m and 11.0 m, respectively. The results of the charge length optimization are shown in Figure 14.

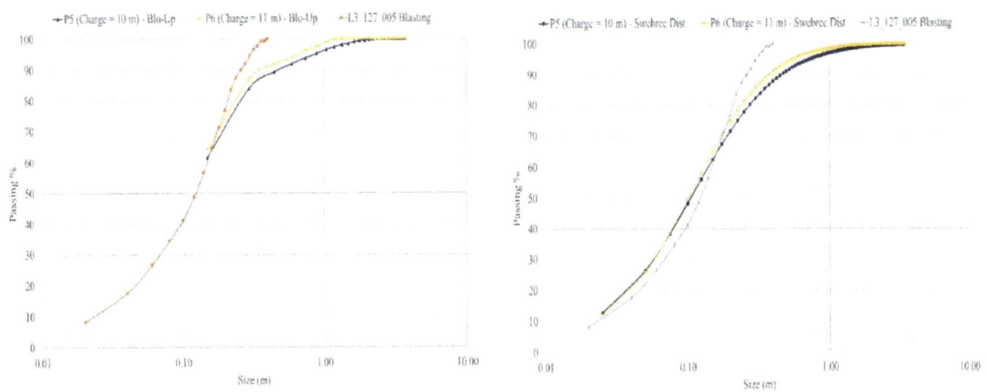

(a) (b)

Figure 14. Charge length optimization results: (**a**) Blo-Up model and (**b**) best fitted Swebrec distribution.

The charge length optimization results show that a finer PSD can be obtained that does not approach the desired target of P_{80}~0.55 m. Compared with the other optimized models, P5 and P6 were not better in terms of performance.

The P_{80} of the optimized model indicates that the best case is model P2, with a P_{80} of ~0.53 m, the closest value to the desired P_{80}. This was optimized with a blasting design with a burden of 6.0 m and spacing of 8.0 m.

5. Conclusions

In this study, a Blo-Up model was constructed to optimize fragmentation from blasting in the BDX lithology of the study area to achieve a desired P_{80} of 0.55 m. Through historical data, observations of a blasting example (L3_127_005), and a predictive analysis, the parameters were calibrated.

Regarding the calibration, the best case had a P_{80} of 0.25 m, and a total error of 4% was obtained when comparing the best fitted Swebrec distribution and the Blo-Up data with the data from the L3_127_005 blasting. For the optimization process, some assumptions were made to better represent the rock mass conditions under vibration and blasting.

Based on the results of the blasting optimization applied for the copper mine in hard rock, the recommended blasting design for BDX must follow the P2 model, with a burden of 6.0 m, spacing of 8.0 m, borehole diameter of 12¼", hole length of 16.0 m, gravel stemming of 5.5 m, charge length of 10.5 m, and subdrill length of 1.0 m, considering the explosive Fortis Advantage 70/100 [25] with a blast sequence of 5.0 ms between holes and 75.0 ms between rows. It is suggested that future works execute this optimal bast design on an industrial scale to validate the PSD of rock fragmentation reached in this work.

Author Contributions: Conceptualization, V.F.N.T., C.C., M.E.V., J.R.F. and L.G.C.S.; Methodology, V.F.N.T., C.C., M.E.V. and L.G.C.S.; Validation, V.F.N.T., J.R.F. and L.G.C.S.; Data Curation, C.C. and M.E.V.; Original Draft Preparation, C.C. and M.E.V.; Writing, Review and Editing, C.C.; M.E.V. and J.R.F.; Supervision V.F.N.T. All authors have read and agreed to the published version of the manuscript.

Funding: This research received no external funding.

Institutional Review Board Statement: Not applicable.

Informed Consent Statement: Not applicable.

Data Availability Statement: Not applicable.

Acknowledgments: The authors would like to thank the Vale Institute of Technological, Vale S.A. and Walm Engenharia e Tecnologia Ambiental for the material resources for the preparation of this study and Itasca Chile for providing the data.

Conflicts of Interest: The authors declare no conflict of interest.

References

1. Ozdemir, B.; Kumral, M. A system-wide approach to minimize the operational cost of bench production in open-cast mining operations. *Int. J. Coal Sci. Tech.* **2019**, *6*, 84–94. [CrossRef]
2. Gadikor, J. Optimization of Drilling and Blasting Practices at a Western US Open Pit Copper Mine. Master's Thesis, University of Montana, Butte, MT, USA, 2018.
3. Park, J.; Kim, P. Use of drilling performance to improve rock-breakage efficiencies: A part of mine-to-mill optimization studies in a hard-rock mine. *Int. J. Min. Sci. Tech.* **2020**, *30*, 179–188. [CrossRef]
4. Zhang, Z.X.; Qiao, Y.; Chi, L.Y.; Hou, D.F. Experimental study of rock fragmentation under different stemming conditions in model blasting. *Int. J. Rock Mech. Min. Sci.* **2021**, *143*, 104797. [CrossRef]
5. Singh, P.K.; Roy, M.P.; Paswan, R.K.; Sarim, M.D.; Kumar, S.; Jha, R.R. Rock fragmentation control in opencast blasting. *J. Rock Mech. Geotech. Eng.* **2016**, *8*, 225–237. [CrossRef]
6. Majid, A.; Gholamreza, L.; Alireza, Y.B.; Mohammada, G. Blastability evaluation for rock mass fragmentation in Iran central iron ore mines. *Int. J. Rock Mech. Min. Sci.* **2015**, *25*, 59–66.
7. Ke, B.; Pan, R.; Zhang, J.; Wang, W.; Hu, Y.; Lei, G.; Chi, X.; Ren, G.; You, Y. Parameter optimization and fragmentation prediction of fan-shaped deep hole blasting in Sanxin gold and copper mine. *Minerals* **2022**, *12*, 788. [CrossRef]
8. Leng, Z.; Fan, Y.; Gao, Q.; Hu, Y. Evaluation and optimization of blasting approaches to reducing oversize boulders and toes in open-pit mine. *Int. J. Min. Sci. Tech.* **2020**, *30*, 373–380. [CrossRef]
9. Amoako, R.; Jha, A.; Zhong, S. Rock fragmentation prediction using an artificial neural network and support vector regression hybrid approach. *Mining* **2022**, *2*, 233–247. [CrossRef]
10. Ouchterlony, F.; Sanchidrián, J.A. A review of development of better prediction equations for blast fragmentation. *J. Rock Mech. Geotech. Eng.* **2019**, *11*, 1094–1109. [CrossRef]
11. Coello-Velázquez, A.L.; Quijano Arteaga, V.; Menéndez-Aguado, J.M.; Pole, F.M.; Llorente, L. Use of the Swebrec function to model particle size distribution in an industrial-scale Ni-Co ore grinding circuit. *Metals* **2019**, *9*, 882. [CrossRef]
12. Tang, J.; Tan, X.; Li, X.; Wei, B.; Wang, Z.; Zheng, C. Research on edge detection algorithm based on line laser scanning. *Procedia* **2019**, *84*, 1101–1106. [CrossRef]
13. Wang, Y.; Tu, W.; Li, H. Fragmentation calculation method for blast muck piles in open-pit copper mines based on three-dimensional laser point cloud data. *Int. J. Appl. Earth Obs. Geoinf.* **2021**, *100*, 102338. [CrossRef]
14. Toriya, H.; Tungol, Z.P.L.; Ikeda, H.; Owada, N.; Jang, H.D.; Adachi, T.; Kitahara, I.; Kawamura, Y. Fragmentation size distribution measurement by GNSS-aided photogrammetry at real mine site. *Mining* **2022**, *2*, 438–448. [CrossRef]

15. Rosin, P.; Rammler, E. The laws governing fineness of powdered coal. *J. Inst. Fuel.* **1933**, *7*, 29–36.
16. Rodríguez, B.A.; García, G.G.; Coello-Velázquez, A.L.; Menéndez-Aguado, J.M. Product size distribution function influence on interpolation calculations in the Bond ball mill grindability test. *Int. J. Miner. Process.* **2016**, *157*, 16–20. [CrossRef]
17. Ouchterlony, F. Fragmentation characterization; the Swebrec function and its use in blast engineering. In *Fragblast 9, Proceedings of the 9th International Symposium on Rock Fragmentation by Blasting, Granada, Spain, 13–17 August 2009*; Taylor and Francis: Oxfordshire, UK; pp. 3–22.
18. Castro, C.; Valencia, M.E. *Optimizing Blast Fragmentation in Salobo Mine, Report ITASCA-669.004.04*; Itasca Chile: Santiago de Chile, Chile, 2020.
19. Braithwaite, M.; Sharpe, G. Simulation of real detonations as an energy source term for the hybrid stress blasting model. In Proceedings of the 9th International Symposium on Rock Fragmentation (Fragblast 9), Grenada, Spain, 13–15 September 2009; Sanchidrian, J.A., Ed.; CRC Press: Boca Raton, FL, USA.
20. Itasca Consulting Group, Inc. *Blo-Up User's Guide, Release 2.7*; Minneapolis: Itasca, IL, USA, 2016.
21. Kapila, A.K.; Bdzil, J.B.; Stewart, B.S. On the structure and accuracy of programmed burn. *Combust. Theory. Modell.* **2006**, *10*, 289–321. [CrossRef]
22. Vale S.A.; WALM, E.T.A. *Estudo Geral—Cava da Mina Elaboração Modelo Geomecânico da Cava Relatório Final do Modelo Geomecânico Relatório Técnico, RL-1000SA-X-00008*; Vale S.A.: Marabá, Brazil, 2019; Unpublished work.
23. Hoek, E.; Diederichs, M.S. Empirical estimation of rock mass modulus. *Int. J. Rock Mech. Min. Sci.* **2006**, *43*, 203–215. [CrossRef]
24. Orica. *Relatório de Desmonte L3_127_005_R00*; Vale S.A.: Marabá, Brazil, 2019; Unpublished work.
25. Orica. Fortis™ Advantage System. Technical Data Sheet. Available online: http://www.oricaminingservices.com/cn/en/product/products_and_services/bulk_systems/page_bulk_systems/fortan_advantage/43 (accessed on 12 July 2022).

mining

MDPI

Review

The Effect of Mining Activities on the Paleokarstic Features, Recent Karstic Features, and Karst Water of the Bakony Region (Hungary)

Márton Veress

Department of Geography, Eötvös Loránd University, Savaria University Centre, 9700 Szombathely, Hungary; veress.marton@sek.elte.hu

Abstract: This study describes the direct and indirect effects of mining on the karst of the Bakony Region. For this, the results of geological and mining research of the last century, the results of hydrological research of fifty years, as well as the investigations of several decades on the karst of the mountain region are used. Direct effects include the exploitation of filling materials (limonite, kaolinite, manganese ore, and bauxite) from paleokarst features, dolomite rubble, activities exploring or destroying cavities, and the pollution of cavity systems with mining waste (dirt). An indirect effect is karst water extraction. Mining activities (coal and quarrying) resulting in the development of pseudokarstic features are also mentioned here. It can be stated that the effects on the karst and karst features may be permanent and even renewing, but the original state may also have returned or can be expected in the near future. Damages may be local or regional. A regional effect is the decrease in karst water level, which has the most significant effect on the environment, but it has already reached its original state by now.

Keywords: Bakony region; mining; karst; karst depression; pseudokarst; karst water

1. Introduction

This study describes the effect of mining on the karst of the Bakony Region, which is the southwestern part of the Transdanubian Mountains. It is surrounded by the Little Hungarian Plain in NW, by Lake Balaton in SE, and the micro-regions of the Balaton Basin (Balaton Riviera, Tapolca Basin), the Great Hungarian Plain (Mezőföld), and the Vértes Mountains in NE (and Mór Graben with graben structure). Its elevation is 150–700 m and its area is 4300 km^2. Its largest area is the Bakony Mountains (2200 km^2). Parts of the Bakony Region are the Keszthely Mountains, Northern Bakony, Southern Bakony, Balaton Uplands and Bakonyalja (Figure 1).

The mountains are of a block structure, where more elevated blocks are mountains, and less elevated blocks are basins. They are mainly built up of Triassic dolomite, which is significantly widespread on the surface too [1]. However, Triassic, Dachstein, and Jurassic limestones frequently occur in great expansion on blocks of higher elevation. Cretaceous and Eocene limestones of small thickness can mainly be found in a larger and smaller expansion on blocks with medium elevation, mostly covered with loess. Blocks of lower elevation (basins) are mostly covered by non-karstic rocks (Middle Oligocene–Lower Miocene gravel, Figure 2). In the mountains, rocks of various ages constitute bands of NE-SW direction. Perpendicular to this, the Bakony Region is of asymmetrical structure. In the SE, older Palaeozoic metamorphic rocks are exposed to the surface. In the middle part, the Triassic floor constitutes a synclinal of NE-SW direction where transgressions (often archipelagic) developed. Its north-western flank is incomplete: although Triassic rocks are exposed, the Palaeozoic features subsided along stepped faults and constitute the floor of the basin of the Little Hungarian Plain [1].

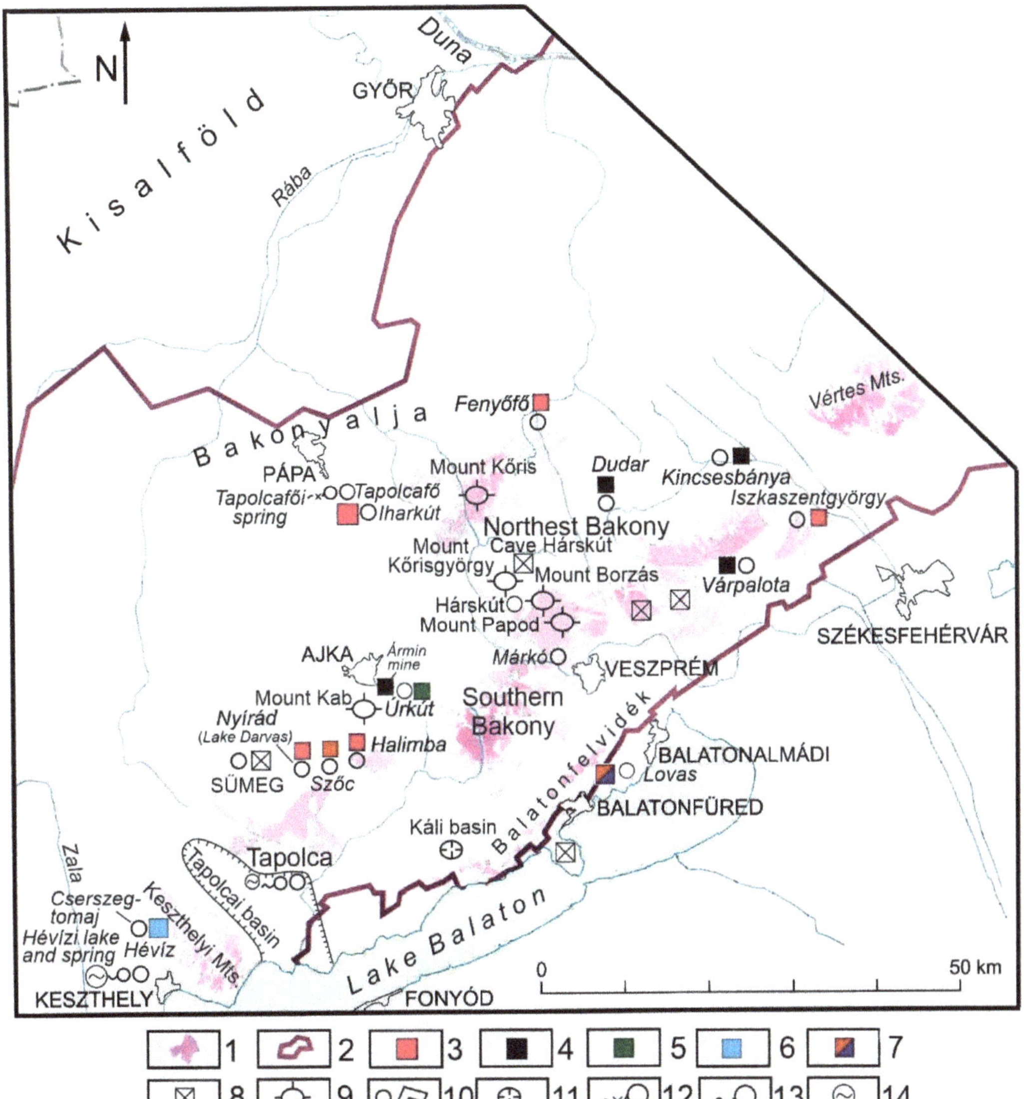

Figure 1. The Bakony Region and main mining sites are mentioned in this study (own figure). Legend: 1. infiltration sites, 2. border of karst water storage, 3. bauxite mine, 4. coal mine, 5. manganese mine, 6. kaolin mine, 7. paint mine, 8. stone and rubble mine, 9. mountain, 10. settlement, 11. basin, 12. karst spring that went dry, 13. active karst spring, 14. spring lake.

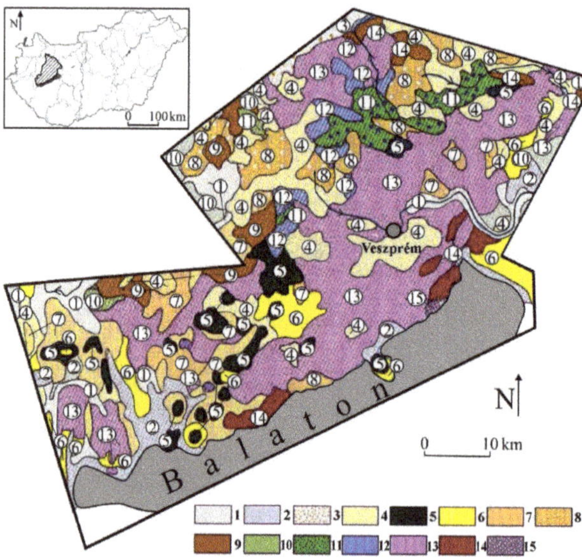

Figure 2. Geology of the Bakony Mountains [2]. Legend: 1. fluvial sediment, 2. Holocene torf, 3. Holocene wind-blown sand, 4. Pleistocene loess, 5. Pliocene basalt, 6. Upper-Miocene freshwater limestone, 7. Middle Miocene limestone, 8. Oligocene gravel, 9. Eocene limestone, 10. Upper Cretaceous limestone, 11. Lower Cretaceous limestone, 12. Jurassic limestone, 13. Triassic limestone, dolomite, 14. Permian sandstone, 15. old Paleozoic phyllite.

Karst develops on soluble rocks (limestone, dolomite, and evaporite). The infiltrating water with carbonic acid creates cavities (caves), fills them, and flows towards the mountain margin (karst water), where it emerges in springs. The infiltrated water constitutes a three-dimensional system whose surface is the karst water level, which undergoes fluctuation of various degrees and periods due to natural and artificial effects, at another time a one-way rise or subsidence [3,4]. Since karst rocks drain water, neither a surface water network nor valleys develop (they are only formed at sites where the valley is inherited from the non-karstic cover or when the karst water level is situated at the valley floor). The dissolved material is transported into the karst with the infiltrating waters; therefore, surface karst features are closed. These are karren, dolines, ponors with blind valleys, and poljes. The material transported in the solution precipitates as freshwater limestone.

Features resembling karst features can also develop during non-karstic processes (e.g., caves and surface depressions). These are pseudokarst features that may also be formed due to mining. Features developing at present are recent karst features, while those that developed earlier and not forming now are paleokarst.

The paleokarst of the Bakony Region is borne by the peneplain that developed during tropical karstification [5]. Its paleokarst is mainly of Cretaceous age [6,7], but others may also occur [6]. The surface karst features mostly developed on Triassic dolomite; the cover is most frequently Eocene limestone and Middle Oligocene–Lower Miocene gravel [6]. During karstification karst depressions (solution dolines), poljes and inselberg karst developed [5,8,9]. Its present surface karst is of an island-like pattern and covered karst that developed on blocks of medium elevation [10]. Features of its covered karst are suffosion dolines, but solution dolines also occur in small numbers at the margin of the mountains [10].

The effects of mining on the karst and karst features are overviewed and classified. This complex overview may have importance in education and environmental protection. The here described approaches might be applied to the qualification of other karst areas.

2. Direct Mining Effects on the Karst

In case of direct effect, the mining activity affects the karst feature: exploitation of minerals or supplying the karst system with materials. The first is some kind of erosion; the latter is accumulation. These are primary effects, but secondary effects may also occur in both cases; for example, during exploitation, the pollution takes place through fuel infiltration.

Direct effects involve the exploitation of the filling material and dolomite rubble, the exploration of karst cavities, the transportation of the mining waste into the karst, as well as feature development and destruction due to mining.

The original feature became exhumed during surface mining when the filling material that accumulated in karst depressions acting as traps was exploited, and the cover was removed. If the mineral resources to be exploited were at a greater depth, exploitation would occur through underground mining operations instead of open-pit mining. The strip pit is either constituted by depressions, or they are parts of the strip pit. Usually, remnants of the mined filling material can still be found in depressions. The original karst feature does not only differ from the original but also because its slope was transformed due to subsequent tectonic stress, and larger and smaller sections of the bearing rock may have been removed.

The exploited material can be material containing limonite, hematite, kaolin (kaolinitic clay), manganese ore, and bauxite.

The limonite–hematite material was exploited from two Palaeolithic so-called paint mines between the settlements of Lovas and Felsőörs in the Balaton Uplands. The paint mines were detected during dolomite rubble mining [11]. According to artifacts found here, they originate from the end of the first third of Würmian $\frac{1}{2}$ interstadial [11,12]. Based on the age of the mining of the paint mine; as a mine, it is probably unique. Unfortunately, its exact site is unknown; thus, geomorphological conditions cannot be studied subsequently. According to published figures, the mined material may have accumulated in the karstic bedrock of the dolomite, or it may have been mixed with this material during rubble formation.

Furthermore, 1–2 km away from Cserszegtomaj (Keszthely Mountains), kaolinitic clay occurrence can be found partly in superficial outcrop and partly covered thinly [13], which are the fills of karst depressions. In his later work, [14] mentions the filling material as bauxite material. The fillings were exploited by hand; thus, deep dolines of 10–50 m with vertical walls were exposed in the Triassic dolomite [8,13]. According to Bárdossy [13], the development age of doline development coincides with the age of Nyirád-Halimba bauxite deposits which is regarded as Lower Cretaceous by [6].

There are several manganese ore occurrences in the mountains. The oxidized manganese ore entered into the karst depressions of Úrkút (Csárda-hegy) by reworking; its cover is of Lower and Middle Eocene, while the bearing karst features deepen into Lower Liassic (Jurassic) limestone [15]. The manganese ore was exploited by hand in the twenties of the 20th century (here, there is a manganese ore mine with an underground mining operation where exploitation still takes place). As a result, a tropical-type doline with steep slopes was exposed with many solution features on its slope, a gentler, rather temperate solution doline, and a terrain with pinnacle karren features (Figure 3).

Figure 3. Exposed paleokarst features of Csárda-hegy (hill) (own figure). Legend: (**A**) tropical doline with steep slopes: (**B**) solution pit of doline slope; (**C**) temperate doline with gentle slopes and pinnacle karren.

Bauxite is widespread in the mountains and may constitute a lenticular colony (diameter is about 20 m, thickness is 30–50 m, the bearing features is shallow doline), a deep-doline colony (diameter is 30–50 m, thickness is 80–100 m, the bearing features are deep doline), a bed-like colony (longitudinal expansion is several hundred meters, the bearing feature is polje or uvala), a structural colony (the bearing feature is tectonic graben), a deep-doline and structural colony (a combination of karst feature and structural feature), and a canyon-like colony when it is several hundred meters long; the karst feature is a blind valley [6,14,16]. Under tropical climates, bauxite is the weathering residue of the rocks of the non-karstic terrain [16], which was transported into the karst depressions, and then accumulated in them but may have formed in situ as well [17], or it may have been transported from below by transgression [18]. The products of lateritic chemical weathering may originate from the peneplain built up of metamorphic rocks surrounding the mountains from S and SE [19]. It cannot have formed in situ since it is separated from the uneven bedrock without transition; however, the iron crust is mostly present on the bedrock, which may have formed during and after bauxite accumulation [16].

The bedrock of karst depressions is Triassic Main Dolomite and Dachstein limestone [6,18], but the material of the depressions of Triassic Main Dolomite may have been reworked and accumulated in Late Cretaceous depressions [6]. Hungarian bauxites—the bauxites of the Bakony Region—can be classified as belonging to various development stages. The second stage (the first stage occurs in the Villány Mountains) involves the colonies of the Albian Lower Perepuszta (near the settlement of Zirc) and those of Bakonyoszlop. Their bedrock is Upper Triassic Dachstein limestone, and their cover is Upper Cretaceous clay marl. The third stage is the Upper Cretaceous Senonian stage. This includes the colonies of Halimba, Iharkút, and Csabpuszta (Sümeg). Their bedrock is Upper Triassic Main Dolomite, and their cover is Upper Cretaceous limestone or fluvial sandstone, clay, and clay marl. The age of the fourth stage is at the Cretaceous-Eocene boundary. The colonies of Fenyőfő, Dudar, Bakonyoszlop, Szőc, Halimba, and Nyirád belong to this. Their bedrock is Upper Triassic Dolomite, and their cover is Eocene limestone [14,16]. However, there are colonies whose cover is Eocene limestone or Middle Oligocene-Lower Miocene gravel and Pleistocene sediment [6,16]. If the caprocks were thinner, their exploitation took place by open-pit mining (Iharkút, Nyírád, Szőc); if they were thicker, they were exploited by underground mining operations (Iszkaszentgyörgy, Halimba, Fenyőfő, Figure 4). Karst depressions, or the terrain dissected by depressions, became exhumed during open-pit mining (Figure 5).

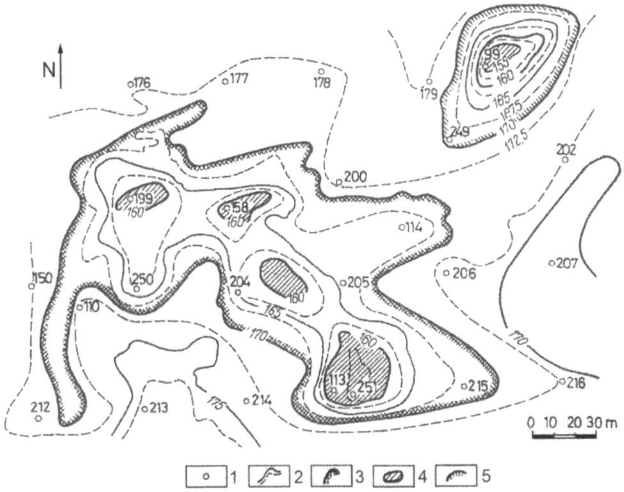

Figure 4. Bedrock map of the Upper Triassic Main Dolomite of the area of Halimba [20]. Legend: 1. drilling and its number, 2. contour line, 3. uvala, 4. the lowest point of doline, 5. doline.

Figure 5. Surface depressions exhumed during bauxite mining that developed Upper Triassic Main Dolomite: (**A**) Iharkút, (**B–D**) Lake Darvas (own figure).

The separation of the manganese ore of Úrkút from mining waste was carried out in the manganese washing plant. The so-formed manganese mud led into one of the depressions of Kab Mountain. The sludge appeared in the spring of the valley located 1.5 km away from this site about three weeks later [21]. The sludge was led into the doline at a rate of 1 m^3/min, which had an annual quantity of 0.5 million km^3 [21], although the duration of supply is unknown.

Dolomite rubble beds are mined on dolomite at several places between the settlement of Márkó and Várpalota in the mountains. The thickness of the rubble ranges from some meters to several ten meters. According to [20], its thickness may also exceed 50 m at Nyirád. The reason for rubble formation is that the calcareous material sticking to dolomite crystals is dissolved, and the dolomite falls into parts [3]. At sites of rubble exploitation, the wavy surface of lower karstic bedrock dolomite is exposed. Inside strip pits, where the rock underwent rubble formation to a lesser degree, the rock was left, and thus, the strip pits are dissected by mounds.

Several caves have been exposed during limestone mining. The longest is the recently discovered several km-long Hárskút Cave, which opens in the quarry of the Kőrisgyörgy Hill. The opened-up caves do not exist for a long time; they become shorter and then decay (e.g., in the quarry of Sümeg).

There was mining activity on the sandstone fields of the Kál Basin. The Pannonian sand became cemented with the amorphous silica that had precipitated from the hot water of post-volcanic activity [22] on which a rich karren landscape developed [23]. In the area of the Szentbékálla block field, the material of blocks was partly exploited (during this, several karren features may have been destroyed) for millstone, while that of Kővágóőrs block field was used to build houses. Similarly, quarrying damaged the (calcareous, quartzic) thermal cones of the Tihany Peninsula. Their rocks were used for construction work. Larger and smaller quarries were created on their sides, but at others, almost half of the cone material was exploited, and thus, the cavity of the inner side of the cone was exposed [24].

3. Indirect Effect on the Karst

An indirect effect develops when the anthropogenic activity does not affect the karst feature, but the alternation of the hydrology of the karst system. The consequences of the changes in hydrological conditions are regional, diverse, and occur fast. The uniform karst water system of the mountains is the main karst water, which primarily developed in the Triassic Main Dolomite. According to drilling and spring data, its former surface was at an elevation of 117.5–154.2 m at the margin of the mountains; it was of an increasingly larger altitude towards the center of the mountains. With the help of the data, a karst water level map was made, which was updated annually. Karst water observation wells were created in order to monitor water level changes; the number of these wells reached 200 within some decades. According to the spring data, the karst water level reached an elevation of 240 m in the inner part of the mountains [25], while it was at an elevation of 280 m between the settlements of Szentgál and Nagyvázsony based on the data of the constructed map of the karst water level [26]. In the latter area, the karst water swelled back because of the impermeable rocks of Southern Bakony. In the interest of the safe exploitation of mineral resources (without karst water flood), an artificial lowering of the karst water level was started in the environs of the mines with pumping. Its intensity reached a value of 460 m^3/min. The water level sank below the mines and reached 100 m at some sites [26]. Depressions developed on its surface. Not only the depth, but also the expansion of depressions increased, and adjacent depressions coalesced. Maximum values of water level subsidence are shown in Figure 6.

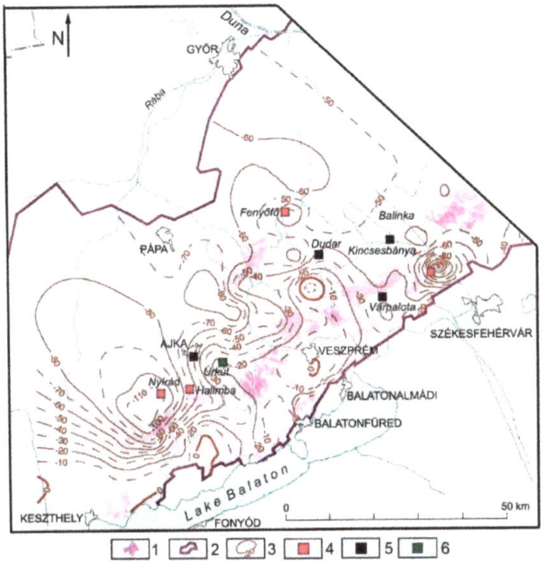

Figure 6. Artificial water level lowering in the Bakony Region [26]. Legend: 1. infiltration areas, 2. limit of karst water storage, 3. artificial karst water level lowering (m), 4. bauxite mine, 5. coal mine, 6. manganese mine.

4. The Artificial Water Level Lowering Resulted in the following Facts

(1) Water level subsidence reached the margin of the mountains. At sites where the level was below the level of karst springs, the discharge of springs decreased (Spring lake of Hévíz), or they went dry (Tapolcafő springs). At the town of Tapolca, not only the spring of Lake Malom (Figure 7) went dry, but the cave providing this site with water went dry too.

Figure 7. Lake Malom of Tapolca which developed by the damming of a stream, but it is also supplied by luke-warm karst springs (therefore, the water temperature is 18–20 °C) (the spring water originates from Sarmatian limestone) (own figure).

(2) Since the karst waters of the mountains flowed into the karstic floor of the surrounding basin (Little Hungarian Plain) and into its sediments, water level lowering could also be observed in the Little Hungarian Plain (Figure 6), and the direction of water flow partly changed (at some places water flow oriented towards the mountains).

5. Anthropogenic Pseudokarst Features Created through Mining

Several coal mines were developed in the area of the coal basin at Ajka. One of them is the Ármin Mine of Bocskor Hill. In the mine, there are seven coal beds below each other in Upper Cretaceous limestone [27]. One of them is shown in Figure 8. Closed depressions were formed at the surface above the mine. Their number exceeds 100, and they are developing at present too. They are areic, aligned in rows, elongated, and grike-like, but there are also features with circular ground plans [23] (Figure 9). Where cover is present, they are inherited onto the cover, and they resemble subsidence dolines (pseudokarstic subsidence doline) in this case.

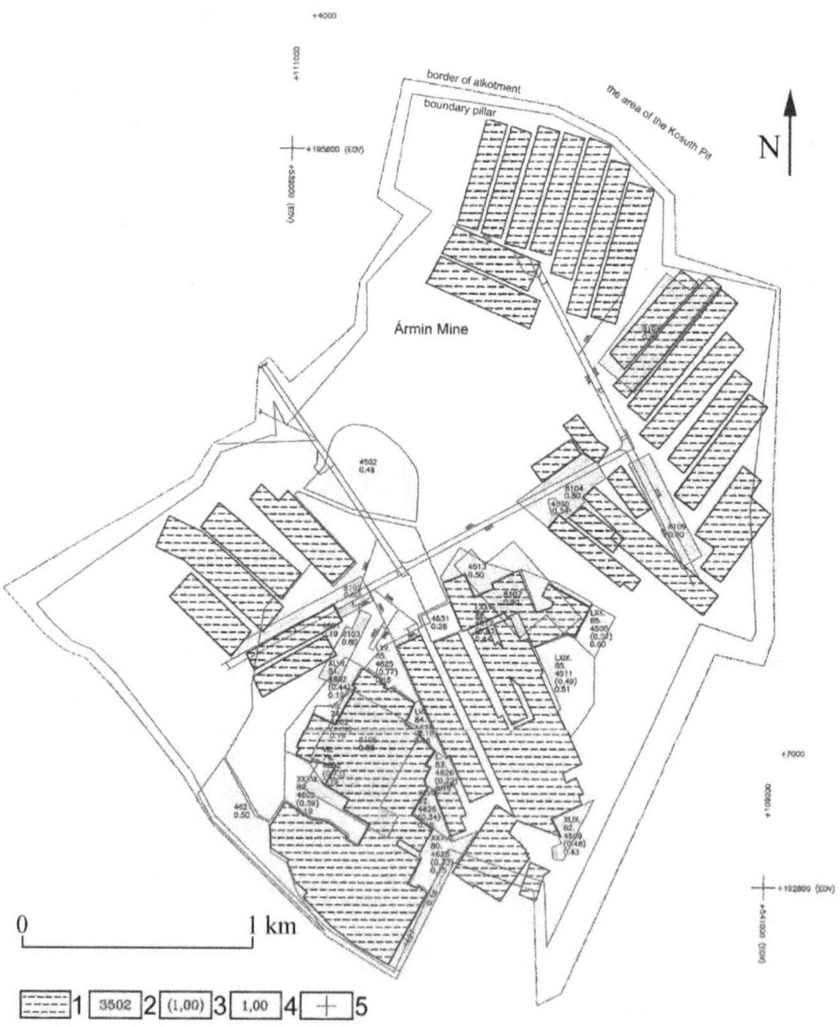

Figure 8. Map of bed IV of Ármin mine [28]. Legend: 1. excavation, 2. block number, 3. abandoned mineral reserves, original mineable value 4. mineable value, 5. map coordinate.

Figure 9. Pseudokarst features from Bocskor Hill: (**A**) graben feature (only partly areic) (Kopecskó); (**B**) circular pseudokarst dropout doline (own figure); (**C**) pseudokarst dropout doline of grike pattern (feature with collapse) (own figure); (**D**) circular pseudokarst suffosion doline (own figure) (the features developed on Middle Eocene limestone).

Their development can be explained by the material equilibrium resulting from the mining of coal beds situated below each other. During this, the covering Eocene limestone warped, tension stress developed at the margin of the warping, which resulted in the development of tension grikes in the rock, then the cover collapsed into some grikes [23] (Figure 10).

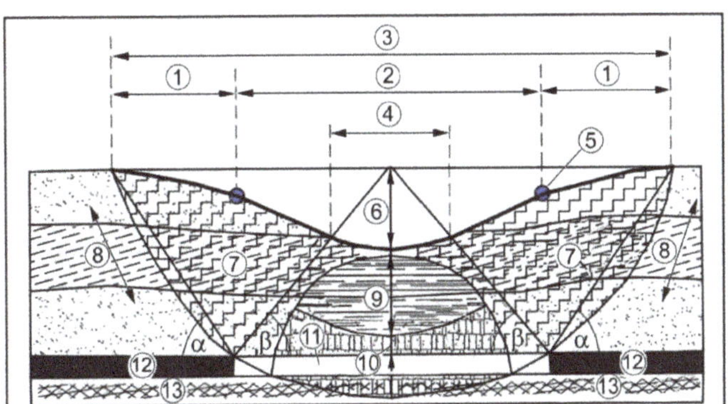

Figure 10. Subsidence trough developed by the sinking of the vault [29–31] (Legend: 1. tension zone, 2. warping zone, 3. original terrain, 4. sole subsidence zone, 5. inflexion point, 6. maximum subsidence, 7. deflection zone, 8. cover layers, 9. fracture zone, 10. collapse zone, 11. excavation field, 12. coal beds, 13. bedrock layers, 14. boundary angle, 15. undercut angle.

From the 18th century to the middle of the 19th century, several glasshouses were operated in various sites in the mountains [32]. For this reason, the rock was exploited at several places creating depressions with a diameter and depth of some meters necessary for the glasshouses. Their number is probably several thousand. They occur in groups (e.g., on the Kőris Hill, Borzás Hill, and Papod Hill). Glass sludge developed during smelting frequently and was found around them (the pieces are porous, and the cavernous coating of the former melt is visible on their surface).

Limestone was exploited at several sites (village, farm) for house building by hand. The exploitation scars are also closed; however, the side margins are straight in the ground plan (but the adjacent margins enclose an angle of almost 90°), their floor and slopes are covered with debris, and there is usually exploited debris around them too. These characteristics show their artificial origin.

6. Conclusions

The relationship between mining and karst in the Bakony Region is described. The main mining effects are classified, and we overviewed what opportunities the karst offered to mine. In the mountains, mining affected the karst of the Bakony Region directly (mining of the material of depressions and of the dolomite rubble, pollution of karst cavities by mining, cavity and cave destruction due to rock mining) and indirectly (karst water extraction). Mining also resulted in the development of pseudokarst features.

Environmental damages caused by mining can be permanent in the karst landscape (rubble mining); others can be solved partly by reconstruction work (e.g., refilling depressions whose material was exploited), but a complete regeneration can also be expected (karst water level). The majority of damage is mostly local; however, karst water extraction affected the whole mountains and their environment, but its degree was different at various sites. Fortunately, due to the reversible character of water balance, the recovery of the original state can be expected after a certain time. Signs of regeneration have been visible at several sites for the past 30 years since the 1990s (there is karst water in the Tapolca Cave again, but several springs have started to operate again).

Funding: This research received no external funding.

Informed Consent Statement: Not applicable.

Data Availability Statement: No new data were created or analyzed in this study. Data sharing is not applicable to this article.

Conflicts of Interest: The authors declare no conflict of interest.

References

1. Kercsmár, Z.; Selmeczi, I.; Budai, T.; Less, G.; Konrád, G. Geology of the karst terrain sin Hungary. In *Cave and Karst Systems of Hungary*; Veress, M., Leél-Őssy, S., Eds.; Springer: Berlin/Heidelberg, Germany, 2022; *in press*.
2. Haas, J. Földtan (Geology). In *Magyarország Térképekben*; Kocsis, K., Schweitzer, F., Eds.; MTA Földrazjtudományi Kutatóin-tézet: Budapest, Hungary, 2011; pp. 35–38. (In Hungarian)
3. Jakucs, L. *Morphogenetics of Karst Regions*; Adam Hilgar: Bristol, UK, 1977; 284p.
4. Ford, D.C.; Williams, P.W. *Karst Hydrogeology and Geomorphology*; John Wiley & Sons, Ltd.: Chichester, UK, 2007; 562p.
5. Pécsi, M. A Pannóniái-medence morfogenetikája (The Morphogenesis of the Pannonian Basin). *Földrajzi Értesítő* **1980**, *29*, 105–127. (In Hungarian)
6. Vadász, E. *Bauxitföldtan (Bauxite Geology)*; Akadémia Kiadó: Budapest, Hungary, 1951; 127p. (In Hungarian)
7. Mindszenty, A.; Sebe, K. Paleokarst in Hungary. In *Cave and Karst Systems of the World*; Veress, M., Leél-Őssy, S.Z., Eds.; Springer: Berlin/Heidelberg, Germany, 2022; *in press*. [CrossRef]
8. Szabó, P. Magyarországi karsztformák klímatörténeti vonatkozásai (Relations between climate and karst features in Hungary). *Földrajzi Közlemények* **1956**, *80*, 183–189. (In Hungarian)
9. Szabó, P.Z. Újabb adatok és megfigyelések a magyarországi őskarszt jelenségek ismeretében (New data and observations to the knowledge of Hungarian paleokarstic phenomena). *Dunántúli Tudományos Gyűjtemény* **1966**, *58*, 1–38. (In Hungarian)
10. Veress, M.; Unger, Z.; Vetési-Foith, S. Anthropogenic pseudokarstic depression on Mount Bocskor (Bakony Region, Hungary). *J. Earth Sci.* **2022**, *in press*.

11. Vértes, L. *Az Őskőkor és az Átmeneti Kőkor Emlékei Magyarországon (Paleolithic and Mesolithic Memories in Hungary)*; Akadémiai Kiadó: Budapest, Hungary, 1965; 385p. (In Hungarian)
12. Mészáros, G.; Vértes, L. A Paint Mine from the Early Palaeolithic Age near Lovas. *Acta Arch.* **1995**, *5*, 1–34.
13. Bárdossy, G. Adatok a Cserszegtomaji kaolinos agyag ismeretéhez (Some contributions to the knowledge of the kaolinitic clay of Cserszegtomaj). *Földtani Közlöny* **1959**, *LXXXIX*, 374–380. (In Hungarian)
14. Bárdossy, G. *Karsztbauxitok (Karst Bauxites)*; Akadémia Kiadó: Budapest, Hungary, 1977; 413p. (In Hungarian)
15. Cseh-Németh, J. Az Úrkúti mangánérc telep kifejlődési típusai (Development types of the manganese ore colony of Úrkút). *Földtani Közlöny* **1958**, *88*, 399–415. (In Hungarian)
16. Juhász, Á. *Évmilliók Emlékei (Memories of Millions of Years)*; Gondolat: Budapest, Hungary, 1983; 512p. (in Hungarian)
17. Komlóssy, G.Y. The Iszkaszentgyörgy bauxite (SE Bakony, Mts. Hungary). Problems of genesis and mineral formation. *Földtani Közlöny* **1970**, *95*, 320–327.
18. Mindszenty, A. A magyarországi bauxitok kutatásának rövid története (1903–2020) (A short history of Hungarian bauxite researches). *Földtani Közlöny* **2020**, *150*, 529–544. [CrossRef]
19. Bulla, B. Néhány megjegyzés a tönkfelszínek kialakulásának kérdéséhez (Some remarks on the question of the development of peneplains). *Földrajzi Értesítő* **1958**, *VII*, 257–274. (In Hungarian)
20. Bíró, B. A halimbai és nyirádi bauxitelőfordulások karsztos fekvője (Karstic bedrock of the bauxite occurrences of Halimba and Nyirád). *Földtani Közlöny* **1969**, *99*, 98–104. (In Hungarian)
21. Kálmán, G.; Pethő, J. Úrkút és Ajka környékének részletes karsztvíz térképe (Detailed kartswater map of the environs of Úrkút and Ajka). *Hidrológiai Közlöny* **1950**, *XXX*, 175–178. (In Hungarian)
22. Győrffy, D. Geomorfológiai tanulmányok a Káli-medencében (Geomorphological studies in the Kál Basin). *Földrajzi Értesítő* **1957**, *VI*, 265–299. (In Hungarian)
23. Veress, M.; Gadányi, P.; Tóth, G. Thermal Spring cones of the Tihany Peninsula. In *Landscapes and Landforms of Hungary*; Lóczy, D., Ed.; Springer: Berlin/Heidelberg, Germany; New York, NY, USA; Dordrecht, The Netherlands; London, UK, 2015; pp. 71–78.
24. Veress, M.; Tóth, G. Kál-Basin: Pseudokarren on Sandstone. In *Landscapes and Landforms of Hungary*; Lóczy, D., Ed.; Springer: Berlin/Heidelberg, Germany; New York, NY, USA; Dordrecht, The Netherlands; London, UK, 2015; pp. 79–87. [CrossRef]
25. Papp, F. Dunántúl karsztvizei és a feltárás lehetősége Budapesten (The karst waters of Transdanubia and the opportunity of exploration in Budapest). *Hidrológiai Közlöny* **1942**, *21*, 146–196. (In Hungarian)
26. Csepregi, A. A karsztvíz kiemelés hatása a Dunántúli-Középhegység vízháztartására (The effect of karst water extraction on the water balance of the Transdanubian Mountains). In *Bányászati Karsztvízszint Süllyesztés a Dunántúli-Középhegységben*; Alföldi, L., Kapolyi, L., Eds.; MTA Földrajztudományi Kutatóintézet: Budapest, Hungary, 2007; pp. 77–112. (In Hungarian)
27. Kozma, K. *Az Ajkai Szénbányászat Története (History of Coal Mining in Ajka)*; Veszprémi Szénbányák: Veszprém, Hungary, 1991; 531p. (In Hungarian)
28. Bakonyi Erőmű ZRT Bányászati Igazgatóság. *Bányavágat Térképe 2004*; Bakonyi Erőmű ZRT: Ajka, Hungary, 2004.
29. Erdősi, F. *A Társadalom Hatása a Felszínre, a Vizekre és az Éghajlatra a Mecsek tágabb Környezetében*; Budapest Akadémia Kiadó: Budapest, Hungary, 1987; p. 228. ISBN 963-05-3885-7.
30. Horányi, L.; Kolozsvári, G. *Eodézia és Bányaméréstan: Bányaméréstan II*; Tankönyvkiadó Budapest: Budapest, Hungary, 1989; p. 267.
31. Sütő, L.; Homoki, E.; Németh, G. Felszínsüllyedési és Bolygatottsági Térkép Készítése a Kelet-Borsodi-szénmedencében. Geoinformatika és Domborzatmodellezés, HunDEM 2009 és GeoInfo Konferencia és Kerekasztal Válogatott Tanulmányai, Hungary, 24 April 2009. Available online: https://www.google.com/url?sa=t&rct=j&q=&esrc=s&source=web&cd=&cad=rja&uact=8&ved=2ahUKEwj4p6eAyKz7AhUV9zgGHanRAHIQFnoECA8QAQ&url=http%3A%2F%2Fwww.uni-miskolc.hu%2F~{}fkt%2FHundem_es_Geoinfo_2009%2FCikkek%2FSutoL_HomokiE_NemethG.pdf&usg=AOvVaw2mZDygeVfeYHJ8AHLCpKTg (accessed on 6 September 2022).
32. Éri, I. Adatok a bakonyi üveghuták történetéhez (Data on the history of the glasshouses of the Bakony Mountains). A Veszprém M. *Múzeumok Közleményei* **1966**, *5*, 143–180. (In Hungarian)

mining

MDPI

Article

A New Methodology for Rockfall Hazard Assessment in Rocky Slopes

Larissa Regina Costa Silveira [1], Milene Sabino Lana [2,*], Pedro Alameda-Hernández [3] and Tatiana Barreto dos Santos [2]

1 Science and Technology of Materials Department, Federal University of Bahia, Salvador 40170-110, BA, Brazil
2 Mining Engineering Department, Federal University of Ouro Preto, Ouro Preto 35400-000, CEP, Brazil
3 Urban Engineering Department, Federal University of Ouro Preto, Ouro Preto 35400-000, CEP, Brazil
* Correspondence: milene@ufop.edu.br

Abstract: This article presents an approach to rockfall hazard assessment for rocky slopes based on a previously published rockfall hazard methodology. The original method is appropriate to high alpine rocky slopes exposed to large scale deformations. It evaluates the parameters related to the geomechanical characterization of rock mass, indications of activity, external influences and event intensity. The original methodology was modified to consider different contexts, including geological, climatic and social environments. Parameters related to external influences were modified; the geometry and characteristics of the slope and the catchment area were introduced. The original methodology and the new proposal were applied to two urban slopes and one railway slope in order to test and compare the methods. The original proposal could not represent the rockfall conditions of these slopes. The new proposal was validated using two mine slopes, whose conditions of stability are known. The results of the analyses with the urban slope and the railway slope were coherent with the situation observed at the field. The validation in the mine slopes showed that this approach is applicable in several situations, being able to determine how hazardous a slope is in relation to rockfall events.

Keywords: rockfall hazard system; probability matrix; hazard matrix; urban slopes; railway slopes; mine slopes

1. Introduction

Several systems of susceptibility, vulnerability, hazard and risk assessment have been proposed because they are easy-to-use and efficient tools for accident prevention and management. Researches on susceptibility, vulnerability, hazard and risk have been carried out in several science fields, such as geology and geotechnics, environmental contamination and ecology [1–7].

Rockfalls are hard to predict because the rock blocks usually do not present previous movement signs and they quickly fail. This condition worsens when monitoring measures are not available or when the rockfall hazard is neglected. Monitoring measures are not available especially in peripheral urban areas and, sometimes, in ecological or adventure tourism areas. A recent example of a serious accident involving a high magnitude rockfall occurred in January, 2022 in Capitólio city, a cliff region of Minas Gerais State of Brazil. In this case, a high quartzite rock block toppled and hit a boat with tourists in a lake located in this region. Another accident involving rockfall, also in Brazil, occurred in January, 2021 at a quarry in the metropolitan region of Salvador, State of Bahia, when a rock block fell from an operational slope onto an excavator, causing the death of one operator.

In populated mountainous regions, rockfalls constitute a major hazard once they can cause damage to properties and personal injuries. Therefore, a geotechnical hazard assessment in these areas is essential, and it consists of the first step of future mitigation planning and risk management. Through geotechnical hazard assessment, it is possible to define the areas with the most urgency of intervention and with the need of control or mitigation measures. These measures can include a constant monitoring plan, block support with bolts and high-resistance screens and removal of overhanging blocks.

In view of the importance of rockfall hazard classification, the main objective of this paper is to propose an easy-to-use approach to rockfall hazard assessment. This approach was adapted from the methodology proposed by [8]. This methodology is easy to use, but it is only suitable for high urban slopes, in alpine regions. Thus, the methodology proposed in this research aims to improve the original proposal by the accurate description of the slope geometry, adaptation of the methodology to rainy regions and inclusion of other seismic situations, like mine blasting and heavy equipment traffic.

The original and the new proposal were applied in three case studies. The first studied slope is located in Mariana town; the second one is located on the railway which connects the towns of Ouro Preto and Mariana; finally, the third studied slope is located in Ouro Preto; all of them in Minas Gerais, Brazil. The locations of the studied slopes are shown in Figure 1. Furthermore, in order to validate the new proposal, the method was applied to two slopes in a quartzite mine, located in São Thomé das Letras town (320 km from Ouro Preto), Brazil (Figure 1), whose stability conditions regarding rockfalls are known.

Figure 1. Location of the slopes in Minas Gerais state, Brazil (modified from nationsonline.org (accessed on 22 October 2022).

2. Rockfall Hazard and Risk Assessment Methodologies

2.1. Concepts

This article applies the concepts defined by the Technical Committee 32 from the International Society of Soil Mechanics and Geotechnical Engineering—ISSMGE [9]. These definitions are internationally accepted and, according to [9], should be used for all zoning, reports and land use planning documents in order to avoid misunderstanding of the terms:

- Susceptibility: a quantitative or qualitative assessment involving rock mass or soil classification, volume (or area) and spatial distribution of mass movement which exists or potentially may occur.
- Hazard: a potential condition that can lead to an undesirable consequence. The hazard description should include the location, volume (or area), classification, velocity of the potential mass movement and the probability of its occurrence within a given time.
- Risk: a measure of the probability of the event occurring and the consequences to health, property or the environment. It is mathematically defined by the multiplication of the failure probability and the consequences of this failure.

2.2. Rockfall Hazard Assessment Methodologies

Several authors proposed methodologies to access the rockfall hazard and risk conditions of slopes. Some of the main proposed methodologies are appropriate for highway slopes. These methodologies consider the traffic parameters, the structural condition and the geometry of the slopes, the catchment area and the previous instability or rockfall frequency. The Rockfall Hazard Rating System (RHRS), ref. [10] and the Modified Colorado Rockfall Hazard Rating System (CRHRS) [11] are examples of hazard assessment systems for highway slopes. RHRS does not establish a hazard (or risk, in the case of the CRHRS) classification, as "low, medium or high hazard/risk". The final result of these methodologies is an index that yields the most problematic regions, associated with the high score values.

Another system of rockfall hazard assessment focused on highway slopes was proposed by [12]. The first step of this approach generates a hazard index from a quick slope data collection. The second step generates an index from detailed field data.

Regarding urban areas, they can be cited a quantitative risk classification proposed by [13], in which risk is obtained by a risk matrix. This system was developed using slopes from a mountainous region in Norway. Nine parameters are used to describe the rockfall slope risk. These criteria can be organized into two groups: the first group is related to the structural conditions of the rock slope, and the second one is related to the displacement rates and activity indications. The consequence is determined by loss of lives. Although it is an effective method, it is limited to populated urban areas; and some parameters, especially those related to mass displacements, require constant monitoring of slopes, which is not always possible, especially in poorly peripheral urban areas.

According to [14], the Slope Mass Rating (SMR) [15] estimates the rockfall hazard. The SMR is calculated by adjustments to the Rock Mass Rating (RMR) [16], multiplying some factors to the basic RMR; for instance, a factor determined by the spatial position of the discontinuities in relation to the slope dip decreases the RMR value. Thus, the SMR classification is not a hazard classification, but a susceptibility index.

Another method focused on urban areas was proposed by [8]. This method evaluates the rockfall hazard for the Bavarian Alps, which takes into account the occurrence probability and the intensity of potential events. It is based on geological-geotechnical data collected in the field and observations. The rockfall probability is related to structural conditions of the slope; to the geomechanical environment, that considers parameters related to the rock mass displacement; and to the activity indications and the external influences, as precipitation and seismic zones. The intensity is related to the volume of failed material is due to the rockfall event. The method is easy-to-use, and it is applicable in urban areas located in mountainous regions.

The methodologies aforementioned are efficient, and some of them are widespread internationally, such as RHRS [10]. Furthermore, they are easy to use, which is an important feature in geotechnical hazard and risk analysis routines. However, they present some limitations regarding applicability. Some of them are suitable only for highway slopes, others only for urban slopes, precisely for alpine regions. Others do not present a hazard analysis. Thus, it is important to propose an easy-to-use methodology, as well as those cited, but one that is more flexible and able to be applied in different contexts. Thus, the described methods were used as a basis for this proposal, considering their strong points and disregarding their weaknesses.

3. Materials and Methods

The main objective of this research is to provide an appropriate methodology for rockfall hazard assessment of rock slopes from urban areas, mines, highways and railways, considering all relevant parameters. Among the methodologies found in the literature, the approach proposed by [8] was selected to be modified, forming the new proposal.

The method was selected because it is an easy-to-use tool based on parameters readily obtained in the field. In addition, the methodology is a preliminary system, which has not been thoroughly tried out and the authors recommended its application and encouraged adaptations. According to [8], adjustments have to be carried out regarding the graduation/scales of disposition and activity as well as of external impacts.

Three slopes susceptible to rockfalls were selected to carry out the rockfall hazard analysis. In case of rockfall occurrence in these slopes, serious consequences are expected, with material losses and damage to the physical integrity of the exposed people. These slopes are: Slope-1, located in the Cabanas neighborhood (Mariana, Minas Gerais District, Brazil); Slope-2, located on the railway which connects the towns of Ouro Preto and Mariana; and Slope-3, located in the Vila Aparecida neighborhood (Ouro Preto, Minas Gerais District, Brazil).

For all slopes, field inspections were made with the purpose of reconnaissance of the area and delimitation of access points to carry out a survey of geotechnical and local characteristics. It was possible to observe and survey both the parameters related to the original methodology developed by [8], as well as other parameters that apparently had an influence on the rockfall hazard in these places, but which were not taken into account in the original methodology of [8]. These steps made possible the proposition of a new system of rockfall hazard analysis, based on the proposal of [8]. In the following sections, the original methodology developed by [8] is presented, as well as the new proposal. Field observations and the application of both methods are presented in the results section.

3.1. *Bauer and Neumann Original Method*

The parameters used in [8] are presented in Tables 1 and 2. Rockfall susceptibility is evaluated by the disposition and the activity. Rock mechanics disposition parameters are discontinuity parameters (orientation, persistence, degree of transection, aperture, roughness and degree of loosening) and the weathering grade [17]. Geomechanical Environment disposition is evaluated by the type of basement, the large-scale deformations and the mass movement in the slope foot. The activity is evaluated by rockfall activity indicators. Weights are attributed to each parameter, and the susceptibility is obtained by their sum.

Following the susceptibility evaluation, the external impact is evaluated by the sum of the weights of precipitation and earthquakes. The rockfall probability matrix is presented in Figure 2a. It is the relationship between susceptibility and external impact. The Figure 2a output is the probability of rockfall occurrence; it is combined with the intensity of the event (consisting of the block volume, shown in Table 2) forming the hazard matrix (Figure 2b), finally obtaining the hazard assessment.

Table 1. Scores of the parameters and rockfall probability values, modified from [8].

<table>
<tr><td colspan="2"></td><td>Parameter</td><td colspan="3">Value</td></tr>
<tr><td colspan="3"></td><td>Unfavorable</td><td>Fair</td><td>Favorable</td></tr>
<tr><td rowspan="10">Susceptibility (Disposition + activity)</td><td rowspan="7">Rock Mechanics (Disposition)</td><td>(1) Discontinuity orientation</td><td>500 (Adverse or slope parallel)</td><td>200 (horizontal)</td><td>0 (Vertical or inward)</td></tr>
<tr><td>(2) Weathering</td><td>200 (W4–W5)</td><td>-</td><td>0 (W1–W3)</td></tr>
<tr><td>(3a) Discontinuity extent</td><td>200 (>10 m)</td><td>100 (1–10 m)</td><td>0 (<1 m)</td></tr>
<tr><td>(3b) Degree of transection of discontinuities</td><td>200 (No mineral bonds)</td><td>-</td><td>0 (Mineral bonds existent)</td></tr>
<tr><td>(3c) Discontinuity aperture</td><td>200 (>1 cm)</td><td>100 (0.5–1 cm)</td><td>0 (<0.5 cm)</td></tr>
<tr><td>(3d) Discontinuity roughness</td><td>200 (Slickensides)</td><td>-</td><td>0 (No slickensides)</td></tr>
<tr><td>(4) Degree of loosening (alternative to category 3)</td><td>800 (Clear indications available)</td><td>400 (Only subordinate indications)</td><td>0 (No indications)</td></tr>
<tr><td rowspan="3">Geomechanical Environment (Disposition)</td><td>(5) Type of basement</td><td>600 (Dissolvable rocks)</td><td>200 (Clayey—marly formations)</td><td>0 (Other formations)</td></tr>
<tr><td>(6) Large-scale, deep-seated deformations</td><td>200 (yes)</td><td>-</td><td>0 (no)</td></tr>
<tr><td>(7) Mass movements in the slope foot</td><td>200 (yes)</td><td>-</td><td>0 (no)</td></tr>
<tr><td colspan="2">(8) Activity</td><td>Initial Activity</td><td>500 (Active)</td><td>-</td><td>0 (Not active)</td></tr>
<tr><td colspan="2"></td><td colspan="4">Total Susceptibility (1 + 2 + (3 or 4) + 5 + 6 + 7 + 8)</td></tr>
<tr><td colspan="2" rowspan="2">External Impact</td><td>(9) Precipitation</td><td>50 (>310 mm)</td><td>20 (230–310 mm)</td><td>0 (<230 mm)</td></tr>
<tr><td>(10) Earthquakes</td><td>50 (Zone 2)</td><td>20 (Zone 1)</td><td>0 (Zone 0)</td></tr>
<tr><td colspan="2"></td><td colspan="4">Total External Impact (9 + 10)</td></tr>
</table>

Table 2. Rockfall intensity, modified from [8].

<table>
<tr><td>Intensity</td><td colspan="2">High Magnitude Rockfalls</td><td colspan="3">Low Magnitude Rockfalls</td></tr>
<tr><td>Volume</td><td>Rock avalanche</td><td>Large rockfalls</td><td>Medium rockfalls</td><td>Small rockfalls</td><td>Single Blocks</td></tr>
<tr><td>Distinction</td><td>Total volume > 1,000,000 m^3</td><td>Total volume close to 1,000,000 m^3</td><td>Total volume close to 10,000 m^3</td><td>Total volume approximately 100 m^3 or fragments with more than 200 mm diameter</td><td>One or few single fragments < 200 mm diameter</td></tr>
</table>

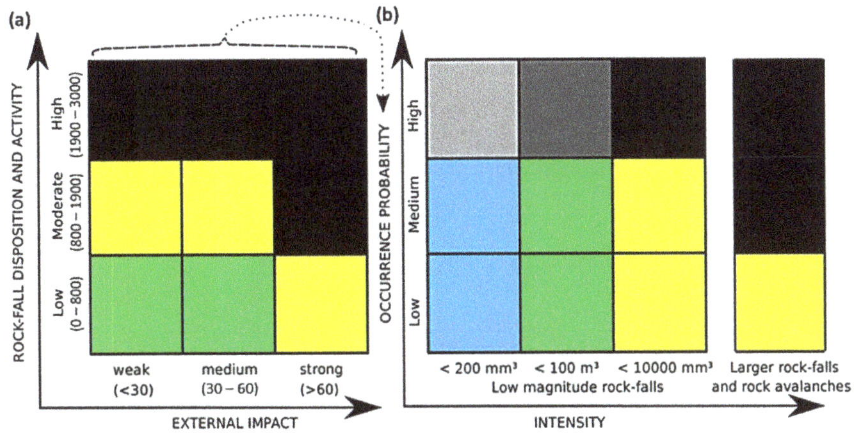

Figure 2. (**a**) Probability Matrix: green, "low"; yellow, "medium"; and black, "high". (**b**) Hazard Matrix: blue, very low hazard; green, low; yellow, medium; gray (with grayscale), high, modified from [8].

3.2. The Proposed Method

The new proposal kept the parameters related to rock mechanics and signs of activity because they are general and applicable to a great variety of situations. The intensity was also kept because it is a measure of block volume, which is related to the energy of the block. Quantifying the energy of a falling block is not an easy task; hence, by considering the block volume it is possible to evaluate this effect indirectly, depicting the intensity of the event.

The first adaptation proposed consisted of the substitution of geomechanical environmental parameters by the geometry and characteristics of the slope and the catchment area, which are important issues regarding the block trajectory in falling movements. In the Bavarian Alps, these parameters were not considered; all the situations were assumed previously hazardous because of the cliff geometry. Anyway, the catchment area is not a concern in the region of Bavarian Alps.

The deformations and mass movement in the slope toe, which were evaluated by the original method in geomechanical environment disposition, are important for high slopes or cliffs. They are not a concern for the slopes analyzed in this research. The geometry of the slope and the catchment area are much more important than those parameters in the context of the slopes analyzed, because failure mechanisms in the studied slopes are not deformation related.

In this adaptation, the type of basement was replaced by the slope height, a crucial parameter, according to [18]. The large-scale deformations item was replaced by the slope dip and surface. Finally, the mass movement in the slope toe was replaced by the characteristics of the catchment area.

The type of basement was a critical parameter in the geomechanical environment item of the original method; thus it received the highest score. In the proposed methodology, the slope height was considered crucial, because of it being directly related to the block trajectory and energy. Furthermore, when it is not possible to build catchment areas or barriers, i.e., in operational open pit slopes, one of the measures is to decrease the height of the slope by building benches.

The score of the slope height is given by Equation (1), where *ISH* is the score and *H* is the slope height (m). This Equation was presented by [19]; it was adapted from [20], based on RHRS.

$$ISH = e^{0.07996H} \tag{1}$$

The slope dip and the roughness-waviness of the slope surface can modify the trajectory of the block (Table 3). The most unfavorable situation is a rough slope with a dip between 30° and almost 75° because, in this range of inclination, the block can roll or jump [21]. Moreover, the block can collide with the irregularities of the slope face and bounce highly and randomly. An intermediary situation is a vertical slope with roughness and overhangs, because a block would behave like a projectile. The favorable situation is a vertical slope, without roughness, like benches, scars or overhangs.

Table 3. Modified scores from [8].

Slope Dip and Surface	Rough SlopeDip 30°–75°	Vertical Slope with Roughness and Overhangs	Vertical Slope, without Roughness and Overhangs
Score	200	100	0
Catchment area	No area; or area with low distance and high inclination. No vegetation.	Moderate distance and inclination. Low or no vegetation.	Larger distance, plane. Or an adequately designed catchment area, like in highways.
Score	200	100	0
External impact	Unfavorable	Fair	Favorable
Precipitation range (mm/days)	>129/6	40–129/6	<40/6
Score	50	20	0
Seismic effects	High seismic zone or damage due to blasting in mines	karst regions, blasting in mines or heavy vehicles and/or medium seismic zone traffic	No seismic effects
Score	50	20	0

If there are vulnerable elements at the slope base, the hazard is directly related to the block path and the available catchment area. If the catchment area is large and approximately flat, the chance of a block reaching the exposed element is low. Another relevant element is vegetation because it can act as a natural barrier, absorbing the block energy (Table 3).

In RHRS, the catchment area is a projected structure with a width and a depth designed according to the height and the slope dip, using the Ritchie abacus [21]. In the new proposal, another approach regarding the catchment area was considered, in order to represent the hazard for urban or mine slopes. Therefore, the catchment area in this proposal is related to the distance between the slope and the exposed element. The inclination, presence of vegetation, talus deposit and the measured distance in relation to the block dimensions are observed (i.e., a block with 6 m of length in an inclined area without vegetation, with around 8 m of distance, is considered an unfavorable situation).

Finally, regarding the weights attributed to the slope dip and the catchment area, for the sake of flexibility, an intermediary situation was proposed between unfavorable and favorable situations.

The second adaptation proposed in this research regards the external impact. The item was kept, but the precipitation values were adapted to cover a variation in precipitation more suitable for rainy regions than the one in which the original method was developed. The earthquake evaluation was replaced by the evaluation of seismic effects, including those related to blasting and traffic. In the area of study, earthquakes are not a concern.

Regarding the external impact due to precipitation, ref. [22] studied the relationship between mass movements and rainfall in the region of Mariana and Ouro Preto; after several statistical analyses, ref. [22] concluded that the accumulated precipitation of 48.2 mm in six consecutive days triggers slope failures. According to the same author, the critical precipitation amount is 129 mm/6 consecutive days, and in rainy seasons (October to March) the critical precipitation is often reached. Ref. [22] analyzed the historical series of precipitation in the region; thus, the author could establish the maximum and minimum precipitation values, which were used for the proposed approach (Table 3). In Table 4 the parameters surveyed in field are presented.

Table 4. Parameters surveyed in field.

Parameters		Unit	Measurement/Classification
Uniaxial Compressive Strength (UCS)		MPa	Geological Hammer/ ISRM Classification [17]
Weathering Degree		-	ISRM Classification [17]
Discontinuities	Orientation	Dip/Dip Direction	Brunton Compass
	Spacing	Metric System	Tape
	Length	Metric System	Tape
	Roughness	-	Barton [23]
	Aperture	Metric System	Tape
Slope and catchment area	Height and inclination	Metric System	Tape and Laser Tape
Block dimension		Metric System	Tape
Vegetation		-	-

4. Results

4.1. Characterization of Rock Masses—Slope-1

Slope-1 (Cabanas neighborhood) consists of a quartzite of the Itacolomi Group. Two areas (PR1 and PR2) were selected for the study due to their geomechanical and structural conditions. These areas have discontinuity planes along the entire slope surface; however, the direct measurement of the parameters at the field is limited to accessible points. These areas are prone to rockfalls, which can reach the buildings below the slope (Figure 3).

Figure 3. Slope-1 overview.

Both areas PR1 and PR2 are very large, 120 m and 75 m in length, respectively. Thus, the geometry of the slope and the characteristics of the catchment area vary considerably along the slope length. Hence, these areas have been split into homogeneous sectors. Variables that define these sectors are the slope height (H), the slope dip (ψs) and the characteristics of the catchment area (Table 5).

Table 5. Geometry of the slope for each sector in Slope-1.

Sector	H (m)	ψ_s (°)	Length (m)
S1A	55	75	60
S1B	30	68	25
S1C	29	57	35
S2A	80	60 from the top to 34.5 m 69 from 34.5 m to the base	30
S2B	59	72	17
S2C	78	73	28

The areas and sectors are shown in Figure 4. In Table 6, the geometry of the catchment areas for each sector is presented, i.e., its inclination, ψca and its distance from buildings, D. Table 6 also describes the general characteristics of the catchment areas like the presence or absence of vegetation and the occurrence of rock debris or rock blocks.

Figure 4. Areas (PR1 and PR2) and sectors of the slope in Cabanas (Google Earth, accessed on 16 December 2016).

Table 6. Catchment area features for each sector in Slope-1.

Sector	ψ_{ca} (°)	D (m)	Characteristics
1A	18	24	Soil and grass.
1B	15	17	Soil, grass and some blocks.
1C	41	9	Soil, debris and blocks.
2A	25 (talus cover); 13 (pasture).	20 (talus cover); 18 (pasture).	Talus cover: small trees, soil, debris and blocks. Pasture: soil and grass.
2B	25	45	Soil and grass
2C	11	50	Soil and grass

Three discontinuity sets were observed in both areas (Table 7). Kinematic conditions for wedge failures were found in both areas. Set 1, developed through foliation, is very persistent in both areas, compared to other sets. Its roughness and persistence are also distinctive, due to its planar surface in the slope face (Table 8). Filled discontinuities for all sectors have a soft sand filling. Weathering degree is low for all sectors; the rock mass is fresh or slightly weathered (Table 8). Uniaxial compressive rock strength is relatively high; it varies from strong (50 to 100 MPa) to very strong (100 to 250 MPa).

Table 7. Geometric characteristics of the discontinuity sets of PR1 and PR2 areas in Slope-1.

Area	Set	Dip/Dip Direction (°)	Spacing (m)	Trace Length (m)	Roughness
PR1	Set 1 (foliation)	21/139	0.42	20	Rough, planar
	Set 2	60/338	1.43	3	Rough, undulating
	Set 3	62/038	0.76	3	Rough, undulating
	Slope Face	67/317	-	-	
PR2	Set 1 (foliation)	27/117	0.24	20	Rough, planar
	Set 2	82/339	0.44	3	Rough, undulating
	Set 3	55/280	0.76	3	Rough, undulating
	Slope Face	70/318	-	-	

Table 8. Characteristics of the sets and rock mass for each sector in Slope-1.

Sector	Aperture (cm)	Filling	Weathering (ISRM 1981)	Strength (ISRM 1981)
S1A	Closed	None	W1	R5
S1B	1–5	Soft sand	W2	R4
S1C	1–10	Soft sand	W2	R4
S2A	Closed	None	W1	R5
S2B	Closed	None	W1	R5
S2C	1–5	Soft sand	W1	R4

4.2. Characterization of Rock Masses—Slope-2

Slope-2 (railway between Ouro Preto and Mariana) consists of schist of the Sabará Group. This slope is homogeneous along its extent of 35 m. During the railway construction, slopes were cut on each one of the two sides of the railway, named A (Figure 5, left side) and B (Figure 5, right side). Slope A is 6 m high, and Slope B is 18 m high. Only Slope B has a catchment area, with 0.80 m of width and 0.40 m of depth (Figure 6). This catchment area was not designed to catch the falling blocks; it is related to the amount of ballast used to build the structure. In the images of the slopes, a level staff (4 m) was used as scale.

Three discontinuity sets were identified in Slopes A and B. In Table 9, the dip and the dip direction of the sets and slopes, the spacing, the trace length and the roughness are shown. Kinematic conditions for wedge failures were found for Faces A and B; block toppling can also be a concern in both faces. All the discontinuities are closed. The rock weathering degree is W2, and the uniaxial compressive strength is R3/R4, i.e., 25–50 MPa/50–100 MPa, which suggests a moderate to hard rock mass.

Figure 5. Railway showing sides A and B of Slope-2.

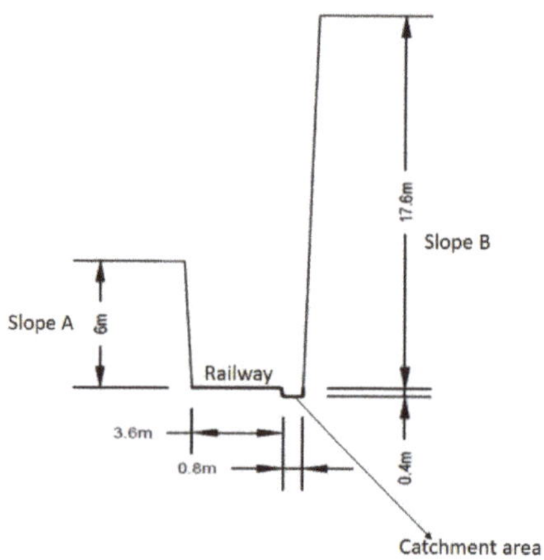

Figure 6. Sketch of the railway slopes.

Table 9. Characteristics of the discontinuity sets in Slope-2 [24].

Set	Dip/Dip Direction (°)	Spacing (m)	Trace Length (m)	Roughness
Set 1	37/194	0.25	15	Smooth
Set 2	44/139	0.42	10	Slightly rough
Set 3	48/040	0.38	10	Slightly rough
Slope A	87/075	-	-	-
Slope B	88/259	-	-	-

4.3. Characterization of Rock Masses—Slope-3

Slope-3 (Vila Aparecida neighborhood) consists of schist of the Sabará Group. The slope height is 6.5 m. It is homogeneous and has no catchment area; the sidewalk is just below the slope toe (Figure 7). In Figure 7 a level staff of 4 m was used as a scale for the slope.

Figure 7. Slope-3.

Three discontinuity sets were observed in this slope. Their characteristics are presented in Table 10. All the discontinuities are closed. Kinematic conditions of planar, wedge and toppling failures were observed.

Table 10. Characteristics of the discontinuity sets in Slope-3 [25].

Set	Dip/Dip Direction (°)	Spacing (m)	Trace Length (m)	Roughness
Set 1	43/211	0.24	3–10	Slightly rough
Set 2	79/298	0.57	1–3	Slightly rough
Set 3	46/044	0.36	1–3	Slightly rough
Slope	76/040	-	-	-

The rock is highly weathered (W4) and very weak, with uniaxial compressive strength in the range 1 to 5 MPa (R1), both conditions classified according to the ISRM (1981) notation.

4.4. Rockfall Hazard Assessment

Rockfall hazard was assessed in the three slopes, according to the Bauer & Neumann [8] method and the new proposal. In both methods, the maximum and minimum rainfalls were considered, in order to assess the hazard in rainy and dry periods.

4.4.1. Bauer & Neumann Original Method

The disposition, activity and external impact scores were obtained according to Table 1; the susceptibility results are presented in Table 11. The rockfall hazard quantification is obtained by the use of the matrix presented in Figure 2b; it is presented in Table 11.

Despite the differences observed in the field, all the slopes presented medium rockfall probability and low hazard, according to the Bauer & Neumann [8] method.

The susceptibility scores of the sectors in Slope-1 are close. All the sectors were classified as low rockfall hazard slopes (Table 11). This result was predictable because the geometry of the slope and the catchment area do not influence the scores in the original method. However, by direct observation of in situ behavior, there are noticeable differences between sectors; hence the original method is not sensitive enough to the actual rockfall hazard on this slope.

Table 11. Rockfall susceptibility, probability and hazard of each slope according to the Bauer & Neumann [8] original method.

Slope	Susceptibility	Maximum/ Minimum Rainfall	Seismic Effects	Maximum/ Minimum Rockfall Probability	Intensity	Hazard
S1A	1400	50/0	0	Medium/Medium	Up to 200 mm	Low
S1B	1600	50/0	0	Medium/Medium	Up to 200 mm	Low
S1C	1600	50/0	0	Medium/Medium	Up to 200 mm	Low
S2A	1400	50/0	0	Medium/Medium	Up to 200 mm	Low
S2B	1400	50/0	0	Medium/Medium	Up to 200 mm	Low
S2C	1600	50/0	0	Medium/Medium	Up to 200 mm	Low
Slope-2 A	900	50/0	0	Medium/Medium	Up to 200 mm	Low
Slope-2 B	900	50/0	0	Medium/Medium	Up to 200 mm	Low
Slope-3	1700	50/0	0	Medium/Medium	Up to 200 mm	Low

The results for Slope-2 are coherent with observations in the field. This slope is stable; no blocks or debris have been observed near the railway trail; no scars suggesting block detachments of the slope were identified in the field.

Slope-3, the one with the worst conditions, has the same classification of the Slope-1 sectors. Once again, the original method is not sensitive to rockfall hazard, when considering field conditions of these slopes, which are very different from those analyzed by Bauer & Neumann [8].

High rainfall can lead to geotechnical problems in Slope-3. Sliding and toppling have been recurrent failures in this slope during rainy seasons.

4.4.2. The New Proposal

The results obtained by the application of the proposal presented in this research are showed in Table 12.

Table 12. Rockfall susceptibility, probability and hazard of each slope according to the proposed method.

Slope	Susceptibility	Maximum /Minimum Rainfall	Seismic Effects	Maximum/Minimum RockfallProbability	Intensity	Maximum/ Minimum Hazard
S1A	1781	50/0	0	Medium/Medium	Up to 200 mm	Low
S1B	2011	50/0	0	High/High	Up to 200 mm	High
S1C	2010	50/0	0	High/High	Up to 200 mm	High
S2A	2200	50/0	0	High/High	Up to 200 mm	High
S2B	1719	50/0	0	Medium/Medium	Up to 200 mm	Low
S2C	2311	50/0	0	High/High	Up to 200 mm	High
Slope-2 A	1102	50/0	20	High/Medium	Up to 200 mm	High/Very Low
Slope-2 B	1104	50/0	20	High/Medium	Up to 200 mm	High/Very Low
Slope-3	1902	50/0	0	High/High	Up to 200 mm	High

Slopes S1B, S1C, S2A, S2C and Slope-3 have a high probability of rockfall occurrence and were classified as high hazard (Table 12). The slopes more susceptible to rockfalls were S2A and S2C due to their height. However, in these sectors, the catchment areas are far from the buildings, contributing to the safety of these areas.

The most problematic slopes were S1B, S1C and Slope-3 due to the high hazard scores and the bad conditions of the catchment area (low distance from the buildings and absence of vegetation), which increase significantly their rockfall potential. Consequently, they require urgent intervention. The other sectors of Slope-1 require monitoring and hazard mitigation.

Slopes 2-A and 2-B have small susceptibility values. However, the rockfall probability for both slopes varies from medium to high because of the consideration of seismic effects due to traffic. These slopes ended up classifying as very low to high hazard. Therefore, the seismic effects should be analyzed deeply because there are no signs of instability in this area, such as the presence of scars or loose blocks.

5. Discussion

5.1. Comparison between Methods and General Comments

The new proposal showed an increase in the susceptibility scores for all the slopes (Table 12), which, in the majority of cases, combined with the external impact (Table 12, Figure 1a) resulted in hazard classification changes (Table 12, Figure 1). This is because the new proposal is more sensitive to parameters, such as: the slope geometry and characteristics of the catchment area. In the original method, instead of analyzing these parameters, the geomechanical environment is analyzed, which does not show any changes among the analyzed slopes, being an irrelevant item for the types of slopes in which this research is focused.

In the original methodology [8], the main problem is related to the geomechanical environment item, because the parameter "type of basement" considered by the authors seems to focus on specific issues concerning the Bavarian Alps. In addition, it considers a wide variety of rock masses as a favorable situation under the label of "other formations". Furthermore, it includes deformation parameters that do not play an important role in the slopes analyzed. The new proposal solves this problem by changing the type of basement by the slope height.

In the new proposal, the impact of maximum rainfall and seismic effects is quite significant. For instance, the same slope is classified as a very low rockfall hazard, when the score for "external impact" is minimal and as a high rockfall hazard, when the score for "external impact" is maximal. This issue appears in Slope-2, which has good structural conditions and no indications of activity. However, considering that this slope is exposed to the vibrations produced by traffic flow, the hazard score changed completely. In this case, the external influences can have more impact on the rockfall hazard than the susceptibility, which is controversial in this particular situation. More research on these slopes would be necessary to evaluate these effects.

Finally, the scores attributed to the characteristics of the slope and the catchment area presented coherent results because the higher slopes had a considerably high susceptibility score. In addition, by considering the slope dip, the surface roughness and the catchment area characteristics, more flexibility in the analyses is introduced.

For the new proposal, the slopes S1C, S2A, S2C and Slope-3 presented high hazard scores. The slope height of S2A and S2C in the Cabanas neighborhood was the most influencing parameter in their high hazard scores. In S1C and Slope-3, the rock mass and the bad conditions of the catchment area influenced their scores. The weathering of the rock mass in Slope-3 is the worst condition.

It is important to observe that the main objective of this work was to propose a new easy-to-use and inexpensive rockfall hazard system, suitable for slopes in several contexts (like urban environments, railways, mines, etc.). The system proposed was based mainly in a pre-existent method proposed by [8]. However, parameters considered in other methodologies of rockfall hazard were incorporated in the proposal, such as the RHRS [10], and the classification methods, such as RMR [16] and SMR [15].

There are other rockfall hazard approaches based on the estimation of rockfall trajectories [26]; other systems consider the frequency of rockfalls or accidents due to rockfalls [27].

The problems of these approaches are the dependence of simulation paths or monitoring data; these are not always available. The system proposed is independent of these issues. Moreover, it indicates the regions that the monitoring and more detailed studies are recommendable.

As mentioned previously, the majority of the empirical methods to determine the rockfall hazard are limited to a single context, i.e., some of them are suitable to urban and alpine rock slopes, for example, the methods proposed by [8] and [13], others only to highways. The results shown in this work proved that appropriate methods for urban alpine slopes are not sensitive and representative for slopes in other contexts, such as the mining industry. The same observation can be extended to methodologies developed for highway slopes, such as the RHRS [10].

The system proposed is flexible to several contexts and can be applied in any type of rock slope. The method was applied in urban and railway rock slopes with consistent results.

The insertion of seismic effects due to blasting was included in the proposal to permit its application to mine slopes. Many previous proposals regarding rockfall assessment are not suitable for mine slopes because they include specific variables only associated to highway slopes or urban environments.

It is important to emphasize that the rockfall is a critical event in mine slopes. This is because, in operational rock slopes, the rock mass is often disturbed by blasting and this can generate overhang blocks. Moreover, it is not common to apply reinforcement in operational mine slopes; therefore it is necessary to know the hazard level of rockfalls to remove these blocks from mine slopes.

5.2. Validation of the Proposed Method

Slopes of a quartzite mine, located in São Thomé das Letras town (Brazil) were chosen in order to validate the proposed method. Its application to mine slopes constitutes a challenge and could indicate if the method can be used in other types of slopes. The mine was selected because it presents slopes prone to rockfalls. Two slopes were chosen for validation; one of them had its access prohibited due to the instability caused by rockfalls (Slope X) and the other presented good structural conditions and no activity indications, being considered stable (Slope Y). Slope X and Slope Y are shown in Figures 8 and 9, respectively.

Figure 8. Slope X.

Figure 9. Slope Y.

Both slopes have rock masses with the same weathering degree condition and intact uniaxial compressive strength degree. The rock masses are fresh to slightly weathered (W1 or W2) and very hard (R5). The main discontinuity set of the rock masses is the foliation, whose average orientation is 11/240 (dip/dip direction), spacing ranging from 3 cm to 30 cm and persistence in the rock mass scale, i.e., more than 10 m when the length of the slope is largest than 10 m.

Slope Y has only the foliation discontinuity set. Slope X has three discontinuity sets: the foliation and two joints with average orientations equal to 64/120 and 76/065. The characteristics of the discontinuity sets are the same for both slopes. The foliation has a planar and smooth surface, and the joints have planar and rough surfaces. The aperture of the discontinuity sets are in the range of 1 mm to 1 cm, and the cracks have apertures around 5 to 10 cm. These cracks were caused by blasting. Sliding planes are noticeable in Slope X.

Slope X and Slope Y have heights equal to 16.78 m and 10 m, respectively. Both slopes are vertical. Slope X presents several indications of activity, such as failed blocks at the slope toe and several scars and cracks at the slope surface. Slope 2 presents a regular surface with no activity indications.

Failed blocks of Slope X are of low magnitude and small-scale, i.e., it has a total volume of about 100 m^3 and/or fragments with a diameter above 200 mm. The slopes did not present a catchment area.

The rockfall hazard conditions of Slopes X and Y were defined according to the method proposed in this research. In the analyses, the maximum and minimum rainfalls were considered. The seismic effects typical of mine slopes are related to blasting and heavy vehicles and/or medium seismic zone traffic. Table 13 shows the susceptibility scores, the maximum and minimum rockfall probabilities and the hazard conditions of Slopes X and Y.

Table 13. Rockfall susceptibility, probability and hazard of Slope X and Slope Y according to the proposed method.

Slope	Susceptibility	Maximum/Minimum Rainfall	Seismic Effects	Maximum/Minimum Rockfall Probability	Intensity	Hazard
X	2003.8	50/0	20	High/High	Up to 200 mm	High
Y	702.2	50/0	20	Low/Low	Up to 200 mm	Low

The results obtained for both slopes were consistent, considering the slope characteristics observed in the field. The structural condition of the rock masses and the signs of activity were the most relevant parameters in the hazard analyses. Slope X presented high susceptibility and a high probability of rockfall and high hazard, considering maximum and minimum rainfall. Slope Y presented low susceptibility and a low probability of rockfall and low hazard for both rainfall conditions.

6. Conclusions

The method for rockfall hazard assessment proposed in this article is a step towards a new comprehensive methodology for rockfall risk assessment in rock slopes.

Strong points of the proposed method can be highlighted:

- The proposal is an easy to use and inexpensive method to evaluate rockfall hazards;
- Slope height, dip and the catchment area are incorporated in the proposal, which are parameters directly related to block path in a falling movement;
- Seismic effects are included in the proposal to evaluate the influence of traffic and blasting in rockfall assessment.
- Important results obtained by the method application are:
- The proposed method was applied successfully to a wide range of slope types, like urban, railway and mine slopes;
- The method proved to be efficient to quantify the most important factors affecting rockfalls.

The majority of current methods for rockfall hazard assessment are not suitable for application in mine slopes. Validation of the proposal in mine slopes showed consistence with the situation observed in the field regarding the tested slopes.

Author Contributions: Conceptualization, M.S.L. and P.A.-H.; methodology, L.R.C.S. and P.A.-H.; validation, L.R.C.S., P.A.-H. and T.B.d.S.; formal analysis, L.R.C.S.; investigation, L.R.C.S.; data curation, L.R.C.S.; writing—original draft preparation, T.B.d.S. and P.A.-H.; writing—review and editing, L.R.C.S., M.S.L., P.A.-H. and T.B.d.S.; supervision, M.S.L.; project administration, M.S.L. All authors have read and agreed to the published version of the manuscript.

Funding: This research received no external funding.

Data Availability Statement: Not applicable.

Acknowledgments: The authors acknowledge CNPq (National Counsel of Technological and Scientific Development), Fapemig (Foundation for Research Support of Minas Gerais) and the Brazilian Federal Agency for Support and Evaluation of Graduate Education (CAPES) for supporting this work and the mining company for the opportunity of the study.

Conflicts of Interest: The authors declare no conflict of interest.

References

1. Luria, P.; Aspinall, P.A. Evaluating a multi-criteria method for hazard assessment in urban design. *Environ. Impact Assess. Rev.* **2003**, *23*, 625–653. [CrossRef]
2. Lara, M.; Sepúlveda, S.A. Landslide susceptibility and hazard assessment in San Ramón Ravine, Santiago de Chile, from an engineering geological approach. *Environ. Earth Sci.* **2010**, *60*, 1227–1243.
3. Filho, A.G. Risk analysis in the spillway of dam Orós by excess of influent flow. *Ann. Braz. Acad. Sci.* **2013**, *85*, 405–414.
4. Lavrentyeva, G.V.; Mirzeabasov, O.A.; Synzynys, B.I. Ecological Risk Assessment for the Terrestrial Ecosystem under Chronic Radioactive Pollution. *Int. J. Environ. Res.* **2014**, *8*, 961–970.
5. Martínez-Graña, A.M.; Goy, J.L.; Zazo, C. Water and Wind Erosion Risk in Natural Parks—A Case Study in "Las Batuecas–Sierra de Francia" and "Quilamas" Protected Parks (Central System, Spain). *Int. J. Environ. Res.* **2014**, *8*, 61–68.
6. Mavrakis, A.; Salvati, L. Analyzing the Behaviour of Selected Risk Indexes During the 2007 Greek Forest Fires. *Int. J. Environ. Res.* **2015**, *9*, 831–840.
7. Cerri, R.I.; Reis, F.A.G.V.; Gramani, M.F.; Giordano, L.C.; Zaine, J.E. Landslides Zonation Hazard: Relation between geological structures and landslides occurrence in hilly tropical regions of Brazil. *Ann. Braz. Acad. Sci.* **2017**, *89*, 2609–2623. [CrossRef]
8. Bauer, M.; Neumann, P. A Guide to Processing Rock-Fall Hazard from Field Data. Geotechnical Safety and Risk. In Proceedings of the 3rd International Symposium on Geotechnical Safety and Risk, Munich, Germany, 1–4 June 2011; pp. 149–156.
9. Fell, R.; Corominas, J.; Bonnard, C.; Cascini, L.; Leroi, E.; Savage, W.Z. Guidelines for landslide susceptibility, hazard and risk zoning for land use planning. *Eng. Geol.* **2008**, *102*, 85–98.

10. Pierson, L.A.; Van Vickle, R. Rockfall Hazard Rating System. In *Transportation Research Record 1343*; National Research Board: Washington, DC, USA, 1993; pp. 6–19.
11. Santi, P.M.; Russel, C.P.; Higgins, J.D.; Spriet, J.I. Modification and statistical analysis of the Colorado Rockfall Hazard Rating System. *Eng. Geol.* **2009**, *104*, 55–65. [CrossRef]
12. McMillan, P.; Matheson, G.D. Rock slope hazard assessment: A new approach. *Eng. Geol. Spec. Publ.* **1998**, *15*, 177–183.
13. Hermanns, R.L.; Oppikofer, T.; Anda, E.; Blikra, L.H.; Böhme, M.; Bunkholt, H.; Crosta, G.B.; Dahle, H.; Devoli, G.; Fischer, L.; et al. Hazard and Risk Classification for Large Unstable Rock Slopes In Norway. *Ital. J. Eng. Geol. Environ.* **2013**, *6*, 245–254.
14. Delonca, A.; Verdel, T.; Gunzburger, Y. Influence of expertise on rockfall hazard assessment using empirical methods. *Nat. Hazards Earth Syst. Sci.* **2016**, *16*, 1657–1672. [CrossRef]
15. Romana, M. *New Adjustment Ratings for Application of Bieniawski Classification to Slopes*; International Society of Rock Mechanics: Zacatecas, Mexico, 1985; pp. 49–53.
16. Bieniawski, Z.T. *Engineering Rock Mass Classifications: A Complete Manual for Engineers and Geologists in Mining, Civil, and Petroleum Engineering*; John Wiley & Sons: New York, NY, USA, 1989; 251p.
17. ISRM. *Rock Characterization Testing and Monitoring*; E.T. Brown: London, UK, 1981; 211p.
18. Youssef, A.M.; Maerz, H. Development, justification, and verification of a rock fall hazard rating system. *Bull. Eng. Geol. Environ.* **2012**, *71*, 171–186.
19. Silveira, L.R.C.; Cunha, L.E.; Hernández, P.A.; Lana, M.S.; Santos, T.B. Rockfall Hazard Rating System Adaptation Proposal for Urban Areas Through a Case Study In Mariana (MG, Brazil). In Proceedings of the 8th Brazilian Symposium of Rock (Cobramseg), Belo Horizonte, Brazil, 19–22 October 2016.
20. Gomes, G.J.C.; Sobreira, F.G.; Lana, M.S. Evaluation of Rockfall Hazard Along Brazil Roads. *Soils Rocks* **2011**, *34*, 163–174.
21. Ritchie, A.M. *Evaluation of Rockfall and Its Control*; U.S. Department of Commerce, Bureau of Public Roads, and the Washington State Highway Commission: Olympia, WA, USA, 1963.
22. Silva, N.L. Correlação Entre Pluviosidade e Movimentos Gravitacionais de Massa no Alto Ribeirão do Carmo/MG. Master's Thesis, Federal University of Ouro Preto, Ouro Preto, Brazil, 2014; p. 114p.
23. Barton, N. Shear Strength Investigations for Surface Mining. In *Stability in Surface Mining, Proceedings of 3rd International Conference (ed. CO Brawner), Vancouver, British Columbia, Society of Mining Engineers*; AIME: New York, NY, USA, 1982; pp. 171–196.
24. Santos, T.B.; Lana, M.S.; Santos, A.E.M.; Silveira, L.R.C. Applicability of geomechanical classifications for estimation of strength properties in Brazilian rock masses. *Ann. Braz. Acad. Sci.* **2017**, *89*, 859–872.
25. Pereira, L.C. Análise Tensão-Deformação e por Equilíbrio-Limite do Mecanismo de Ruptura por Tombamento de Blocos. Master's Thesis, Federal University of Ouro Preto, Ouro Preto, Brazil, 2012; p. 179p.
26. Alvioli, M.; Santangelo, M.; Fiorucci, F.; Cardinali, M.; Marchesini, I.; Reichenbach, P.; Rossi, M.; Guzzetti, F.; Peruccacci, S. Rockfall susceptibility and network-ranked susceptibility along the Italian railway. *Eng. Geol.* **2021**, *293*, 106301. Available online: www.elsevier.com/locate/enggeo (accessed on 19 October 2021). [CrossRef]
27. Budetta, P.; De Luca, C.; Nappi, M. Quantitative rockfall risk assessment for an important road by means of the rockfall risk management (RO.MA.) method. *Bull. Eng. Geol. Environ.* **2016**, *75*, 1377–1397. [CrossRef]

mining

MDPI

Article

Modeling of Coalmine Methane Flows to Estimate the Spacing of Primary Roof Breaks

Vladimir Sergeevich Brigida [1,2,*], Vladimir Ivanovich Golik [3,4] and Boris Vitalievich Dzeranov [5,6]

1 Federal Research Centre the Subtropical Scientific Centre of the Russian Academy of Sciences, 354002 Sochi, Russia
2 Department of Biomedical, Veterinary and Ecological Directions, Peoples' Friendship University of Russia (RUDN University), 117198 Moscow, Russia
3 Department "Technique and Technology of Mining and Oil and Gas Production", Moscow Polytechnic University, 107023 Moscow, Russia
4 Department "Mining", North Caucasian Institute of Mining and Metallurgy, 362021 Vladikavkaz, Russia
5 Geophysical Center of the Russian Academy of Sciences (GC RAS), 119296 Moscow, Russia
6 Geophysical Institute, Vladikavkaz Scientific Center of the Russian Academy of Sciences (GPI VSC RAS), 362002 Vladikavkaz, Russia
* Correspondence: 1z011@inbox.ru; Tel.: +7-901-105-55-63

Abstract: The global challenge for the mining sector is the problem of "decarbonization" of coal mining. The modeling of emission flows of coalmine methane is stipulated by the need to prove the environmental effect of the implemented technological changes. For longwall geotechnology, the dynamics of methane concentration in the gas–air mixture extracted by the degassing system reflects the complex relationship between emission and geomechanical processes in the rock mass. In this regard, the aim of the work was to improve the methods for assessing the steps of caving the main roof when mining gas-coal seams. The method of work consisted of processing experimental data (smoothing—Loess, three-dimensional interpolation, regression—the method of least squares) to obtain reliable response functions in three-dimensional space. When developing algorithms in the Python language, the Vi Improved text editor was used. Graphical representation of the results was carried out in "Gnuplot". As a result of modeling, it was found that the increase in the span of the main roof from 83 to 220 m (S = 1340–1120 m) in the distance range of 120 m in front of the stoping face line and up to 50 m behind it (L = −120–50 m) leads to an alternating cyclicity of local extrema of the dynamics of methane release, according to a polynomial dependence. This fact is a consequence of the implementation of deformation-wave processes in geo-environments, which produce cyclic nonlinearities in the nature of the aero-gas regime of mine methane emissions into anthropogenically disturbed rock masses. In addition, the influence of the situational geomechanical conditions of the excavation area in the goaf was clarified. This makes it possible to reliably identify the caving steps of the main roof.

Keywords: coalbed methane; longwall; emission zone; roof stability

1. Introduction

According to a number of researchers, in 2016, the share of the Russian Federation in the global emissions of greenhouse gases was 7%. Up to 50% of this volume was caused by the extractive sector of the economy [1]. From a global perspective, anthropogenic methane emissions can be up to 19% of all greenhouse gas emissions [2]. The share of mine methane out of this volume ranges from 11 to 13% [3]. Based on the foregoing facts, it follows that ensuring the environmental transformation of the extractive sector in the aspect of "decarbonization" of the extraction of anthropogenic georesources is not just an urgent scientific problem, but rather a challenge for the sustainable development of the industry as a whole.

A number of studies [4–7] have been devoted to the formation of a sustainable state of the mining industry, while the climatic aspects, in most cases, were left largely unaddressed. For example, in previous studies [8–10], more emphasis was placed on practical issues of tailings disposal as elements of goaf backfilling. Other studies aimed at ecological modernization of *in situ* leaching of uranium and only addressed the issues of carbon dioxide capture and disposal during mining [11,12]. More significant, in our opinion, are attempts to improve geoecological monitoring based on unmanned aerial vehicles in combination with advanced GIS technologies [13]. The need to model the emission flows of climatically active gases is stipulated by the need for a reliable assessment of the effect of "decarbonization" from the implemented technological changes.

Using the example of coalmine methane, emission flows are formed not only by the difference in the gas balance in the extraction area and the efficiency of degassing, but also by the processes of anthropogenic disturbance of the massif up to the surface. Modern approaches to predicting the caving steps of the main roof are far from perfect. It is known that the process of rock layers collapsing is a complex dynamic process with a significant share of the stochastic component [14]. The existing approaches to estimating the main caving span are based on processing empirical data; five categorical Q-Q models based on the RSCi index [15]; the application of Cosser at theory for 3D finite element modeling using COSFLOW [16] or ANSYS [17]; two-dimensional synthetic rock mass models (SRM) [18]; and genetic programming or GEP methods [19]. Moreover, for calculations of the stress–strain state of the massif (even for the classical theory of "pressure arch"), the cyclicity of loads on the racks of powered support is taken into account [19]. The non-linearity of rock pressure manifestations can also be associated with the difference in the stratigraphic structure of the roof rocks. For example, in [20], it was found that greater rigidity of the roof layers leads to an expansion of the support pressure zone by 5 m with an increase in stresses by 10.4 MPa. From the analysis of this work, it is obvious that only a change in mining and geological factors during the development of an extraction column can cause the migration of local maxima of the support pressure (with a constant advance rate of stop). The most striking empirical evidence of the manifestations of periodic nonlinear manifestations of the "pressure arch" is the work of Polevshchikov G.Ya. [21]. This paper states "In the inelastic zone, there are two "elements" with the same "wave" characteristics", while the author himself points to ""deformation-wave" nature of rock mass response to the change of external conditions". These assumptions are apparently based on the confirmed fact of the presence of the phenomenon of zonal disintegration [22] around the development workings (the theory itself was fully formed by academician Oparin V.N. only by 2008 [23].

It is worth noting that this direction began to develop as early as the works of Spackeler G. (1930) who, based on field studies of Weber H. (1916), proposed a hypothesis about the existence of a nonlinear ("wave") component in the theory of rock pressure. Further, only by the 1990s ("dynamic waves" I.P. Chernyak) did this direction find its continuation. In addition, it should be noted that in fundamental works on the disclosure of the features of destruction of unequally loaded rock samples, both in situ [24] and in laboratory studies [25,26], similar effects are encountered. The evolution of induced micro fracturing in shear faults is most clearly reflected by wave-like graphs of changes in acoustic emission parameters (Q-profiles) under the influence of dynamic stresses (see Figure 2 in [25]).

The theory of degassing is known as the leading role of "destressing" in the decomposition of a solid gas-coal solution. This process causes an increase in methane release (methane concentration) in wells. The possibility of a qualitative assessment of rock pressure parameters in a degassed massif based on the dynamics of methane release into underground wells is indicated in many papers [27–29]. At the same time, their main indicator maybe the distribution of the methane concentration in the gas–air mixture extracted by underground wells. The issues of describing the features of aerogas processes caused by stopped excavation have been solved, both by Russian [30] and foreign researchers [31–33]. At the same time, a number of phenomena remain not fully understood. In this regard, the purpose of the present paper is to improve approaches to assessing the caving step of the main roof during longwall mining of gas-coal seams.

2. Materials and Methods

The task of geoecological modeling in the form of estimating an approximating function to a set of experimental data (data set) is equivalent to establishing a regression model of the studied process given in an implicit form. The time factor introduces its own difficulties, creating an additional dimension, but also provides a potential way out—reformatting the primary data into an invariant form, using dimensionality reduction methods, etc., "expansion" of the problem and the optimal selection of influencing components in determining the type of response function (CH_4).

One of the difficulties is taking into account the mining and geological features of the extraction area and, in connection with this, the "individuality" of the spatial distribution of gas for each coal seam [34]. In this regard, it is necessary to develop models of several types of wells to form a "portrait" of emission processes in different conditions of mining the same seam.

The drilling of underground degassing wells was carried out on the seam m_3 CJSC "Mine named after. A.F. Zasyadko" (development depth exceeded 1300 m, longwall length of 300–305 m) during the mining of the 18th eastern longwall. A series of 3 wells with different parameters of spatial orientation ("pads") were sunk at a frequency of 20–25 m, while in each pad, there were two wells (No. 2 and No. 3) oriented to the overlying massif (with re-drilling of gas-bearing sandstone) and one "axial" (No. 4, with a zero turn), or "on the goaf" (No. 4). To assess the features of methane emission flows, two types of wells were selected—type No. 2 and No. 4 due to the fact that the angle of their turn to the bottom of the longwall was the same ($\beta = 60°$). Moreover, the ascent angles relative to the horizon were also equal ($\alpha = 60°$, Figure 1), which determines the similarity in the development of deformation processes in the conditions of the previously undermined massif and the beginning development of reserves.

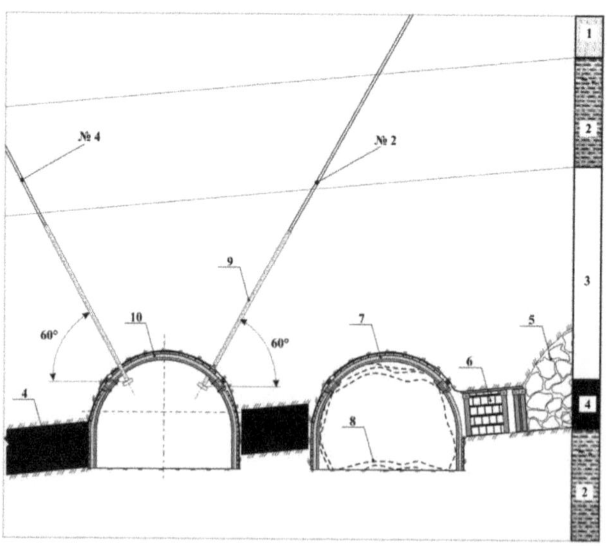

Figure 1. The sink scheme of wells of various types in the excavation area: 1—gray sandstone; 2—dark gray siltstone; 3—mudstone; 4—seam m_3; 5—mined-out space of previously worked out longwall; 6—bi-support; 7—belt road of 17th eastern longwall; 8—contour deformations; 9—casing; 10—air drift of 18th eastern longwall.

In addition, the choice of "goodness-of-fit" criteria (an indicator of assessing the reliability of modeling the initial data) for three-dimensional models remains open. The most commonly used are correlation tightness indicators [35,36], special correlation functions (for additive random processes [37]), relative errors, coefficient of determination (for example, in "wavelet analysis") [38], or approximations of the thermodynamic equilibrium of carbon dioxide [11].

In our study, the proportion of the model variance, which is based on experimental data (R^2), was used in the first stage—the selection of filtering/smoothing parameters according to the method given in [39]. In contrast to the criteria approach [40], artificial neural networks (ANN) [41], the multivariate regression method (with the application of SPSS software) [42], or the finite element method (FEM) [43], further, already smoothed data (on a distributed grid) were interpolated similarly to the study [39], after which the parameters of high-degree polynomials were selected (regression) by the method of least squares. The algorithms were first written in "Vi Improved" (version 9.0, open-source software from Bram Moolenaar, Holland) in Python (version 2.7.10., Python Software Foundation, DE, USA). To improve the quality of modeling at each stage, the residuals on the projections ($Z = f(X;Y)$) were analyzed, and the presentation of graphical results (vectorization) was programmed in the "gnuplot" program (version 5.4, Thomas Williams & Colin Kelley). To assess the quality of the final models, at the last stage, quantile–quantile (Q-Q) plots were additionally built in MS Excel.

The basis for the data set was mine data (obtained by the standard method) of methane concentration measurements for 151 points (L, S, CH_4)—well type No. 2—and 261 points—type No. 2 (due to the fact that wells in goaf were drilled through one picket). The "data set" fragment is presented in Table 1.

Table 1. Experimental data on methane concentration in degassing wells.

N	Wells No. 4			Wells No. 2		
	L, m	S, m	Measurements CH_4, %	L, m	S, m	Measurements CH_4, %
1	−20	1330	2	−30	1330	60
2	−18	1330	10	−28	1330	40
3	−6	1330	100	−26	1330	27
4	8	1330	88	−24	1330	30
5	10	1330	86	−22	1330	13
6	16	1330	86	−20	1330	25
7	28	1330	65	−18	1330	24
8	46	1330	48	−6	1330	44
9	−26	1310	15	8	1330	30
10	−12	1310	72	10	1330	27
11	−10	1310	100	16	1330	15
12	−4	1310	95	28	1330	15
13	8	1310	90	46	1330	15
14	26	1310	80	50	1330	14
15	34	1310	72	−50	1310	90

3. Results

An example of a graphical representation of experimental data (data set vectorization) for comparison with the resulting response surface (after all stages of processing with the formation of a regression model) for the example of type 2 wells is shown in Figure 2.

Wells No	1	2	3	4	5	6
Piket No	133	131	123	121	117	113
S, m	1330	1310	1230	1210	1170	1130

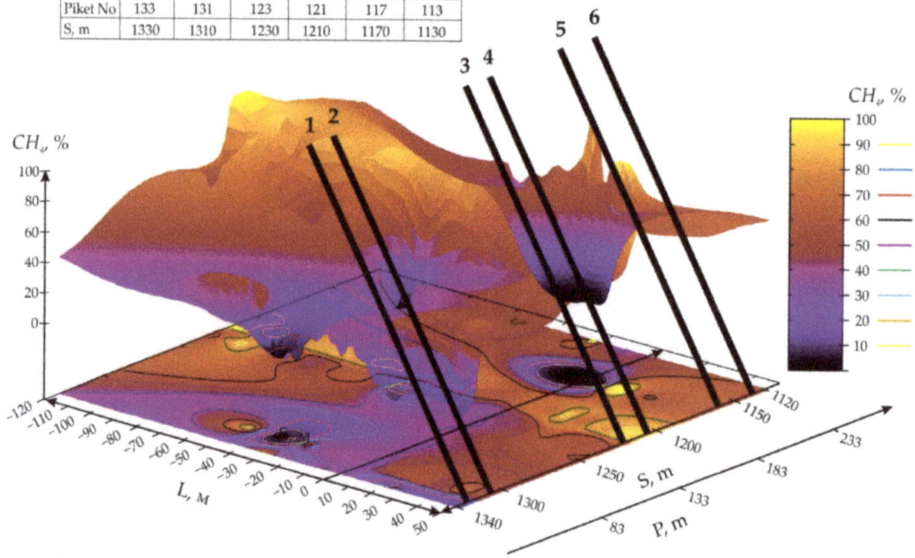

Figure 2. The distribution of methane concentration in the previously mined rock mass during reverse development of the m_3 seam (according to well type No. 2): L is the distance to the stopping face line of the 18th eastern longwall, m; S is remoteness of the face of the 18th eastern longwall relative to the beginning of the extraction column (pickets), m; P is main roof span, at a different point in time; CH_4 is the concentration of methane in the extracted mixture, %.

The spatial distribution of gas flows (after processing the experimental data according to the accepted methodology) from wells drilled into the mined-out space of a previously worked longwall during longwall mining of the m_3 seam is shown in Figure 3.

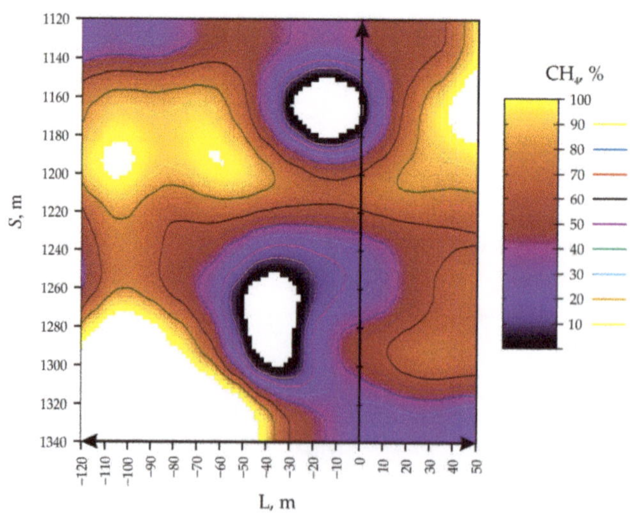

Figure 3. Evolution of the intensity of emission flows of coalmine methane during reverse mining of the m_3 seam (according to data from wells type No. 4).

From the analysis of Figure 3, it follows that at the beginning of reserves development (at S = 1340 m) for wells drilled into the goaf of a previously worked out longwall, a "high content" of methane in the extracted mixture is characteristic—$CH_4 \geq 60\%$ (the area limited by the black curve) from −120 to −30 m before the stope.

Ahead of the longwall, the general trend of the "local maximum" concentration (CH_4 = 100–90%, the area bounded by the dark yellow curve) can be traced from −120 to −40 m, followed by its sharp decrease from −38 to −14 m (from 80 to 40%), which is replaced by a more gentle decline to −4 m (from 40 to 30%). Stabilization and constancy of the methane concentration = 30%, which began −4 m ahead of the longwall, continues up to 50 m behind it, while this trend does not change during the first 63 m (S = 1320 m) of the longwall. When the stoping face line approaches the mark S = 1320 m, up to S = 1268 m, the area of the local maximum decreases from the area −120 to −43 m to the area −106 to −98 m. Further, the area of the maximum concentration disappears and begins to be traced only from the range S = 1218–1165 m, at distances from −120 to −110 m to −60 to −43 m ahead of the longwall. Behind the longwall of the studied area, it does not exist up to S = 1197, while it is localized at the level L = 50 m. With subsequent mining of the reserves, the area of the local maximum gradually increases in size. It reaches its maximum expansion (L = from 30 to 50 m) at S = from 1180 to 1160 m relative to the beginning of the extraction column. Further, from S = 1160, the studied area sharply decreases and shifts again to L = 50 m at S = 1140 m.

The zone of high methane content ($CH_4 \geq 60\%$), starting from S = 1320 m, is removed relative to the distance to the bottom hole (L = −110 to −85 m) and decreases in size, after which, up to S = 1220 m, the size and localization of this zone remain stable. A very interesting mining area is S = 1218–1200 m, which is characterized by the presence of $CH_4 \geq 60\%$ for any distance from the stope, both in front and behind the longwall. After S = 1200 m, the zone is removed relative to the distance to the face L = from −120 to −10 m and decreases in size L = from −120 to −53 m in front of the longwall at S = 1140 m.

The area of a dangerously "local minimum" concentration ($CH_4 \leq 20\%$, limited by the orange curve in Figure 3) begins to form from 1310 m from the beginning of the extraction column at L = −28 m in front of the longwall and gradually expands in size. The maximum width of the region (L = from −60 to −8 m) is traced at S = 1250 m, after which it sharply narrows to a point (S = 1240 m L = −35 m). Further, the area of the minimum concentration disappears and begins to be traced only from S = 1183 m, at a distance of −11 m in front of the longwall. The maximum width of the region (L = from −37 to 7 m) is traced at S = 1162 m, after which it sharply narrows to a point (S = 1140 m L = −15 m). Behind the longwall, this area is essentially absent.

As a result of the studies (the graphical solution of which is shown in Figure 3), a polynomial dependence of the methane concentration on the distance from the stoping face line was established with a decrease in the distance to the beginning of the extraction area (R^2 = 0.97):

- The definition domain of the points (orthogonality interval of the approximating polynomials) corresponds to $S = S'$ for all $S' \in [0, \pi]$ and $L = L'$ for all corresponds to $S = S'$ for all $S' \in [0, \pi]$ and $L = L'$ for all $L' \in [0, \pi]$.

$$
\begin{gathered}
CH_4 = 68.3 - 6.6\cos(S) + 12.5\cos(L) + 15.2\cos(2S) - 43.3\cos(S)\cos(L) + 30.8\cos(2L) - \\
-23.7\cos(3S) + 21.7\cos(2S)\cos(L) - 3.79\cos(S)\cos(2L) - 12.59\cos(3L) + 3.69\cos(4S) - \\
-31.28\cos(3S)\cos(L) + 17.4\cos(2S)\cos(2L) - 14.2\cos(S)\cos(3L) - 14.4\cos(4L) - \\
-0.29\cos(5S) + 10.8\cos(4S)\cos(L) + 1.1\cos(3S)\cos(2L) - 8.0\cos(2S)\cos(3L) + \\
+10.5\cos(S)\cos(4L) - 4.6\cos(5L) + 4.06\cos(6S) - 12.5\cos(5S)\cos(L) - \\
-8.3\cos(4S)\cos(2L) + 2.5\cos(3S)\cos(3L) - 0.4\cos(2S)\cos(4L) + 10.2\cos(S) + \\
+\cos(5L) - 0.8\cos(6L) - 3.6\cos(7S) + 6.6\cos(6S)\cos(L) - 8.5\cos(5S)\cos(2L) - \\
-8.5\cos(4S)\cos(3L) + 1.2\cos(3S)\cos(4L) - 0.5\cos(2S)\cos(5L) + 3.5\cos(S)\cos(6L) - \\
-3.8\cos(7L) + 0.4\cos(8S) - 5.8\cos(7S)\cos(L) - 4.7\cos(6S)\cos(2L) + \\
+6.9\cos(5S)\cos(3L) + 7.1\cos(4S)\cos(4L) - 6.4\cos(3S)\cos(5L) + \\
+5.5\cos(2S)\cos(6L) - 3.8\cos(S)\cos(7L) - 3.4\cos(8L) - 2.6\cos(9S) + \\
+4.5\cos(8S)\cos(L) - 1.2\cos(7S)\cos(2L) - 0.2\cos(6S)\cos(3L) + 4.4\cos(5S)\cos(4L) - \\
-0.8\cos(4S)\cos(5L) - 2.3\cos(3S)\cos(6L) + 1.0\cos(2S)\cos(7L) + 0.6\cos(S)\cos(8L) - \\
-3.9\cos(9L) + 1.1\cos(10S) - 3.9\cos(9S)\cos(L) - 0.6\cos(8S)\cos(2L) + \\
+1.1\cos(7S)\cos(3L) + 3.5\cos(6S)\cos(4L) - 4.6\cos(5S)\cos(5L) + \\
+1.6\cos(4S)\cos(6L) + 1.3\cos(3S)\cos(7L) + 1.3\cos(2S)\cos(8L) + \\
+3.1\cos(S)\cos(9L) - 1.8\cos(10L).
\end{gathered} \tag{1}
$$

The results of the distribution of gas flow from wells drilled into the overlying adjacent seams during the longwall development of the m_3 seam are shown in Figure 4.

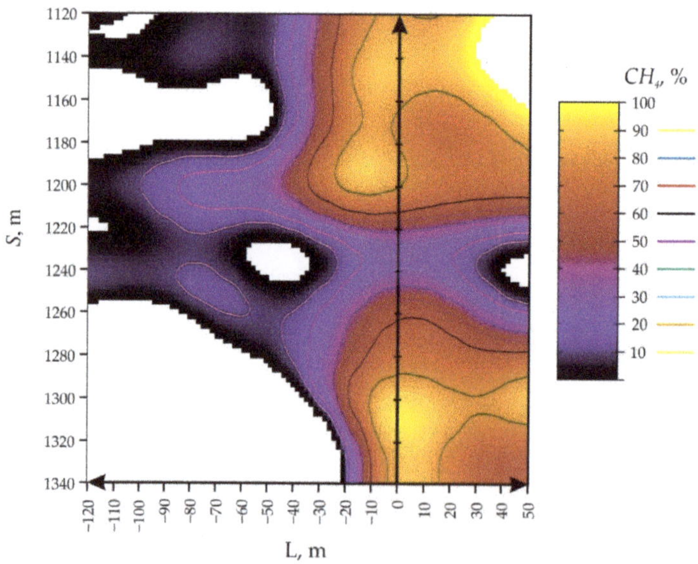

Figure 4. Spatial distribution of methane release dynamics in well No. 4.

From the analysis of Figure 4, it follows that at the beginning of the development of reserves for wells drilled above the developed seam to create a barrier (cross-measure boreholes), it is characteristic that the zone of the "local maximum" concentration ($CH_4 \geq 60\%$) begins from points S = 1323 m and L = 4 m behind the longwall. The maximum width of the region (L = from −6 to 15 m) is traced at S = 1305 m, after which sharp narrowing to a point (S = 1298 m L = from 2 m) is observed. Further, insignificant dimensions of the studied zone begin to be traced only from the range S = 1197–1194 m, at distances from −12 to −9 m ahead of the longwall. Subsequently, behind the longwall, the zone begins to form from 1172 m from the beginning of the site at L = −50 m in front of the longwall and gradually expands in size. The maximum width of the region (L = from 20 to 50 m) is observed at S = 1140 m, after which it sharply narrows to a width of L = from 33 to 50 m at S = 1120 m.

The area of high methane content ($CH_4 \geq 60\%$), starting from S = 1340 m, is stably traced at a distance L = −10 m in front of the longwall and in the entire range L = 0–50 m behind it (which means the width of the area is 60 m) to the picket S = 1298 m. Subsequent mining of the reserves leads to a decrease in the width of the area to the point S = 1261 m L = 5 m. Of particular interest is the area of the extraction column S = 1260–1218 m, which is characterized by the presence of $CH_4 \leq 60\%$ for any distance from the stoping face, both in front of and behind the longwall. Moreover, several zones of the "local minimum" are dispersed on one line in this range. Further, the area of high productivity of degassing begins from the point (S = 1218 m; L = −13 m), expanding sharply up to 50 m behind the longwall (starting from S = 1208 m) and up to −34 m in front of the longwall (S = 1200 m). Further, ahead of the longwall, there is a slight decrease in the width of the zone of high methane content, followed by stabilization at a value of L = −25 m, from S = 1168 m to S = 1120 m.

The area of a dangerous "local minimum" of concentration ($CH_4 \leq 20\%$) in front of the longwall can be traced from a picket of 1340 m, at a distance L = from −120 to −18 m and gradually decreasing (at the first stage) in size to S = 1278 m (L = from −120 to −35 m). Further, the area expands to maximum values of −120 to −13 m at picket 1239 m, after which the width of the area again begins to decrease to minimum values (L = from −120 to −84 m) at S = 1200 m. In the last stage, the growth and stabilization of the zone width up to L = from −120 to −39 m after the picket = 1168 m is typical. Behind the longwall, the area of the local minimum begins to form from 1260 m from the beginning of the excavation column to 50 m ahead of the longwall and gradually expands in size. The maximum width of the region (L = 20 to 50 m) is traced at S = 1239 m, after which it sharply narrows to a point (S = 1220 m L = 50 m).

Numerical values (analysis of residuals) for the formation of ideas for assessing the quality of the obtained models are given in Table 2.

Table 2. An example of a set of actual and model data.

N	Wells No. 4		Wells No. 2	
	Model, %	**Measurements, %**	**Model, %**	**Measurements, %**
1	6.51	2	60.77	60
2	20.19	10	56.72	40
3	76.19	100	53.30	27
4	87.34	88	50.37	30
5	85.77	86	47.82	13
6	79.08	86	45.51	25
7	65.99	65	43.36	24
8	60.70	48	31.43	44
9	5.62	15	24.47	30
10	73.84	72	24.57	27
11	79.95	100	25.89	15
12	92.23	95	27.30	15
13	94.23	90	20.46	15
14	79.06	80	20.29	14
15	77.16	72	72.92	90

The results of constructing Q-Q graphs for two types of wells, based on the data in Table 2, are shown in Figures 5 and 6.

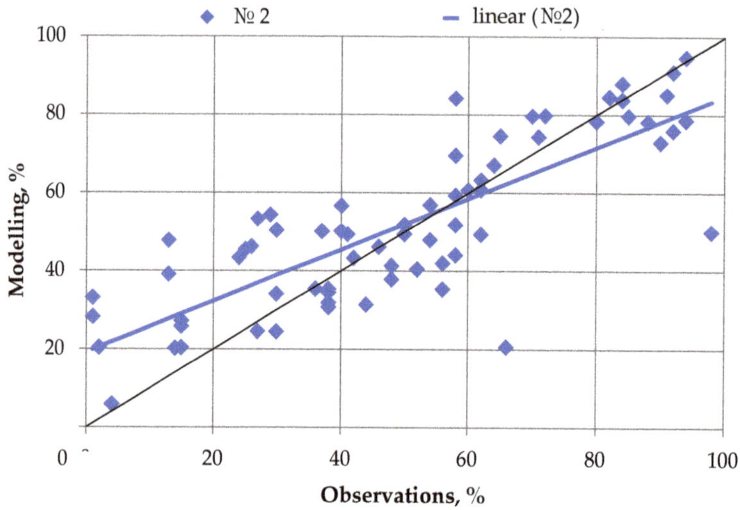

Figure 5. Scatter of modeling residuals (M) in relation to experimental data (O) for well type No. 2.

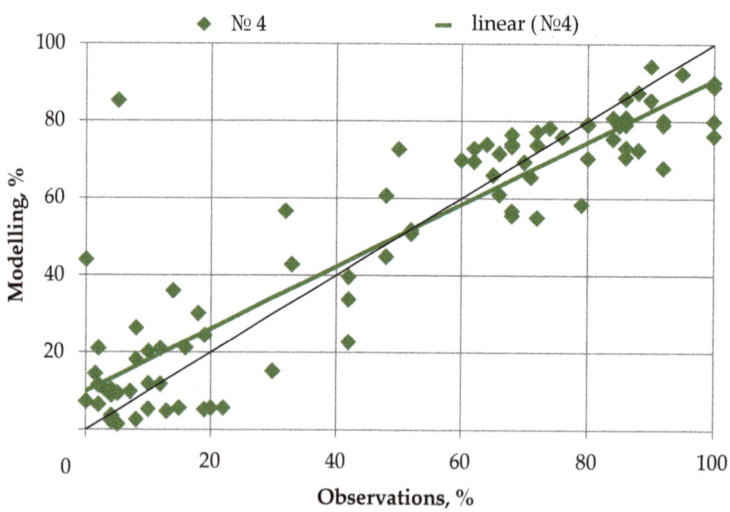

Figure 6. Scatter of modeling residuals in relation to experimental data for wells type No. 4.

The analysis of Figure 5 shows that the linear trend for the model data (high-order polynomials) deviates slightly from the reference black line (maximum likelihood), while the presence of "releases" resulted in (R^2 = only 0.66). The equation of the studied straight line has the following form:

$$M = 0.654O + 19.27 \quad (2)$$

The analysis of Figure 6 shows that the linear trend for model data (high-order polynomials) deviates minimally from the reference trend (maximum likelihood), while $R^2 = 0.82$). The equation of the studied straight line has the following form:

$$M = 0.806O + 99.77. \quad (3)$$

From the analysis of the angles of deviations of linear trends for different types of wells, it follows that for them, with the same approach to data processing, insignificant discrepancies were obtained in modeling quality.

This can be explained by a significant discrepancy in sample sizes (sets of experimental data)—151 for type No. 2 and 261 for type No. 4, which causes a 50–70% larger slope of linear trends. At the same time, the modeling accuracy in both cases is sufficient to establish the general patterns of the studied process.

4. Discussion

A striking feature of the operation of undermined wells (type No. 4, see Figure 4) is the complete loss of performance, based on the proposed mechanism of well collapse [44,45] throughout the entire length of the local degassing network. In general, the results confirm the well-known fact of "overflow" of gas flows in the zone of advanced fracturing (enveloping the surface of partial unloading around the stoping face line). In addition, it confirms a number of the manifestation features (see Figure 9 of the paper [46]) of deformation-wave processes in geo-environments [21,46]. The projections of the simulated response surfaces on the S-L axis (the length of the extraction column of the 18th eastern longwall–the distance to the longwall) make it possible to identify significant isogases for interpreting the relationship between deformation and aerogas processes during the stopped excavation. The absence of parallelism of the curve in the zone of advanced fracturing (the area of high methane content is the black isoline in Figure 4), as in [47,48], confirms the assumption about the manifestation features of situational geomechanical conditions in the extraction area. The presence of such wave manifestations is indicated in the works of researchers from China [49,50].Based on this, it should be assumed that (for well type No. 4) in the area S = from 1240 to 1220 m, the main roof was caved (failure of wells due to cutting of the boreholes). Prior to this, S = from 1250 to 1240 m, a decrease in the productivity of degassing due to the collapse of casing strings (in zones of ultimate stress state) occurred. In wells drilled into goaf, a different pattern is observed. Obviously, at a minimum distance (−40 to −30 m) ahead of the longwall, the concentration of methane in the extracted mixture should not be high due to the interception of gas flows (by more productive wells No. 3 and No. 4, see the diagram in Figure 1 [29]). Moreover, the formation and expansion of the area of the local minimum of methane concentration from −60 to −20 m and, accordingly, the narrowing of the zone of maximum productivity in the zone S = from 1310 to 1240 m, is a response to a change in the productivity of wells drilled above the developed formation and an increase in the stress concentration in the support pressure zone.

The caving of the main roof (according to Figures 3 and 4, it occurred in the area of picket No. 1240 m, which corresponds to the displacement of the longwall from the installation pass at 163 m), based on the simulation results, does not lead to a complete failure of the degassing section networks consisting of wells drilled into goaf [51]. Wells "recover" faster due to less damage to stability caused by extreme operating conditions. Starting from section S = from 1218 to 1220 m, an atypical phenomenon of high methane emission is observed in all operating wells in the section from −120 to 50 m (in wells No. 2 both in front of and behind a longwall). This can be explained by the transitional period of "gassing after caving" when the volumes of the new goaf above the mined seam do not yet form an anthropogenic gas collector [52,53]. In subsequent periods, the balances of distribution of gas flow in wells drilled into the goaf of the previously worked out longwall and into the mined massif stabilize, which indirectly confirms the possibility of obtaining no less reliable models than, for example, those based on "machine learning" [53]. As a result of applying the author's approach [54] to overcome the drawbacks of deterministic methods of three-dimensional interpolation of data, the resulting models are not inferior in reliability to models based on kriging. The results obtained can be used to improve the coal and gas co-mining methodology [55].

5. Conclusions

The development of anthropogenic gas-coal deposits while reducing the emission of greenhouse gases into the atmosphere requires significantlycomplexapproaches to modeling gas flows. Along with the common stochastic methods of geostatistics, the improvement of deterministic interpolation methods in combination with three-dimensional regression models has made it possible to obtain the following results:

- Deformation-wave processes in geo-environments produce cyclic non-linearities in the nature of the air–gas regime of mine methane emissions into anthropogenic rock masses, while only a part of the gas flows is captured by the degassing network in the extraction area.
- It has been established for the first time that a decrease in the distance of the stoping face line from the start of mining of the extraction column S = from 1340 to 1120 m and the distance in front of the longwall L = from −120 to 0 m leads to undulating changes in gas release (in wells No. 2), according to a polynomial dependence.
- The influence of situational geomechanical conditions of reserve mining on the area and the shape of the local extrema of the models was clarified. These models are transformed in proportion to the development of the stoping front and are displaced at certain angles to the alignment with the longwall.

Further research should be focused on improving the methodology of geoecological monitoring of emission flows of climatically active gases during the transformation of natural and technical systems. Conclusions are only valid for the lower technical limit of the m_3 seam mining in Donbass. The main constraints include a development depth of 1250–1350 m, natural methane content of 23 m^3/t dry ash-free mass and above, a coal seam thickness of 1.3–1.55 m, (the main limiting mining and technical factor) a rock temperature that exceeds 41 °C.

Author Contributions: Conceptualization, V.S.B. and V.I.G.; methodology, V.S.B.; software, V.S.B.; validation, V.S.B. and B.V.D.; formal analysis, B.V.D.; investigation, V.S.B.; resources, V.I.G.; data curation, V.S.B.; writing—original draft preparation, B.V.D.; writing—review and editing, B.V.D.; visualization, B.V.D.; supervision, V.S.B.; project administration, V.I.G. All authors have read and agreed to the published version of the manuscript.

Funding: This research received no external funding.

Data Availability Statement: Not applicable.

Conflicts of Interest: The authors declare no conflict of interest.

References

1. Saunois, M.; Jackson, R.B.; Bousquet, P.; Poulter, B.; Canadell, J.G. The growing role of methane in anthropogenic climate change. *Environ. Res. Lett.* **2016**, *11*, 120207. [CrossRef]
2. Shukla, P.R.; Skea, J.; Slade, R.; Khourdajie, A.A.; van Diemen, R.; McCollum, D.; Pathak, M.; Some, S.; Vyas, P.; Fradera, R.; et al. *IPCC, 2022: Climate Change 2022: Mitigation of Climate Change. Contribution of Working Group III to the Sixth Assessment Report of the Intergovernmental Panel on Climate Change*; Cambridge University Press: Cambridge, UK; New York, NY, USA, 2022; p. 2913. [CrossRef]
3. Silvia, F.; Talia, V.; Di Matteo, M. Coal mining and policy responses: Are externalities appropriately addressed? A meta-analysis. *Environ. Sci. Policy* **2021**, *126*, 39–47. [CrossRef]
4. Rylnikova, M.V. The Harmonious Development of Mining, Science and Higher Education—A Guarantee of Stable State of Mining Regions in Russia. *Sustain. Dev. Mt. Territ.* **2020**, *12*, 154–161. (In Russian) [CrossRef]
5. Golik, V.I.; Razorenov, Y.I.; Dmitrak, Y.V.; Gabaraev, O.Z. Safety improvement of the underground ore extraction considering mass geodynamics. *Occup. Saf. industry.* **2019**, *8*, 36–42. (In Russian) [CrossRef]
6. Rybak, J.; Tyulyaeva, Y.; Kongar-Syuryun, C.; Khayrutdinov, A.M.; Akinshin, I. Geomechanical substantiation of parameters of technology for mining salt deposits with a backfill. *Min. Sci.* **2021**, *28*, 19–32. [CrossRef]
7. Amrani, M.; Taha, Y.; El Haloui, Y.; Benzaazoua, M.; Hakkou, R. Sustainable Reuse of Coal Mine Waste: Experimental and Economic Assessments for Embankments and Pavement Layer Applications in Morocco. *Minerals* **2020**, *10*, 851. [CrossRef]
8. Kongar-Syuryun, C.B.; Faradzhov, V.V.; Tyulyaeva, Y.S.; Khayrutdinov, A.M. Effect of activating treatment of halite flotation waste in backfill mixture preparation. *Min. Inf. Anal. Bull.* **2021**, *1*, 43–57. (In Russian) [CrossRef]

9. Gabaraev, O.Z.; Konovalova, T.G.; Turluev, R.R. Ecology and utilization of ore dressing tailings. *IOP Conf. Ser. EarthEnviron. Sci.* **2022**, 1021. [CrossRef]
10. Khayrutdinov, A.; Paleev, I.; Artemov, S. Replacement of traditional components of the backfill mixture with man-made waste. *IOP Conf. Ser. Earth Environ. Sci.* **2021**, *904*, 012005. [CrossRef]
11. Dzhioeva, A.K. Prospects for mining ecologization to reduce harmful emissions into the atmosphere. *Ugol'* **2022**, *10*, 29–32. (In Russian) [CrossRef]
12. Rybak, J.; Khayrutdinov, M.M.; Kuziev, D.A.; Kongar-Syuryun, C.B.; Babyr, N.V. Prediction of the geomechanical state of the rock mass when mining salt deposits with stowing. *J. Min. Inst.* **2022**, *253*, 61–70. (In Russian) [CrossRef]
13. Adero, N.J.; Drebenstedt, C.; Prokofeva, E.N.; Vostrikov, A.V. Spatial data and technologies for geomonitoring of land use under aspect of mineral resource sector development. *Eurasian Min.* **2020**, *1*, 69–74. [CrossRef]
14. Mohammadi, S.; Ataei, M.; Kakaie, R.; Mirzaghorbanali, A.; Aziz, N. A Probabilistic Model to Determine Main Caving Span by Evaluating Cavability of Immediate Roof Strata in Longwall Mining. *Geotech. Geol. Eng.* **2021**, *39*, 2221–2237. [CrossRef]
15. Mohammadi, S.; Ataei, M.; Kakaie, R.; Mirzaghorbanali, A. A New Roof Strata Cavability Index (RSCi) for Longwall Mining Incorporating New Rating System. *Geotech. Geol. Eng.* **2019**, *37*, 3619–3636. [CrossRef]
16. Khanal, M.; Qu, Q.; Zhu, Y.; Xie, J.; Zhu, W.; Hou, T.; Song, S. Characterization of Overburden Deformation and Subsidence Behavior in a Kilometer Deep Longwall Mine. *Minerals.* **2022**, *12*, 543. [CrossRef]
17. Islavath, S.; Deb, D. Interaction of a Shield Structure with Surrounding Rock Strata Under Geo-static and Fatigue Loadings. *Geotech. Geol. Eng.* **2022**, *40*, 2949–2965. [CrossRef]
18. Arasteh, H.; Esmaeili, K.; Saeedi, G.; Farsangi, M.A.E. Discontinuous Modeling of Roof Strata Caving in a Mechanized Longwall Mine in Tabas Coal Mine. *Int. J. Geomech.* **2022**, *22*, 04022040. [CrossRef]
19. Rasouli, H.; Shahriar, K.; Madani, S.H. Prediction of the Height of Fracturing via Gene Expression Programming in Australian Longwall Panels: A Comparative Study. *Rud. GeoloskoNaft. Zb.* **2022**, *37*, 91–104. [CrossRef]
20. Zhu, Z.; Wu, Y.; Liang, Z. Mining-Induced Stress and Ground Pressure Behavior Characteristics in Mining a Thick Coal Seam With Hard Roofs. *Front. Earth Sci.* **2022**, *10*, 843191. [CrossRef]
21. Polevshchikov, G.Y. Deformation-wave processes under production face advance in coal and rocks. *J. Min. Sci.* **2013**, *49*, 715–723. [CrossRef]
22. Shemyakin, E.I.; Fisenko, G.L.; Kurlenya, M.V.; Oparin, V.N.; Reva, V.N.; Glushikhin, F.P.; Rosenbaum, M.A.; Tropp, E.A.; Kuznetsov, Y.S. The effect of zonal disintegration of rocks around underground workings. *Doklady Akademii Nauk* **1986**, *1*, 289.
23. Shemyakin, E.I.; Fisenko, G.L.; Kurlenya, M.V.; Oparin, V.N.; Reva, V.N.; Glushikhin, F.P.; Rozenbaum, M.A.; Tropp, E.A.; Kuznetsov, Y.S. *Zonal Disintegration of Rocks and Underground Workings Stability*; Siberian Branch of the Russian Academy of Sciences: Novosibirsk, Russia, 2008; p. 276.
24. Young, R.P.; Nasseri, M.H.B.; Sehizadeh, M. Mechanical and seismic anisotropy of rocks from the ONKALO underground rock characterization facility. *Int. J. Rock Mech. Min.* **2020**, *126*, 104190. [CrossRef]
25. Ghaffari, H.; Nasseri, M.; Young, R. Faulting of Rocks in a Three-Dimensional Stress Field by Micro-Anticracks. *Sci. Rep.* **2014**, *4*, 5011. [CrossRef] [PubMed]
26. Zhang, Z.; Zhang, R.; Cao, Z.; Gao, M.; Zhang, Y.; Xie, J. Mechanical Behavior and Permeability Evolution of Coal under Different Mining Induced Stress Conditions and Gas Pressures. *Energies* **2020**, *13*, 2677. [CrossRef]
27. Karimpouli, S.; Tahmasebi, P.; Ramandi, H.L. A review of experimental and numerical modeling of digital coalbed methane: Imaging, segmentation, fractures modeling and permeability prediction. *Int. J. Coal Geol.* **2020**, *228*, 103552. [CrossRef]
28. Szlazak, N.; Obracaj, D.; Korzec, M. Estimation of Gas Loss in Methodology for Determining Methane Content of Coal Seams. *Energies* **2021**, *14*, 982. [CrossRef]
29. Brigida, V.S.; Zinchenko, N.N. Methane Release in Drainage Holes Ahead of Coal Face. *J. Min. Sci.* **2014**, *50*, 60–64. [CrossRef]
30. Gryazev, M.V.; Kachurin, N.M.; Vorob'ev, S.A. Mathematical Models of Gas-Dynamic and Thermophysical Processes in Underground Coal Mining at Different Stages of Mine Development. *J. Min. Inst.* **2017**, *223*, 99–108. [CrossRef]
31. Qu, Q.; Shi, J.; Wilkins, A. A Numerical Evaluation of Coal Seam Permeability Derived from Borehole Gas Flow Rate. *Energies* **2022**, *15*, 3828. [CrossRef]
32. Cheng, Y.P.; Wang, L.; Zhang, X.L. Environmental impact of coal mine methane emissions and responding strategies in China. *Int. J. Greenh. Gas Control.* **2011**, *5*, 157–166. [CrossRef]
33. Jiangfu, H.; Wenchao, H.; Chengpeng, Z.; Zhongguang, S.; Xiaoyi, S. Numerical simulation on the deformation characteristics of borehole failure in gas-bearing coal seams considering the effective stress principle under complicated stress path conditions. *Geomech. Geophys. Geo-Energy Geo-Resour.* **2022**, *8*, 95. [CrossRef]
34. Hou, X.; Liu, S.; Zhu, Y.; Yang, Y. Evaluation of gas contents for a multi-seam deep coalbed methane reservoir and their geo-logical controls: In situ direct method versus indirect method. *Fuel* **2020**, *265*, 116917. [CrossRef]
35. Bosikov, I.I.; Klyuev, R.V.; Khetagurov, V.N.; Azhmukhamedov, I.M. Development of methods and controls for aerogasodynamic processes at mining sites. *Sustain. Dev. Mt. Territ.* **2021**, *1*, 77–83. (In Russian) [CrossRef]
36. Kadochnikov, V.G.; Dvoynikov, M.V. Development of Technology for Hydromechanical Breakdown of Mud Plugs and Improvement of Well Cleaning by Controlled Buckling of the Drill String. *Appl. Sci.* **2022**, *12*, 6460. [CrossRef]
37. Bosikov, I.I.; Klyuev, R.V.; Azhmukhamedov, I.M.; Revazov, V.C. Assessment of coal mine ventilation control using statistical dynamics methods. *Min. Inf. Anal. Bull.* **2021**, *11*, 123–135. (In Russian) [CrossRef]

38. Yao, H.; Peng, H.; Hong, B.; Guo, Q.; Ding, H.; Hong, Y.; Zhu, Y.; Cai, C.; Chi, J. Environmental Controls on Multi-Scale Dynamics of Net Carbon Dioxide Exchange From an Alpine Peatland on the Eastern Qinghai-Tibet Plateau. *Front. Plant Sci.* **2022**, *12*, 791343. [CrossRef]
39. Dzhioeva, A.K.; Brigida, V.S. Spatial non-linearity of methane release dynamics in underground boreholes for sustainable mining. *J. Min. Inst.* **2020**, *245*, 522–530. [CrossRef]
40. Qu, Q.; Balusu, R.; Belle, B. Specific gas emissions in Bowen Basin longwall mines, Australia. International. *Int. J. Coal Geol.* **2022**, *261*, 104076. [CrossRef]
41. Borowski, M.; Zyczkowski, P.; Cheng, J.; Luczak, R.; Zwolinska, K. The Combustion of Methane from Hard Coal Seams in Gas Engines as a Technology Leading to Reducing Greenhouse Gas Emissions—Electricity Prediction Using ANN. *Energies* **2020**, *13*, 4429. [CrossRef]
42. Alipenhani, B.; Majdi, A.; Amnieh, H.B. Determination of Caving Hydraulic Radius of Rock Mass in Block Caving Method using Numerical Modeling and Multivariate Regression. *J. Min. Environ.* **2022**, *13*, 217–233. [CrossRef]
43. Khanal, M.; Adhikary, D.; Balusu, R.; Wilkins, A.; Belle, B. Mechanical study of shear failure of vertical goaf drainage hole. *Geotech. Geol. Eng.* **2022**, *40*, 1899–1920. [CrossRef]
44. Niu, Y.; Zhang, X.; Wang, E.; Li, Z.; Cheng, Z.; Duan, X.; Li, H.; Wei, Y.; Qian, J.; Cai, G.; et al. A new method of monitoring the stability of boreholes for methane drainage from coal seams. *Meas. J. Int. Meas. Confed.* **2020**, *15*, 107521. [CrossRef]
45. Brigida, V.S.; Golik, V.I.; Dmitrak, Y.V.; Gabaraev, O.Z. Ensuring Stability of Undermining Inclined Drainage Holes During Intensive Development of Multiple Gas-Bearing Coal Layers. *J. Min. Inst.* **2019**, *239*, 497–501. [CrossRef]
46. Qu, Q.; Guo, H.; Balusu, R. Methane emissions and dynamics from adjacent coal seams in a high permeability multi-seam mining environment. *Int. J. Coal Geol.* **2022**, *253*, 103969. [CrossRef]
47. Li, Y.; Wu, S.; Nie, B.; Ma, Y. A new pattern of underground space-time Tridimensional gas drainage: A case study in Yuwu coal mine, China. *Energy Sci. Eng.* **2019**, *7*, 399–410. [CrossRef]
48. Dzhioeva, A.K.; Tekhov, A.V. Reduction of mine methane emissions for ensuring sustainable development of geotechnologies in the transition to Industry 3.0. *IOP Conf. Ser. Mater. Sci. Eng.* **2021**, *1064*, 012008. [CrossRef]
49. Qin, B.; Shi, Z.S.; Hao, J.F.; Ye, D.L.; Liang, B.; Sun, W.J. Analysis of the Space–Time Synergy of Coal and Gas Co-mining. *ACS Omega* **2022**, *7*, 13737–13749. [CrossRef]
50. Shi, Z.; Ye, D.; Hao, J.; Qin, B.; Li, G. Research on Gas Extraction and Cut Flow Technology for Lower Slice Pressure Relief Gas under Slice Mining of Extra-thick Coal Seam. *ACS Omega* **2022**, *7*, 24531–24550. [CrossRef]
51. Zhang, C.; Tu, S.; Bai, Q.; Yang, G.; Zhang, L. Evaluating pressure-relief mining performances based on surface gas vent hole extraction data in longwall coal mines. *J. Nat. Gas Eng.* **2015**, *24*, 431–440. [CrossRef]
52. Liu, T.; Zhao, Y.; Kong, X.; Lin, B.; Zou, Q. Dynamics of coalbed methane emission from coal cores under various stress paths and its application in gas extraction in mining-disturbed coal seam. *J. Nat. Gas Eng.* **2022**, *104*, 104677. [CrossRef]
53. Khanal, M.; Poulsen, B.; Adhikary, D.; Balusu, R.; Wilkins, A.; Belle, B. Numerical study of stability and connectivity of vertical goaf drainage holes. *Geotech. Geol. Eng.* **2021**, *39*, 2669–2679. [CrossRef]
54. Zhou, J.; Lin, H.; Jin, H.; Li, S.; Yan, Z.; Huang, S. Cooperative prediction method of gas emission from mining face based on feature selection and machine learning. *Int. J. Coal Sci. Technol.* **2022**, *9*, 51. [CrossRef]
55. Zhong, S.; Lin, D. Evaluation of the Coordination Degree of Coal and Gas Co-Mining System Based on System Dynamics. *Sustainability.* **2022**, *14*, 16434. [CrossRef]

MDPI AG
Grosspeteranlage 5
4052 Basel
Switzerland
Tel.: +41 61 683 77 34

Mining Editorial Office
E-mail: mining@mdpi.com
www.mdpi.com/journal/mining

www.ingramcontent.com/pod-product-compliance
Lightning Source LLC
Chambersburg PA
CBHW061836210926
51850CB00006B/908
9783725866663